A Look into Mirrors

Work at Sea: The Evolution of Shipboard Technology (2026)

Airships: Their Science, History and Future (2025)

*But Will It Fly?: The History and Science of Unconventional
Aerial Power and Propulsion* (2025)

Poseidon's Progress: The Quest to Improve Life at Sea (2024)

*Arming the Warship: Naval Weapons Technology and Gunnery
from the Spanish Armada to the Cold War* (2024)

A Look into Mirrors

Their Making and Use Throughout History

IVER P. COOPER

McFarland & Company, Inc., Publishers

Jefferson, North Carolina

ISBN (print) 978-1-4766-9855-7
ISBN (ebook) 978-1-4766-5629-8

LIBRARY OF CONGRESS CATALOGING DATA ARE AVAILABLE

Front cover images: *left to right* bronze mirror with a support in the form of a draped woman, Greek, mid–5th century BCE (The Met); James Webb Space Telescope Mirror (NASA/Chris Gunn).

Printed in the United States of America

McFarland & Company, Inc., Publishers
Box 611, Jefferson, North Carolina 28640
www.mcfarlandpub.com

Table of Contents

Introduction

Sometime in the distant past, a human being noticed his or her reflection in a body of water. And unlike almost all other animals, that person perceived that the reflection was of his or her own body.

It was, perháps, love at first sight.

Over the course of several millennia, humans learned how to make that reflection portable, first by polishing stone and metal, then by bonding a metal of some kind to the back of a clear glass or plastic.

To see a reflection, one needed a light source to illuminate the subject. The first light source was the sun. Artificial lighting was gradually improved, with candles being replaced with Argand (oil) and kerosene lamps, and later gas lights, limelights, and electric lights.

The famous "Sphinx" illusion (1865) inspired the phrase "it's all done with mirrors." While the reference was to stage magic, the use of mirrors in civilization has been far more pervasive than most people realize.

As the reflections became clearer and brighter, people discovered that the mirror could be used for more than just self-inspection. For extension of our vision, in periscopes, telescopes, microscopes, and cameras. For communication, in heliographs, photophones, and the mirror galvanometers of undersea telegraph cables. For amusement, in kaleidoscopes and stage illusions. For energy-related purposes, in solar furnaces and sunlight-propelled spacecraft. And for other uses, in many additional ingenious devices.

This is the story of how mirrors were made and how they have been used, for work and play.

PART I

Mirrors and Mirror-Making

CHAPTER 1

Ancient Mirrors

Water, Stone, and Metal

Water Mirrors

Undoubtedly, the first mirrors were puddles, ponds, lakes, and rivers. The *Speculum Dianae*, the "Mirror of Diana," was not a manufactured mirror, but a lake, *Nemorensis Lacus* (Lake of the Woods, Lake Nemi), over 100 feet deep and 3½ miles in diameter. It lies inside an extinct volcano. The lake is shown in a painting by Richard Wilson (1713–1782), "Landscape with Diana and Callisto" (circa 1757). In his day, the *Speculum Dianae* was a stopping point for upper class Europeans making the "Grand Tour." Lord Byron wrote of the "oval mirror" of Lake Nemi in "Childe Harold's Pilgrimage."[1]

Sir James George Frazer, in *The Golden Bough*, remarked: "No one who has seen that calm water, lapped in a green hollow of the Alban hills, can ever forget it. The two characteristic Italian villages which slumber on its banks, and the equally Italian palace whose terraced gardens descend steeply to the lake, hardly break the stillness and even the solitariness of the scene. Diana herself might still linger by this lonely shore, still haunt these woodlands wild."[2]

On the northeast shore was a temple to Diana. Diana's bynames included *Limnaea* ("of the lake"), *Phoebe* ("bright moon"), *Aria* ("of the oaks"), *Pitanatis* ("of the pines"), and *Pheraea* ("lover of hunting"). Whether the lake was called "Diana's Mirror" because of this temple, or the temple was built at the lake because the reflection of the moon could be seen in it,[3] is unclear. The woods nearby were hiding places for runaway slaves (her temple was an asylum for them). One such, the leader of her cult, was the *Rex Nemorensis*, the King of the Woods. His rule was a precarious one, as each *Rex* obtained his office by slaying his predecessor. The "Golden Bough" of Frazer's book was the branch, possibly of an oak tree, broken off by a challenger to prove his right to contest the office.

According to the United States Geological Survey, as of September 5, 2024, there were 154 American bodies of water which included "Mirror Lake" in their name.[4] To this, one may add the occasional "Mirror Creek," "Mirror Cove," "Mirror Pool," "Mirror Pond," "Mirror Spring," "Mirror Harbor," and "Mirror Bay," bringing the total to 186. There are also a score of geographical locations with "Looking Glass" (or a variant form) in their names.

Lakes had one serious disadvantage for practical use as mirrors: They weren't necessarily located where you needed them to be. To see yourself, you had to walk to the lake. And even this option was absent if the purpose wasn't to see yourself, but to enhance the appearance of a monument of some kind. The problem was solved in three different ways.

One was to excavate your own pool of water, just in front of the edifice to be reflected. For Americans, the best-known example of this may be the Lincoln Memorial Reflecting Pool, which provides a beautiful view of the Washington Monument to the east. The Washington Monument was completed in 1888. However, there then was no reflecting pool nearby. That architectural enhancement was proposed, together with the nearby Lincoln Memorial, by the McMillan Commission in 1902. Both were completed in 1922, and Robert Todd Lincoln, the oldest son of Abraham Lincoln, was present for the dedication ceremony. The pool measures 2,000 feet long by 160 feet wide and is about 3 feet deep.[5] Perhaps the greatest moment in the history of this pool was when it witnessed Martin Luther King Jr.'s "I Have a Dream" speech, in August 1963.

Similarly, a commemorative reflecting pool, 1,750 feet long and 200 feet wide, mirrors the shaft of the 570-foot-tall San Jacinto monument in Texas. The monument, crowned with the Lone Star of Texas, commemorates the 1836 battle in which the Mexicans were decisively defeated and their President-General, Santa Anna, was captured.[6]

A much older reflecting pool is the one outside the walled temple of Angkor Wat, in present-day Cambodia, built by Suryavarman II in the early twelfth century. Angkor was the capital of the Khmer Empire, which dominated southeast Asia. Another famous watery reflection is the one available outside the Taj Mahal. This was a white marble tomb built in 1631–1648 in Agra, seat of the Mughal Empire, by Shah Jehan for his wife, Arjumand Banu Begum.[7] The great dome is reflected in the central canal.[8] Perhaps the oldest reflecting pool is the small one outside the Parthenon. This lies in front of the cult statue of Athena Parthenos, toward the west end of the colonnade.[9]

Another approach was to carry the water in a bowl. Now the mirror could go anywhere, and quickly, too. "In China, the first tool to serve as mirror is a shallow bronze jar (called Jian?) which holds water to provide reflection of one's face. However, this was a very expensive item, available only to the rulers."[10] According to Li Xueqin of the Chinese Academy of Sciences, the jian was "a large basin used for three purposes: first, as a bath tub; second, as a mirror when filled with water; and third, for keeping ice…. The base inside … is often embellished with a turtle, fish, coiled dragon or other designs."[11]

Finally, mankind identified certain stones and metals that were reflective when polished. These were the first true mirrors, that is, reflective objects specifically manufactured for that purpose. An Egyptian sage of the First Intermediate Period (circa 2181–2055 BCE) wrote, "Lo, she that beheld her face in water is now the possessor of a mirror."[12]

Stone Mirrors

Some Old-World Stone Mirrors

It is presently believed that the first manufactured mirror was created in the seventh millennium BCE. It was unearthed in the Çatalhöyük excavations, 52 kilometers south of Konya. This mirror was made of obsidian.[13]

"A roughly rectangular polished flake of selenite (a crystalline form of gypsum), found in a Badarian grave (c.4400–4000 BC), has been interpreted as one of the earliest Egyptian mirrors. This item, now in the British Museum, was found alongside other cosmetic equipment."[14]

Theophrastus (371–287 BCE) refers to the use of "anthrakion from Orchomenos in Arcadia" to make mirrors.[15] This might be obsidian, or some other dark stone.[16] Thoresen warns that "anthracite" is a "false cognate" and argues in favor of "Etruscan garnet," which "looks opaque black in reflected light, but viewed under a strong transmitted light or sunlight, its transparency is apparent and its color resembles that of a glowing coal."[17]

"Smaragdus" is the name the Romans gave to the emerald and to several other green stones. Pliny writes in *Natural History*:

> When the surface of the smaragdus is flat, it reflects the images of objects in the same manner as a mirror. The Emperor Nero used to view combats of the gladiators upon a smaragdus.[18]

Pliny said that there were twelve different varieties of "smaragdus"; these probably included dioptase, diopside, peridot, epidote, fluorite (fluorspar), malachite, chalcedony, and green jasper. We do not know the size of Nero's viewing stone, but if it were as large as a normal Roman hand mirror (6–10" diameter), it is not very likely that this smaragdus was a true emerald.

More old-world stone mirrors are discussed later in this section.

Some New-World Stone Mirrors

The most ancient American mirrors are those of coastal Peru. In village 1 of the late pre-ceramic complex of Asia, 60 miles south of Lima, Frédéric Engel found "a small piece of very shiny stone cemented to a clay tablet," which he identified as a mirror. The tablet is decorated with shell inlays. The village was abandoned in 1400 BCE.[19] "Mirrors are widely spread in South America, the oldest being found at Huaca de los Reyes and Gramalote, on the Peruvian coast and in the Andes at Shillacoto and Kotosh."[20]

The oldest North American mirrors are those of the Olmecs, and, according to radiocarbon dating, the earliest specimens studied are between 2,100 and 3,100 years old. The Olmecs made mirrors out of mica, obsidian, magnetite, ilmenite, hematite, and pyrite.[21] The refractive indexes of mica (muscovite 1.552–1.618) and obsidian (1.450–1.550)[22] are similar to that of glass (~1.5), which would imply a normal (perpendicular) reflectivity similar to glass (~4% per surface). The polished iron ores have higher reflectivities: magnetite (21%), hematite (28%), and pyrite (55%).[23] Ilmenite is about 20 percent.[24]

The mirrors could have been ground with emery (aluminum oxide), hematite powder, or perhaps sand. Ochre (hematite jeweler's rouge) could have been used for polishing. The Olmecs mostly used concave mirrors with focal lengths of 5–80 centimeters. In Olmec art, these mirrors are depicted as pectorals, that is, as ornaments worn on the chest. Their use in fire-starting and in divination is conjectural. There are a few Olmec convex mirrors, and it has been suggested that they were used for cosmetic purposes, that is, to show the whole face in a small mirror.

The Maya, in the Yucatan peninsula, also used mirrors: typically "flat ... polished iron-ore polygons fitted in a mosaic pattern onto a slate backing."[25] Mirrors are depicted on several Mayan vases. One polychrome vase shows a ruler looking into a mirror.[26] Another shows a monkey dancing with, and gazing into, a mirror.[27] Some scholars have speculated that the Chay Abah, the oracular obsidian stone of the sixteenth-century Cakchiquel Maya, was actually a mirror.[28]

The Aztecs used both obsidian and pyrite, as discussed below.

Obsidian

Obsidian is the best known of the stone mirror materials. Obsidian is a black volcanic glass with a high sheen. It forms when molten lava cools quickly—usually when it meets a body of water. It is usually black in color and takes a fine polish. Its edges are often translucent, giving them a smoky appearance. Obsidian is usually found in small outcrops, but the Glass Buttes of Oregon are made entirely of obsidian. It breaks with a very sharp, conchoidal (shell-shaped) fracture, leading to its use in the Old Stone Age to make edged tools.[29] "The edge of obsidian can be flaked to one molecule in thickness, much sharper than any knife or razor edge."[30]

"Obsidian is extremely hard to work, and the skill required for these prehistoric townsfolk to transform it into perfect, scratch-free mirrors astounded the British archaeologists who excavated Çatal Hüyük in the 1960s."[31]

At least eight mirrors have been found at this site, at levels IV–VI. Archaeologist James Vedder decided to determine, experimentally, how hard it was to achieve a good obsidian mirror.

> In December 1999, I started to fashion a small one by hand, using various readily available materials (some artificial) to grind and polish a broken obsidian nodule.... The fine grained stone gave a fairly good polish, while the others were used briefly to verify that a wide range of materials are effective in the process. I observed the sun's reflection and examined the obsidian surface with a 10x hand lens to evaluate the progress of the grinding and polishing and took great care to keep the surface clear of extraneous material to reduce the possibility of

generating scratches.... Materials for grinding and polishing comparable to those listed above can be found in Turkey.

The largest mirror surface I produced is about 4 by 6 centimeters (1.6 by 2.4 inches). With experience, one could probably select an optimum set of materials to expedite the creation of a mirror surface. Since achieving the final polish may take the longest time, I can not estimate now the time required to make a mirror.

All of the mirrors produced good images, and all were slightly convex as expected from manual grinding in which linear and rotary motions result in greater pressure being applied around the perimeter of the surface. The only technical reference that I have seen on an obsidian mirror from Çatalhöyük states that it is slightly convex.

With special preparation of a core and great care during the grinding process, one could probably make a nearly flat mirror with no obvious distortions in the image.[32]

During the Protoclassic (1–300 CE) and Classic (300–900 CE) periods of Mesoamerica, the flaked obsidian was used "as is." However, during the Late Post Classic period (1200–1519 CE) the reflective properties of obsidian mirrors were improved by grinding and polishing.[33]

Obsidian is most closely associated with the Aztecs. They used it in jewelry, in weapons, and, of course, in mirrors. The edge of a newly chipped flake of obsidian was sharper than surgical steel, which is why obsidian blades have been used in eye surgery. The Aztecs used obsidian arrowheads, and edged their melee weapons with obsidian shards, as in their double-edged sword, *macuahuitl*. While it is too brittle to be resharpened, obsidian is plentiful in Mexico and Guatemala, so the Aztec warriors could readily re-edge their swords after a battle.[34]

The name of the god Tezcatlipoca was used as a synonym for obsidian, and some scholars consider him to be a personification of that stone. Obsidian mirrors were said to be "smoking" because of their dark color. "Both reflective and translucent, the obsidian mirror was seen as a threshold between two worlds, with the obsidian conceptualized as a membrane or tissue separating this earthly world from the beyond. Many ancient depictions survive showing mirrors being worn as part of ceremonial and military costumes, especially by the ruler."[35] According to Bernardino de Sahagún (circa 1499–1590), Aztec mirrors were made by specialists (*tecachiuhqui*).[36]

John Dee (1527–1608) was Queen Elizabeth I's court astrologer. He used an obsidian mirror for scrying. Shortly after her thirty-fifth birthday, the Queen came to Dee to look into his "magic glass."[37] After his death, it passed to the Mordaunts, the earls of Peterborough, and then through two further owners to Horace Walpole (1717–1797), the author of the first Gothic novel (*The Castle of Otranto*). A 1784 inventory of his estate listed it as a "speculum of kennel-coal," which was "curious for having been used to deceive the mob by Dr. Dee." It then passed through the hands of John Hugh Smyth-Pigott and on to Lord Londesborough; Albert Londesborough characterized it as a "polished oval slab of black stone."[38]

The British Museum acquired this unusual artifact in 1966. Inspection has confirmed that Dee's "shew-stone" is made of obsidian, and of a roughly circular shape, about 7.5 inches in diameter and one-half inch thick, with a short handle.[39]

Benjamin Goldberg suggests that Dee obtained his "shewstone" from the

Spanish, who in turn imported it from Mexico.[40] This is possible; Cortes arrived in Mexico in 1519. And chemical analysis shows that it is from Pachuca, Hidalgo, Mexico,[41] near Mexico City.

Pyrite

Pyrite (from the Greek *pyr*, "fire"), also known as "fool's gold," was also used in mirrors. Chemically, it is iron disulfide, often contaminated with nickel or cobalt. It is quite abundant in North America. Pyrite mirrors were used by the Maya,[42] the Olmecs (Zoque),[43] the Toltecs, and the Aztecs. In Classic Mesoamerica, iron pyrite was the most popular mirror material. Iron pyrite could be cut and laid, mosaic fashion, on a slate backing. The latter was often circular. These Classical pyrite mirrors could reach diameters of 30 centimeters, or even larger. "The Tomb of the Jade Jaguar at Tikal (Burial 196, Structure 5D-73) included the largest pyrite mosaic mirror yet found in all of Mesoamerica."[44]

Besides these large mosaic mirrors, there were also small, circular pyrite mirrors which the Mayan and Toltec nobility wore as decorations, on the small of the back, or on the back of a belt.[45]

Pyrite is more reflective than obsidian but unfortunately deteriorates; the iron sulfide reacts with water to form sulfuric acid. It also oxidizes readily. Today a Classic Mesoamerican mirror might be just a reddish or yellowish cast on the slate.[46]

Pyrite mirrors have been found in a tomb in the Pyramid of the Moon, at Teotihuacan.[47] In this tomb, discovered by Arizona State University (ASU) archaeologist Saburo Sugiyama, a skeleton, thought to be that of an adult male who was bound and sacrificed, was buried in a square chamber 11.3 feet on each side and five feet deep. He was surrounded by more than 150 burial offerings, including pyrite mirrors. This mirror was not of Aztec origin; Teotihuacan burned down around 750 CE, and the Aztec tribes entered the Valley of Mexico about 600 years later. Nonetheless, Valliant suggests that the Aztecs made more extensive use of pyrites than of obsidian.[48]

The Aztecs wore back mirrors, which they called *tezcacuitlapilli*. Their presence in a grave site is an indication that the deceased were soldiers.[49] In an Early Postclassic (Toltec) *tezcacuitlapilli*, a central pyrite mirror was surrounded by a turquoise mosaic with representations of Xiuhcoatl, the serpent of fire. It is associated with the god of fire, Xiuhcuhtli, also known as the Turquoise Lord, and with Huitzilopochtli, the Sun God. The flash of the mirror might well have contributed to this association. Examples of these *tezcacuitlapilli* have been found at Tula and at Chichen Itza; the "warrior" columns from Mound B at Tula wear this kind of back mirror.[50]

"[A] consensus has been reached that pyrite mirrors had been used mainly for ritual divination and magical-civic activities, such as communicating with the ancestors, serving as portals to alternate realities, to start fires or reflect light beams, as part of clothing, or a social symbols or prestige objects used in ceremonies."[51]

Gallaga attempted to produce a pyrite mirror using pre–Columbian materials and tools (stone hammers). It took him "more than 40 person-hours" to shape a "single

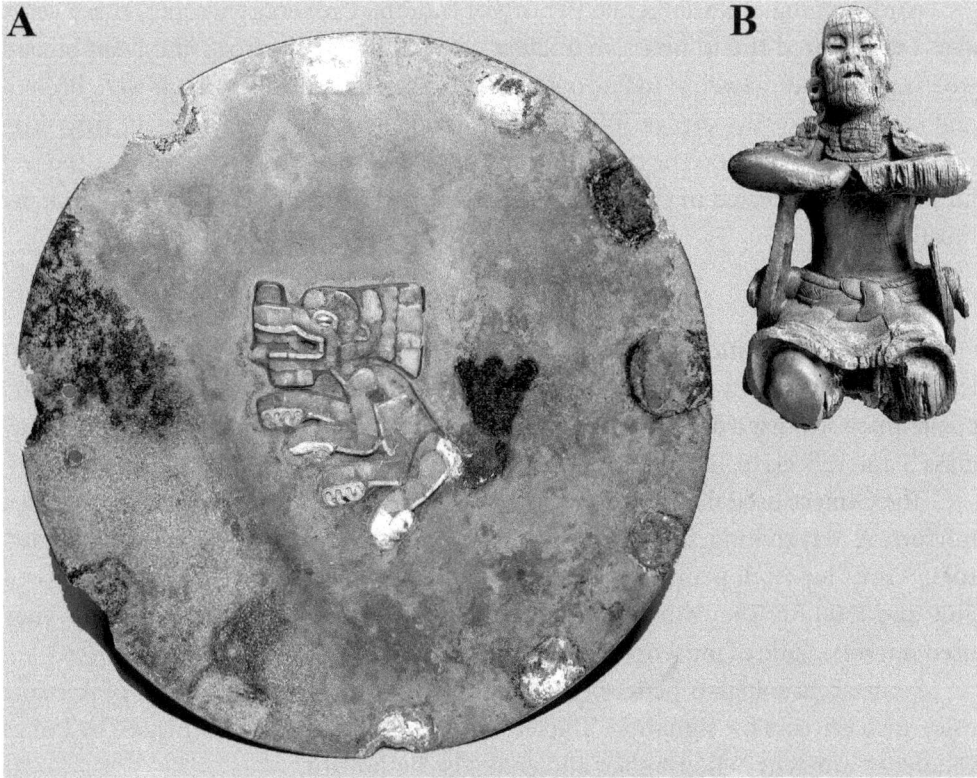

A. A mirror from Teotihuacan, Mexico (500–600 CE). It is "made of a single sheet of polished pyrite stone and includes a jade jaguar mosaic at its center." 7.5 inches diameter. (Art Institute of Chicago, S. DeWitt Clough and Ada Turnbull Hertle endowments, Ref. 1994.313.) B. A Mayan wood sculpture of a mirror-bearer (410–650 CE; "Early Classic"). "Clearly defined notches in the skirt and under the arms would have held a removable plaque approximately 5 inches square, probably covered in a mosaic mirror of pyrite or obsidian. The plaque would have been inserted under the arms and then hooked into the skirt notches.... The mirror-bearer to the ruler was an important role, sometimes filled by a woman, but more often by courtly dwarves. Their primary function was to reflect the image of Maya lords and ladies as those dignitaries preened in self-regard." (The Michael C. Rockefeller Memorial Collection, Bequest of Nelson A. Rockefeller, 1979, Accession 1979.206.1063. The Met.)

pyrite plaque … with a surface of 1 cm square … an average mirror used between 20 and 30 pyrite plaques of similar dimensions." Gallaga suggests that "the pre–Hispanic artisans who worked the pyrite were full-time specialists," and that the skill and time investment required meant that the mirrors "they were probably restricted to nobles who had the power to commission such objects from royal artisans."[52]

Hematite

Hematite was another Mesoamerican mirror material. It is iron oxide (i.e., chemically the same as rust), and is red, brown, gray or black in color. It is found throughout North America. Some field specimens are highly specular, and pieces of hematite can be polished.[53]

Anthropologist Karen Anne Pyburn of Indiana University wrote: "Three years ago, I discovered the tomb of an ancient Maya king…. The king had been buried wearing his finest jewelry, including hematite mirror earrings, which were broken and came out in tiny pieces. When we cleaned them they were perfect tiny mirrors and as I looked into them and saw my own eyes reflected I realized that the last human visage reflected in these mirrors had been an ancient Maya priest. It took my breath away."[54]

Mica

The micas are silicate minerals, found in pegmatites throughout North America. They cleave easily into thin, flexible flakes, which are both transparent (muscovite mica was used as a window substitute in Russia, hence the name, "Muscovy glass") and reflective, like glass (itself a silica).

The Olmecs used mica mirrors as early as 1600 BCE.[55] The Mayans also had mica mirrors. A Mayan king, buried about 350 CE, was found with an apron of mica mirrors.[56] The Hopewell people inhabited much of the Eastern United States between 2100 and 1500 BP. They are famous as mound builders. There are reports that they used mirrors made of muscovite mica.[57]

Some archaeologists believe that Indians in the Chaco Canyon area of Arizona used mica mirrors for signaling. The feasibility of such use was confirmed by James Riddle, an amateur heliographer, and geologist Beth Boyd:

> Surprisingly good results were obtained with the mica sample composed of numerous flakes varying from 1/16" to 1/8" thickness, and the piece measured an average of 2.5" by 3.25". The longest range tested with the mica was 3.92 miles between the "scar" on the east side of Thumb Butte, and the parking lot at the north end of the "Prescott Resort," ironically owned by the Yavapai Indian Tribe. Both sites are adjacent to Prescott, Arizona…. Interestingly, the mica outperformed a good quality 3" × 4" hand mirror, not quite so brilliant, but much easier to get "on target." This was due to the mica diffusing the reflected light providing a much wider field of reflection. It would have been much more difficult to reflect intelligible signals with a handheld flat glass mirror without mechanical aid such as is found in "modern" heliographs.

Riddle speculated that a mica mirror could be used at a range of ten or more miles but commented that mica is fragile. (Of course, the mica could be mounted on a backing of some kind to protect it.) Interestingly, the other rock and mineral specimens tested (schist, quartz, galena, and copper pyrite crystals) did not come close to the effectiveness of mica.[58]

Mica is very common in India, so one would expect India, too, to have known mica mirrors. It has been reported that mica flakes were used before metal or glass mirrors in the traditional Gujarat "sheeshedar," a form of embroidery in which small mirror discs were sewed onto fabric.[59] Judy Shorten writes:

> Originating in India, shi sha mirror embroidered embellishments can be seen on clothing, household items and other decorative and/or ceremonial items, and is quite commonly observed in areas such as Bali, Thailand, Singapore, etc. The Rabari women of the Kutch desert area in India seem to have been the first to do shi sha embroidery. They live in a

monotonously hot, dry climate, so the women gather together and embroider in much the same way American women gather for quilting bees, etc, making it a social event.... [T]he women usually use square or triangular shaped mica mirrors. Another group of women, the Sindhis, in southern Pakistan, more commonly use round mica mirrors. Both groups use strong cotton or silk thread with an interwoven stitch to hold the mirrors onto the fabric.[60]

Mica mirrors were also known to the ancient Nubians (present-day Sudan) of 3100–2800 BCE.[61] In Lower Nubia, the "A Group" people, discovered in 1907 by Boston archaeologist George Reisner and known to us only through their grave goods, buried their dead with pottery, stone palettes for grinding cosmetics, food jars, linen, copper tools, and mica mirrors.[62] It is unlikely that we will ever learn more about them, as the Aswan Dam flooded the remains of their settlements.

It is uncertain whether the Romans used mica in any mirrors. Pliny, in his *Natural History*, refers to what is translated as "specular stone" or "mirror stone."[63] Translator John Bostock assumes that this is "transparent selenite or gypsum." However, since Pliny says that it "can be split into leaves as thin as may be desired," it seems more likely that it was mica. Pliny implies that "mirror stone" is used for building purposes but does not say how. It could be as decoration, in view of its reflective properties, or in thin enough form it might be used as a window.

Pliny even mentions a use for the "shavings" and "scales" of the mirror stone: "the Circus Maximus having been strewed with them at the celebration of the games, with the object of producing an agreeable whiteness."[64] The modern equivalent is the incorporation of mica powder into the sidewalks of Hollywood, so they sparkle.

Anthracite and Jet

While obsidian is forever associated with the Aztec god "Burning Mirror," if you really wanted a mirror that burns, you would need to make it out of polished coal. Coal is formed by the compression and heating of decomposed plant matter. It first becomes peat, then lignite, then bituminous coal, and finally anthracite. Anthracite, the hardest form of coal, is glassy in appearance. "It burns slowly, with a pale blue flame and very little smoke."[65] Most coal is derived from terrestrial deposits, especially plant matter in the swamps of the Carboniferous period. The term "jet" usually refers to a form of coal found in marine deposits and thought to have been derived from driftwood.[66] Soft jet is formed under freshwater and hard jet under seawater.[67] However, it is also used loosely to refer to polished coal.

In Peru, the Chavins polished anthracite mirrors.[68] The name "Chavin" comes from Chavin de Huantar, in the highlands, where a great pre–Incan temple is located. The Chavin culture, which had a distinctive decorative style, flourished between 1200 and 300 BCE.[69] One Chavin mirror was described in a 1997 auction catalogue as follows: "ca. 1200 BC Anthracite flat mirror, of trapezoidal form with finished sides and beveled edges and high polish."[70] At Shillacoto, a 1200 BCE (Kotosh period) site, a burial of unusual richness, denoting the importance of the deceased in his society, was found to include two jet mirrors.[71] Another Shillacoto find was of a semicircular jet mirror.

Use of anthracite was not limited to the Chavin. There was also the Cupinisque culture, which prospered on the north coast from 1200 to 200 BCE.[72] A Cupisnique anthracite mirror was found by Junius Bord at Huaca Prieta in the Chicama Valley. The mirror is unframed and stands 7.3 centimeters high.[73]

Burger viewed the Cupinisques as a people with an unusually high interest in personal adornment: "For example, rings of carved bone were sometimes found on two, three, or even five fingers of buried individuals." There were ear pendants of carved bone with shell or turquoise inlays, and shell ornaments, once sewn on clothing. Rollers, sometimes found still covered with red pigment, were used for skin painting. "Not surprisingly, one of the most common non-ceramic artifacts at Cupinisque sites are anthracite mirrors highly polished to reflect an image."[74]

Why was coal, rather than obsidian, the mirror material of choice? Undoubtedly, this was because obsidian was not readily available locally in significant quantities. Obsidian was found mostly at Quispisisa in the south central highlands. This source was 470 kilometers south of Chavin de Huantar, a month's travel in either direction by llama caravan.[75]

Other Stones

In British Columbia, the Tsimshian people polished flat pieces of slate. "Some archaeologists call them mirrors because when the flat slate is wet it reflects." However, their actual purpose is unknown. If they were used as mirrors, some think that they were used for purposes of divination rather than self-inspection. George T. Emmons wrote in 1921 that the Tsimshian mirror "was the property of the women of higher class, and was worn suspended around the neck by a cord of hide or of twisted root."

Emmons had been a lieutenant in the U.S. Navy and wrote of his experiences and observations in Alaska and British Columbia. It is not clear from his writings how he came to conclude that women of high rank used the mirror. Was it mere speculation or did he see them being used? In southern British Columbia, J.A. Teit, a Scottish colonist who observed and wrote about its First Nations peoples, wrote to Emmons that he had only heard of dark stone being used for mirrors in the past. He noted that sheets of mica had been more commonly used as mirrors by the interior Salish people.[76]

Marble takes a fine polish and therefore can give a reflection. In Dante's *Purgatorio*, Canto IX, the first step of the stairway to St. Peter's Gate, symbolic of confession, was made of "white marble, so polished and so clear that I was mirrored there as I appear in life."[77]

Marcus Vitruvius Pollo (90–20 BCE), author of the first engineering handbook, said that plaster can be used in a similar way: "that which is well covered with plaster and stucco, and closely laid on, when well polished, not only shines, but reflects to the spectators the images falling on it."[78] In fifth-century Sri Lanka, King Kasyapa ordered the construction of the pleasure gardens of Sigiriya on top of a monolith rearing more than 1,000 feet above the Ceylonese jungle. The charms of this garden include a three-meter-tall brick Mirror Wall (*Kat Bitha*), so called because it was plastered with lime that clearly reflected the courtiers who walked along it. The wall is adorned with at least 1,500 graffiti from the seventh to thirteenth centuries, most dealing with the maidens, heavenly or otherwise, depicted in Sigiriya's famous frescoes.[79] Sigiriya is a UNESCO World Heritage Site.

In Christopher Marlowe's poem "Hero and Leander," the "crystal shining fair" pavement of Venus's temple near the town of Sestos was called "Venus' glass," because it reflected the decorations on the ceiling of the building.

Mirror-Cut Gems

A "mirror-cut" gem could serve as a very small (and very expensive) mirror. "The mirror-cut diamond was an early 16th century phenomenon. It had an extraordinarily large table consisting of as much as 90 percent of the width of the stone. This made the gem highly refractive and literally gave it the properties of a mirror."[80]

Several famous gems were mirror-cut—for example, "The Mirror of Portugal." This was a 30-carat diamond; there is a picture by van Dyke in the Hermitage of Henrietta Maria (the wife of Charles I of Great Britain) wearing it. It first came into history as one of the Portuguese crown jewels. An unsuccessful usurper of the Portuguese throne fled with it to England and gave it to Queen Elizabeth. It was passed down to Charles I, whose queen sold it to Cardinal Mazarin. It became one of the French crown jewels and disappeared after the French Revolution.[81]

Rock Crystal (Quartz)

At the Louvre, there is a "mirror composed of a rectangular plate of rock crystal"; the frame is 40 centimeters tall, but the plate is shorter, less than half that height. The museum dates it to 1630/1635.[82] However, Beale says that it was a Venetian gift to Marie de' Medici on her marriage to Henri IV (1600).[83] (I do wonder whether it is actually *cristallo*, a clear Venetian glass mimicking crystal.)

Metal Mirrors

Only a few metals are naturally found in native (elemental) form: iron, copper, silver, and gold.[84] They were the first metals to be exploited. Of these, copper was by far the most readily available.

Initially, copper was cold-worked (pounded with a hard rock to change its shape). This could make it more brittle, but its ductility could be restored by annealing it (heating it briefly). This led naturally to forging the metal (working while it was hot). Once copper could be heated to its melting point (1083°C), it could be cast into a shape established by a mold it was poured into.

Metal use expanded greatly once the smiths learned how to smelt the native metal from its ores. The smelting of copper may have occurred as early as 7000–6000 BCE, at Çatalhöyük in Turkey. The large-scale production of copper began during the fourth millennium BCE, which hence was called the Chalcolithic (copper-stone) Period.

Alloying—the combination of known proportions of pure metals into metallic mixes—increased the variety of materials available to craftsmen, as the alloys had properties different from those of their constituent elements. The first alloy was arsenical copper, also known as arsenical bronze. However, alloys of tin and copper, known as tin bronze or even just as bronze, became more important. The earliest tin bronze objects had "only small amounts of tin (1–4%)"; tin content increased as metalworkers experimented with different formulas. Bronzes rich (~10%) in tin are known to have been used in Mesopotamia (2800 BCE), Egypt (2600 BCE), Minoan Crete (1700 BCE), the Indus Valley (2500 BCE), China (2000 BCE), and the southern Andes (1000 BCE). Tin ore was available from Iberia, Brittany, Cornwall, and Erzgebirge in Europe, from China, and from Bolivia in South America.

Lead and zinc were also added to copper. The alloy of copper and zinc is called brass, and it was popular at an early date on the Indian subcontinent.

The first iron objects were made from meteoritic iron, obviously a limited source. The Iron Age was made possible by the smelting of iron ore. In Europe and the Near East, the general use of iron for weapons and tools is considered to have begun with the Hittites at about 1200 BCE. The superior martial qualities of weapons made from steel, an alloy of iron and carbon, increased the importance of iron.

Copper, silver, gold, and iron, and various metal alloys have been used by many different civilizations to make mirrors.

Nonferrous Mirrors in the Ancient Near East

The Near East was the birthplace of the Sumerian, Akkadian, Hittite, Assyrian, and Babylonian civilizations. "At Tello [ancient Sumerian Girsu], in a level dated to the late Uruk period (ca. 3200 BC), tombs containing … small copper disks identified as mirrors were found…. Several copper mirrors of varying size and consisting of disks with short handles were found in Jamdat Nasr (circa 3000 BC) tombs at Kish." A bronze mirror was found in a thirteenth–fourteenth century BCE Assyrian tomb in Mari, and "Tushratta, king of Mitanni (circa 1380 BC) … sent the pharaoh two silver mirrors."[85]

For the late second millennium BCE Hittites, "a mirror paired with a spindle symbolized womanhood." The goddess Kubaba (not to be confused with Queen Kubaba of Kish) was depicted (in northern Anatolian and Iranian finds) holding a mirror in one hand, but very little is known about her.[86]

Bronze mirrors were used by the Jews. The Book of Exodus (Revised Standard Version, 1:23) contains this passage: "And he made the laver of bronze and its base of bronze, from the mirrors of the ministering women who ministered at the door of the tent of meeting."

Electrum—an alloy of gold and silver—was also used as a mirror metal. Such a mirror was found in the tomb said by historian Stephanie Dalley to be that of two Israelite princesses, Yabaa and Atal-ya, "sent to be wives of King Sargon of Assyria by their Jewish fathers Kings Uzziah and Jotham. The Biblical Kings Uzziah and Jotham both ruled over Judah in the late 8th century BC, when Assyrian power was reaching its peak under Sargon II of Assyria…. The tomb in which the princesses were found was located in the ancient Assyrian city of Nimrud."[87]

EGYPT

A First Dynasty (circa 3100–2900 BCE) Egyptian grave was found to contain a pear-shaped copper mirror, with a handle.[88]

Copper was the main metal used in ancient Egypt. "The earliest copper finds are dated at approximately 5000 BC and contain a high level of impurity. By the time of the 1st Dynasty circa 3000 BC arsenic was alloyed with the copper to improve the metal. This development of the alloys advanced through bronze (a copper–tin alloy), to brass [a copper–zinc alloy,] which only became available during the Roman period (30 BC)."[89] Bronze has a lower melting point and greater hardness than that of pure copper.

An arsenic content of 4–6 wt%, which was typical for arsenical copper Egyptian mirrors, would have made the metal "golden in colour." However, some Egyptian mirrors exhibit surface enrichment for arsenic (to ~30 wt%), although the bulk metal did not exceed 6 wt%. Thomas argues that "silver in Ancient Egypt was more valuable than gold … so silver-appearing mirrors, produced through surface enrichment, may have been fashioned for the owners to look as if they were of high status."[90]

The most familiar of all hieroglyphics, the ankh (eyed cross), is the Egyptian symbol for life and for the mirror;[91] there is a mirror case, from the tomb of Tutankhamun, in the shape of an ankh.[92] Calcutt asserts that it was also the hieroglyphic for copper,[93] but Davey says that in the Old Kingdom, the "hieroglyphic symbol for copper" was the "silhouette" of a "crucible."[94] In any event, the ankh came to be the astrological symbol for the planet Venus, and the alchemical symbol for copper.[95]

From the Fourth Dynasty on, Egyptian mirrors often had a slightly flattened circular shape. It is thought that this was to represent the apparent flattening of the rising and setting sun.[96]

A copper mirror was found at Kahun, a Middle Kingdom (Twelfth Dynasty, circa 1895 BCE) workers' village.[97] This shows that mirrors were used by the common people, not just the upper classes. In a Middle Kingdom grave of a woman of "minor position and limited means," we find a copper mirror lying on her breast. Its ivory handle represents a lotus flower, a symbol of the sun and of rebirth.[98] Other handle designs included papyrus stalks, and minor gods and goddesses.

According to archaeologist Elizabeth Eaton, "the mirror had both magical and utilitarian value, and is everywhere represented, in reliefs and coffins and on stelae, as the primary requisite of women of any pretensions of wealth whatsoever."[99] Mirrors appear often in female graves, but not in male graves. This is curious, as in ancient Egypt, both men and women wore cosmetics,[100] and would have used hand mirrors to guide their application.

Despite the importance of copper in Egyptian life, most Egyptian mirrors that have survived to the present day are bronze.[101] Bronze is more easily worked, and also harder, than copper. The Metropolitan Museum of Art has a bronze Egyptian hand mirror that dates back to the Eighteenth Dynasty (about 1500 BCE).[102]

Despite its greater hardness, polished bronze scratches easily,[103] so it is not surprising that hand mirrors from Egypt and elsewhere often had protective covers. These covers also offered an opportunity for artistic embellishment.

Silver "generally appears less frequently in the Egyptian archaeological record than gold or cupreous metals." It had to be imported to Egypt, it tarnishes rapidly, and it is also "highly susceptible to the corrosive salts found in most Egyptian burial environments." Silver mirrors were nonetheless "associated with the so-called foreign wives of Thutmose III (R) (circa 1479–1425 BC)."[104] Another ancient silver mirror is that of the sixth century BCE Nubian king Amaninatakilebte; its handle shows four different Nubian gods.[105]

Artistic representations of silver and gold mirrors were found on stelae in Egyptian tombs, so women could at least enjoy them in the afterlife even if they did not own one while alive.[106]

The Greeks, the Etruscans, and the Romans

The Greeks also used metal mirrors. Aeschylus (525–456 BCE) wrote, "bronze is the mirror of the form, wine, of the heart."[107]

"The most common toilet article appearing on vases is the mirror, usually made of polished silver or bronze. These and a wide variety of cosmetic implements are often excavated in tombs, sanctuaries dedicated to female divinities, and in the domestic quarters of ancient towns."[108] According to a fifth century BCE Attic vase, even when Andromeda was being staked down as a sacrifice, a servant carried her mirror for her.[109]

Charles Panati asserts that "in 328 BC, the Greeks established a school for mirror craftsmanship. A student learned the delicate art of sand polishing a metal without scratching its reflective surface."[110] A metallurgist claimed that "the goddess of love, Aphrodite, is said to have emerged out of the sea near Cyprus looking at her image in a copper mirror," and joked, "this shows that the oldest profession is therefore metallurgy."[111]

The Greeks had three basic types of mirrors. The "hand mirror" had a disk-shaped reflective surface, a decorated back, and a handle. "A ring at the top allowed [it] to be hung from the wall." It has been found in Mycenaean sites (about 1400 BCE).[112]

In the second type, the handle was given a base, so the mirror could stand

A. "Caryatid mirror" (bronze, New Kingdom Egypt ca. 1540–1069 BCE). "Mirrors with handles in the form of naked young girls were the height of fashion in mid–Dynasty 18; numerous examples exist.... The retinue of Hathor consisted precisely of such beauties, called *nefrut* in Egyptian." Note the flattened circle shape of the mirror. The mirror is supported by a "papyrus umbel" (shoots coming from the papyrus stalk and drooping down). (Leonard C. Hanna Jr. Fund 1983.196, Cleveland Museum of Art.) B. "Caryatid mirror with Aphrodite" (bronze, ca. 460 BCE Greece). The winged figures are representations of Eros and there is a "siren at the top of the disk." (Acquired by Henry Walters, Accession 54.769, The Walters Art Museum.)

upright. The handle was often in a human shape (caryatid), although most likely representing a goddess. Both of these types were known in Ancient Egypt, too.

The box mirror was akin to a modern woman's pocket mirror. It had a clamshell design, with the upper disk protecting the reflective surface on the lower disk. This is the kind of mirror alluded to in Aristophanes' *The Clouds* (419 BCE), where Strepsiades says, "if I purchased a Thessalian witch, I could make the moon descend during the night and shut it, like a mirror, into a round box and there keep it carefully."[113] The box mirror, apparently a Greek invention, appeared in the fifth century BCE and was the "most popular type of mirror from the Hellenistic period onwards."[114]

A. Bronze box mirror, 6.25 inches diameter, Greek, mid-fourth century BCE. The cover shows, in relief, the head of a woman. (Rogers Fund, 1907, Accession 07.258a-c, The Met.) B. "Terracotta statuette of a woman looking into a box mirror," Greek, third to second century BCE. (Rogers Fund, 1912, Accession 12.229.19, The Met.)

Among the Etruscans, mirrors were luxury objects, as is shown by their prevalence as grave goods. (Like the Egyptians, they believed that "you *can* take it with you.") So far, Etruscan mirrors have only been found in women's tombs.

One Etruscan mirror carries a message of filial piety: "tite cale: atial: turce: malstria: cver" ("Tite Cale to his mother gave this mirror as a gift"). The use of *cver*, signifying a sacred object, suggests that the mirror was given to her posthumously. To make sure that the living did not steal and openly reuse the mirrors of the dead, some were marked *suthina* ("funeral offering").[115]

Mythological scenes were common decorations; a bronze mirror from Perugia, circa 300 BCE, shows the hero Pherse (Perseus) facing a seated Turms (Hermes). Beside Turms is Menerva (Minerva; Athena). All three are staring down at the reflected image of the severed head of Medusa, which Menerva is holding. Mirror backs are also an important source of Etruscan inscriptions.

Like the Greeks, the Romans used both bronze and silver mirrors. Pliny the Elder (CE 23–79) said that in the "times of our ancestors," the best mirrors were made at Brundisium, out of an alloy of "stannum" (tin) and copper—i.e., bronze. He also claimed that the first silver mirrors were made during the time of Pompey the Great (106–48 BCE), by Pasiteles, but of course he was mistaken. Possibly, he was thinking only of Roman mirrors.[116] By Pliny's time, the silver mirrors had displaced the bronze ones: "everybody, our maid servants even, … use silver ones."[117] However, the architect Vitruvius (first century BCE) warned that "a silver mirror, made from a thin plate, reflects the image confusedly and weakly, whilst from a thick solid plate it takes a high polish, and reflects the image brilliantly and strongly."[118]

Hand mirrors, similar to those used by the Greeks and Etruscans, were the most common type. One such find, a bronze mirror dating to 200 BCE, is decorated on the back with an illustration of a young man and a young woman playing tabula, a race-type game played with dice and a backgammon-like board. According to the inscription, the lady is saying, "I believe I've won."[119]

There were also larger mirrors. Seneca (4 BCE–65 CE) wrote, "We think ourselves poor and mean if our walls [of the baths] are not resplendent with large and costly mirrors."[120]

China and Japan

In 1976, in a tomb at Xiaotun, China, four bronze mirrors were discovered. "All four were round, with a small central boss on the back encircled by geometric designs." These mirrors date back to the Shang Dynasty (circa sixteenth to eleventh century BCE). In the Warring States period (475–221 BCE), Chinese bronze mirrors were square or round and had no handles. Mirrors with handles did not appear until after the Tang Dynasty. Most of the Warring States mirrors had four staggered Shan (a Chinese character, looking like a "T" with pronounced downward barbs on the crosspiece, and slanted) characters on their backs. Others were decorated with dragons, phoenixes or other animals, sometimes interspersed with floral patterns.[121]

A modern Chinese source writes, "Bronze was too expensive for all but high-ranking aristocrats in the ancient times, so bronze mirrors were found only in the homes of the well-to-do. For the ordinary people, water was probably the best mirror."[122] However, during the Han Dynasty, iron mirrors appeared, no doubt because they were less expensive.[123]

In China, the Han dynasty was followed by the "Six Dynasties." "From the Eastern Jin through the Liu Song (317–479), iron mirrors seem to have been favored in the north." This was attributable, at least in part, to "a shortage of copper" in northern China.[124]

There is evidence that the Japanese cast bronze mirrors as early as 200 BCE.[125] "In the sixteenth century, Jesuit missionaries observed that Japanese noblemen dressed in front of a mirror, and the late seventeenth century *Hagakure* told samurai to check their appearance before going out."[126] In Japan, mirrors were found not only in the home, but also in Shinto shrines and Buddhist temples.[127]

The most famous Chinese and Japanese mirrors are the "magic" or "light penetrating" mirrors, whose optics are discussed in Chapter 6.

Central Asia

The use of mirrors was not limited to the high civilizations of the ancient world. Scythians—a warlike nomadic people of central Asia—carried them as they wandered. The Scythians were noteworthy for their woman warriors, thought by some to be the basis of the Amazon legends of the Greeks. "A team of archaeologists investigating 2,400-year-old burial mounds built by the Scythian people on the upper River Don has found that five of twenty-one graves contained the bodies of young women with their weapons…. They are buried with womanly things—mirrors of silver and bronze; necklaces of gold, glass or clay; earrings; and sometimes a symbolic spindle," Dr. Gulyayev said. "But alongside these are weapons—a quiver, bow and arrows, and often two throwing spears."[128] Indeed, it has been asserted that all of the tribesmen, not just women, "carried a small mystical mirror on their belt at all times, and were buried with it upon death."[129]

Among another nomadic people, the Pazyryk, all "had mirrors—men, women, and children. The mirrors were made of wood, [and] bronze, or silver. They were always at their side, carried in a bag and hung from the belt. There is even a hole for it there. And when they died it was placed in the grave."[130]

Iron Mirrors in Europe and the Near East

According to Giles and Joy, mirrors were "prestige objects" in the "middle-late Iron Age (ca. 400 BC–43 AD)":

> The earliest mirrors from Britain and Ireland were made of iron, and date between the 4th and 1st centuries BC. Five examples are known from East Yorkshire: two from Arras, two from Wetwang Slack and one from Garton Slack. There are also three mirrors thought to date to the 1st century AD including an iron mirror handle from the Carlingwark Loch hoard, Kirkcudbrightshire and a fragment of iron mirror from a layer behind the hillfort ramparts at Maiden Castle. An iron mirror found in an inhumation on Lambay Island, Co. Dublin is also likely to be of 1st century AD date. Most of these examples consist of a roughly circular iron plate, once polished, varying between 165mm to 198mm diameter.[131]

The authors point out that we do not know whether these mirrors were used by the deceased in life or were postmortem gifts.

Despite the availability of iron, the use of bronze mirrors continued into the late Iron Age. Tsujita described sixty-two bronze mirrors found in "Britain, Ireland, the Netherlands and France dating from the Iron Age to the Early Roman Period."[132]

"Ferrous mirrors seem to have been reasonably common in the medieval and post-medieval Islamic world until superseded by glass mirrors imported from Europe from the 17th century AD on." At least some of these ferrous mirrors were made from crucible steel rather than wrought iron.[133]

Gutenberg's Pilgrim Mirrors

A 1487 woodcut shows two women and a child holding pilgrim mirrors.[134] Before making moveable type, Johann Gutenberg sought to corner the market on pilgrim's mirrors for the pilgrimage to Aachen in 1439. These small, convex metal mirrors were thought capable of collecting the holy emanations from the Aachen relics, and the Aachen goldsmiths knew from their experience during the last pilgrimage, in 1432, that they could not meet the demand: "over 10,000 people every day for a fortnight." Gutenberg's plan was to mass produce 32,000 mirrors, to be sold for half a gulden each. His costs would have been about 600 gulden, so he was looking at a profit of more than 2,600 percent.[135]

Unfortunately, Gutenberg didn't have the seed capital. In 1438, Hans Riffe agreed to put up the money, in return for two-thirds of the profits. Then two more craftsmen talked their way into the partnership; these promised to pay Gutenberg for instruction in the mysterious technique by which Gutenberg intended to mass-produce the mirrors. In the summer of 1438, the plague invaded Aachen, and the authorities decided to postpone the pilgrimage until 1440.[136] We do not know whether Gutenberg ever did make the pilgrim mirrors, but it hardly matters.

It appears that pilgrim badges were made of a lead-tin alloy. The "lead type" that he was to use with his printing press was a lead–tin–antimony alloy.[137]

"Speculum" Bronze

The word "speculum" comes from the Latin *specere*, to look at. The word was first applied to a surgical instrument (1598), then to a "burning mirror" (1650), and next to a telescope mirror (1704) (*Oxford English Dictionary*).

Adding tin to bronze increases its reflectivity.[138] Speculum metal is a "high-tin" bronze. Roman Britain mirrors with tin content as high as 27 percent are known.[139] Biringuccio in 1540 said that the "ancient" composition for concave "burning" mirrors was the same as that for bells: "three-quarters of copper and one of tin, and in order to make it somewhat lighter in color they added an eighteenth part of antimony." The tin also made the metal softer (thus easier to work into the required shape), but too much tin made it brittle.[140]

Newton's notebook from 1661 to 1665 gives a protocol for "metall for reflection."[141] His ratio of copper to tin is the same as in the "ancient" formula provided by Biringuccio. However, he also used arsenic (some of which was lost as fumes), as well as antimony.

Mills believes that the "Newton's Telescope" in the possession of the Royal Society is his first telescope, constructed in 1668, based on the diameter of the primary mirror. Chemical analysis of the mirror shows it to be a 3:1 copper:tin alloy, with 10 percent arsenic.[142]

In 1777, Mudge advocated a 2:1 copper:tin alloy, but without arsenic or antimony, for telescope mirrors.[143] (That is the usual modern proportion.[144])

Mills prepared a mirror with a "Newtonian" alloy (30 g copper, 10 g tin, 5 g arsenic); it had a reflectance of 50 percent for white light at 45 degrees incidence, as compared to 59 percent for a 2:1 copper–tin alloy. While the arsenic thus reduced reflectivity, it rendered the alloy less brittle.[145]

Ideally, the metal is "perfectly white." Increasing the copper content makes the metal yellower, and increasing the tin content, bluer. However, if there is too much tin, the alloy "becomes brittle and cannot be worked further."[146]

Vickers, surveying speculum compositions, commented, "It has been frequently attempted to increase the hardness of this metal by adding arsenic, antimony, and nickel; but with the exception of nickel, these additions have, however, an injurious effect, the specula readily losing their high luster, this being especially the case with a larger quantity of arsenic." He suggested a standard composition of 68.21 percent copper, 31.78 percent tin, without additives, but he acknowledged compositions with as much as 2 percent arsenic. The composition of the six-foot-diameter Ross telescope mirror was 70.24 percent copper, 29.11 percent tin, 0.38 percent zinc, 0.1 percent iron, 0.01 percent lead, and 0.01 percent nickel.[147]

For telescope mirrors, speculum metal was replaced by silvered and later aluminized glass. However, "at the village of Aranmula in Kerala in southern India, a unique mirror making tradition survives. Here, a cast high-tin bronze mirror of 33 percent tin ... is made which is comparable to, if not better than, modern mercury [sic] glass coated mirrors."[148]

Metal Mirror Casting

In *Pirotechnica* (1540), Vannoccio Biringuccio describes how to cast and polish a metal mirror. To make a flat mirror, he melted the alloy and poured the molten metal into a mold, typically three dita (inches) thick. The metal piece was removed from the mold and fastened to a board with plaster of Paris, pitch, or glue. Next, the metal was polished, using a millstone, or sand and water. Biringuccio warned the reader not to "continue to rub long in one direction." Scratches made by the coarse materials were removed with very fine emery or powdered pumice, placed on a woolen cloth. Then the mirror was dusted with "tripoli," ochre, or "calcined tin," and rubbed some more. Finally, the mirror was detached from the board and framed.

The manufacture of a concave mirror was similar, but one started with a concave mold, and the mirror was polished while still in the mold, which was turned on an axle like a potter's wheel.[149]

As Giambattista della Porta (1535–1615) pointed out, to make a concave mirror, you actually need two "forms," one defining the outside and the other the inside, with the thickness of the mirror being defined by the space between them. You would need some way to hold the two forms apart, too. Della Porta assumed use of a clay mold.[150]

The Science Museum in London "had a go at making speculum metal telescope mirrors." The metal was melted "in a chimney furnace at over 1000 degrees Celsius before pouring it into disc-shaped moulds." Their mirror was cast six inches in diameter, with "an almost imperceptibly shallow concave face—only about a millimetre deep in the centre." Hand-grinding and polishing the mirror to give it the proper shape took "over 20 hours."[151]

Chapter 2

Modern Mirrors

Glass and Beyond

"Silvered" Glass Mirrors

Introduction

In the mid-thirteenth-century *De Proprietatibus rerum*, Bartholomew Anglicus wrote, "no matter is more apt to make mirrors than is glass."[1]

In general, glass is used by itself as a mirror only when the goal is "beam splitting," that is, dividing the incident light into reflected and transmitted rays. That's because it's not very reflective; with thick glass, there is about 4 percent reflection from each surface. Hence, in Bartholomew's time, "mirrors be tempered with tin."[2] The glass had to be flat and, if the metal were on the back of the glass, as clear as possible.

In most modern mirrors, a reflective metal coating is applied to the front or rear surface of a substrate (glass or plastic). Household mirrors are rear-surface mirrors; the substrate protects the metal. However, there will be a "ghost" reflection from the substrate surface, and if light strikes the surface obliquely, it will be offset from the metallic reflection. Hence, front-surface mirrors are used in scientific applications.

The metal coating of a front-surface mirror may be overcoated with a thin layer of a transparent material (thereby protecting the metal surface from scratching and chemical action). Because the protective coating (usually a silicon oxide, or magnesium fluoride) is thin, ghosting is minimized. However, the overcoat does alter the reflective properties of the mirror. Depending on the thickness of the overcoat and its optical properties, it may enhance reflection at some wavelengths and reduce it at others.

While glassware has been made for millennia, it was only relatively late in the history of the craft that it became possible to make large panes of flat, clear glass. Once that was achieved, it was then necessary to develop techniques for covering it with metal. We look first at glassmaking.

Glass in Antiquity

Glass is formed when sand (silica), soda (alkali), and lime are fused at high temperatures. Glass can occur in nature; the strike of lightning on a sandy beach can create a glasslike rock, fulgurite.[3]

It is believed that glass was invented in Mesopotamia (present-day Iraq), as long ago as 2500 BCE. Glass was probably first made accidentally; perhaps, while firing pottery, the ceramic became glazed.[4] Glass vessels appeared in Mesopotamia and Egypt after 1500 BCE. In Egypt, only the pharaoh, the high priests, and the nobility owned glassware.[5]

Roman Glass and Mirror Making

Glassblowing was invented around 50 BCE, in the Roman Empire. It made possible the mass production of glass products. Seneca said, "a person finds himself poor and base ... unless his vaulted ceiling is covered with glass."[6] However, glass remained expensive. But by the time of the Emperor Gallienus (253–268 CE), glass was so common that he scorned it, drinking only from cups of gold.[7]

Lucius Annaeus Seneca wrote, in 65 CE, that "certain devices have come to light only within our own memory—such as the use of windows which admit the clear light through transparent tiles."[8] It is unclear whether he was thinking of glass or some other transparent material, such as thin sheets of mica or of alabaster.[9] Butti and Perlin are persuaded that he meant glass because the exact word he used for the material, *testa*, refers to a baked material.[10] "The Romans were capable of making fairly large panes of glass—at Pompeii, archaeologists found a slab that was 32" × 44", and ½" thick."[11]

The Romans had two techniques of making window glass. One method produced "panes of uneven thickness that are fire polished, or 'glossy' on one side, and pitted, with a matt finish on the other." While this type of glass has been termed "cast glass," it probably was not made by pouring glass into a mold because "it would not reproduce both the forms of the edges and corners and the tool marks seen on original Roman glass." Modern glassmakers have been able to reproduce the characteristics of the old Roman window glass by pouring molten glass onto a damp surface, flattening it with a large block of damp wood into a disc, and then pulling out corners so that it becomes square.[12]

The Romans ceased to use the cast glass method in the third century CE; it was superseded by the "cylindrical" method, which is discussed later, in the section on medieval glassmaking and mirror-making. The newer method "produces panes of even thickness which are glossy on both sides."

Silicon dioxide is the principal constituent of premodern glasses. A pure SiO_2 glass ("fused silica") is colorless and transparent. Impurities, such as iron oxide, can cause it to become colored. In the sixth century BCE, the eastern Mediterranean civilizations discovered that antimony decolorized the glass. Later, perhaps in the first century CE, the Romans discovered that manganese had the same effect. (The decolorization is not perfect; if you look at a modern windowpane edge on, it has a green cast, the residual effect of the iron ions.)[13] For a mirror glass, decolorizing was certainly advantageous; a colored glass would give a cast to the reflection. (This had been a problem with gold and copper mirrors, too.)

From clear window glass, of course, it is but a short step to mirror glass; all one needs is a method of applying metal to the surface. One problem with making glass mirrors from Roman window glass was that at least some of it "had a bluish-green tint and was not completely transparent."[14] However, it is not clear whether the tinted glass in question was the best the Romans could achieve. Pliny declared that "the most highly valued glass is colorless and transparent."[15] Examples can be found of transparent Roman glassware.[16]

Care must be taken in judging the transparency of the Roman glass from that of the surviving pieces. It is conceivable that, over two millennia, the glass that was once acceptably transparent has become discolored. A mid-first-century CE wall painting from Oplontis shows a glass bowl filled with fruit; some pieces can be seen both through the glass and directly. If the painting is true to its subject, then the glass, although a little frosted, was free of a color cast.[17]

Even if the Romans could not avoid a color cast, the issue was not whether their glass was equal in quality to modern window glass, but whether it was clear and flat enough so that a metal-backed glass would be competitive with the traditional metal mirrors for some purposes.

Some believe that the first reference to a metal-backed glass mirror is by Pliny the Elder (23–79 CE).[18] After discussing bronze and silver mirrors, Pliny adds, "more recently, a notion has arisen that the object is reflected with greater distinctness, by the application to the back of the mirror of a layer of gold."[19]

It was certainly within the capabilities of the Romans to apply a layer of gold to the back of a transparent glass, for example, by fastening gold foil to the glass, or by painting the glass with a mercury–gold amalgam such as that used in gilding. However, in the chapter in question, which is specific to mirrors, Pliny says nothing about glass.

Certainly, the modern mind recognizes that there would be no advantage to placing a layer of gold on the back of a silver (or bronze) mirror. However, it is a mistake to attribute modern scientific knowledge to the ancients. They had no sure idea of how light was reflected (Pliny attributed it to a "repercussion of the air") and, with no way of quantifying the reflection from a mirror, could well have fancied that a gold-backed silver mirror was superior to an "ordinary" silver one.

Nonetheless, there is a second, provocative passage in Pliny's *Natural History*. In discussing the various kinds of glass, and how they are made, he says, "Sidon was formerly famous for its glass-houses, for it was this place that first invented mirrors."[20] The implication is certainly that the mirrors were of glass, and if that was the case, a layer of gold could certainly have been applied to their back to increase reflectivity. However, Beckmann argued that the word *excogitaverat*, which Bostock translates as "invented," actually means just "thought of," and Beckmann "gives it as his opinion that attempts were made at Sidon to form glass mirrors, but that the experiments had not completely succeeded."[21]

Werner Ehlich insists that "glass mirrors were known in antiquity":

They were manufactured probably for use in fairly thicker [*sic*] glass panels in the glass factories of Sidon. Because a mercury layer was unknown at the time, they attached thin leaves of gold, silver, copper or tin to one side of the glass. Occasionally, these leaves were laid between two pieces of glass. Because the glass was not polished, it was very uneven so that the mirror certainly gave distorted reflections. Neither in the Greek nor in the Italian floors did the glass mirror appear to be in common use. All the same, there have been found however quite a few fragments in the Roman camps of Salzburg and in other cities, such as in Regensburg.[22]

Similarly, anthropologist Alan MacFarlane and glass historian Gerry Martin say:

The Romans knew how to make glass mirrors, yet metal mirrors were preferred. Archaeological investigations have uncovered only a few examples of the former. The glass was generally coated with tin or, more rarely, silver. It was used in hand mirrors but larger mirrors in which a man could see himself from head to toe have also been found.[23]

The last statement is remarkable if true; to show a whole person, the mirror would have to be at least half that individual's height, say 2½ to 3 feet.

In any event, glassware, glassworkers, and glassmaking techniques spread across the Roman Empire. "The glass industry of Northern Europe was established by Syrian glassmakers who settled in an area between the Seine and Rhine."[24]

Medieval Glassmaking and Mirror-Making

There were two medieval methods of making of large panes of glass. In the Crown glass ("Normandy") method, a bubble of glass was blown, cut open, and spun about. The spinning resulted in a circular pane. The glass was frequently reheated during this process, giving it a high polish (fire polish). The glass was cooled, and the workers first cut out the center (bulls eye), and then cut out straight pieces. A painting by Van Eyck (1385–1441), *The Marriage of Arnolfini* (1434), features a small convex mirror that was probably made in this way. Another convex mirror appears in Petrus Christus's *A Goldsmith in His Shop* (Bruges, 1449). Curiously, the mirror is showing an image of two young men outside, not the family indoors.[25]

This technique possibly dates back to late Roman times. Werner Ehrlich writes:

In the Roman Gaulic graves of Reims, mirrors were also found which originated in the 3rd or 4th centuries AD and were manufactured with a clearly different technique. They consisted of an hour-glass like structure, and arched and round glass pieces of 5 to 3 cm, with lead poured behind. The glass piece was probably cut from a glass balloon which was warmed in order to avoid breakage and then the lead was poured in. This mirror naturally also gave a distorted and miniaturized picture.

Generally, this mirror, (pic. 114) due to its small size was regarded as a toiletry article and was not a piece of furniture.[26]

Goldberg says that the Roman glass mirrors were convex and made from glass "blown as large bubbles ranging from ten to twenty-eight inches in diameter." The glass was allowed to cool partway, down to the melting point of lead, and then molten lead was poured in, coating the inside of the bubble.[27]

Another source says that crown glass was first produced in Ravenna after the fifth century CE.[28] Yet another theory is that it was developed in Syria around 650 CE, or even in Rouen, France, in 1330.[29] The last seems unlikely to this author.

In the Broadsheet ("Lorrainer") method, the glass was blown, then swung to form a long cylinder, a "sausage." The craftsmen cut off the ends, opened the cylinder lengthwise ("muffing"), placed it in a flattening oven, and polished it.[30] (It may have been cut once and unrolled, or cut twice, leaving two pieces.) The first written description of the process is in the treatise *On Divers Arts* (early twelfth century)[31] However, it, too, may have been of Roman origin.

In the late twentieth century, the window glass for the White House was still made from mouth-blown glass by the Lorrainer method. Each "sausage" of glass, when slit and ironed out, made an 18-inch by 25-inch pane. They contained "air bubbles and wavy bands," but the White House treasured the antique look.[32]

The Rise of Venice to Dominance of the Mirror Glass Industry

> It is a Venetian mirror from Murano, … compared with which the finest mirror of steel or silver is mere darkness.—George Eliot, *Romola*[33]

Glasshouses may have existed in Venice as early as the seventh century CE,[34] but the industry received several boosts in the thirteenth century. First, there was the fall of Constantinople (to errant Crusaders under Venetian influence) in 1204. This resulted in the decline of Byzantine competition, and in the migration of glassworkers from the Near East to Venice. Second, there was the development of the glassmakers guild, first documented in 1224.[35]

In 1271, the *fiolarii* (makers of glass vessels) signed a formal agreement with the Venetian Republic. This *Capitolare Fiiolariis* provided, significantly, that no glass would be imported into Venice and that no foreign masters would be allowed to make glass. Plainly, the Republic wanted to foster a homegrown glass industry and to avoid industrial espionage. In 1291, the Venetian government pressed the glassmakers' guild to move to island of Murano, ostensibly to eliminate the risk that a fire from their furnaces would burn down Venice, but probably also to heighten security.

Two centuries later, the Venetian industrial policy paid off handsomely. The key experiment began on February 21, 1457, when Angelo Barovier (Beroviero?) and Niccolo Mozetto were given permission to work on "crystal glass" outside the normal production season.[36] In 1460, they began commercial production of a colorless soda lime glass, *cristallo*, named after rock crystal. It was used in many products, notably lenses and—once the problem of applying a metal coating was resolved—mirrors.

This *cristallo* was far superior in clarity to the older glasses, and it sold well. What was its secret?

One possible factor was the source of the "glass former," silica. Usually, glassmakers obtained it from either quarried sand or crushed siliceous stones. These sands and stones varied in composition, depending on where they were collected. Ancient Egyptian glass, for example, was made from a sand with a high iron content, giving its glass a greenish cast.[37]

"Analysis of river pebbles from the Ticino River indicate that it was a source of almost pure silica making it ideal for glassmaking." Venetians were collecting river bed pebbles from the Ticino and the Adige at least by 1332, and Antonio Neri's *L'Arte Vetraria* (1612) confirms that Murano glassmakers preferred pebbles from the river Ticino.[38]

Another important glass ingredient is the modifier or flux. This was a salt that served to lower the melting point of the silica. The modifiers were usually salts of sodium, potassium or lead. The sodium salts were usually obtained from certain minerals (e.g., natron) or by burning seaweed (yielding "soda ash," sodium carbonate). The potassium salts were derived by combustion of hardwoods such as beech (yielding "potash," potassium carbonate).[39] (There is a reference to this aspect of glassmaking in Chaucer's "The Squire's Tale": "some said that it was / Wondrous to make fern-ashes into glass / Since glass is nothing like the ash of fern.")

One also needed a stabilizer, such as a calcium, magnesium, barium, or aluminum salt, to protect the glass from water. If this was not already present as an impurity in the silica (sand can be calcium-rich) or flux source (some plants are rich in magnesium), it had to be added. Roman glasses were "soda lime glasses," in which the silica was combined with sodium oxide (derived from soda ash) and calcium oxide (derived from lime).[40]

After the fall of Rome, northern Europe switched to potash as the flux, while the Venetians continued to use soda ash. William S. Ellis says that "the breakthrough was achieved with the use of a vegetable ash rich in potassium oxide and magnesium. The sale di vetro, or glass salts, extracted from the ash were used as the fluxing agent in the melting process, mixed with the sand to produce a sodium-potassium-based crystal-like glass."[41]

The particular vegetable ash used by the Venetians was *alume catino*, imported "by the end of the thirteenth century." It was derived from small bushes of the Levantine coast, especially Syria. The raw ash was subjected to purification by repeated crystallization, resulting in a material (*sal di cristallo*) richer in sodium and of much reduced iron content.[42]

Manganese was called "glassmaker's soap" because it "'washed' the glass of undesired tints caused by iron," and it was used in Venice at least by the thirteenth century.[43] Melchior-Bonnet says that the Venetian masters knew that ashes of kali, an Egyptian herb, "acted as a bleaching agent because of its low phosphorous content and richness in manganese."[44] However, Neri's treatise taught that the "manganese of Piedmont" was "the best."[45]

If glass was to be used to make *cristallo*, steps were taken to purify the glass melt (by removing "a sort of salt called sandever" that contained "chloride and sulfate impurities") and to homogenize it (by "maintaining the furnace at the highest possible temperature" for "4 to 6 days" and "stirring continually").[46]

The next major achievement came in 1507, when Andrea and Domenico de'Anzolo del Gallo realized that the *cristallo* could be given a highly reflective surface by hammering tin into thin sheets, amalgamating it with mercury, and then laying the sheets of *cristallo* onto the amalgam. The critical advantage of this tin amalgam process was that it did not require heat and therefore was unlikely to crack the glass through thermal shock. They petitioned the Venetian Council of Ten, claiming to be the possessors of "the secret of making mirrors of crystalline glass, a most valuable and singular thing … unknown to the whole world, except for one house in Germany and one in Flanders, who sell their mirrors at excessive prices."[47] (The reference to Germany is consistent with the report that Nuremberg craftsman had metallized mirrors in the twelfth century.[48]) They asked for 25 years of exclusivity but were awarded 20. This was enough to provide a formidable jump start to mirror-making in Venice, and the *specchiai*, or mirror makers, became so important that in 1569 they organized their own guild.[49]

Venetian mirrors, besides being products in their own right, were incorporated into Venetian wall lights. By the late seventeenth century, these usually had a mirror glass back plate to enhance illumination.[50]

At least from the fifteenth century on,[51] the Venetian Republic sought to keep glassworkers from taking their secrets elsewhere. The Venetians used both the carrot and the stick. On the one hand, the nobility and the glassmakers were allowed to intermarry without penalty.[52] (This was true, elsewhere, too; the coat of arms of the Counts Spiegel zum Desenberg was "Gules, three round mirrors argent in square frames or."[53])

On the other hand, the Venetians were adamant that glassmakers could not leave the Republic: "If any worker or artist should transport his talents to a foreign country and if he doesn't obey the order to return, all of his closest relatives will be put to prison and if, in spite of the imprisonment of his relatives, he stubbornly insists on remaining in the foreign country, an emissary will be sent to kill him and, after his death, his family will be set free."[54] There were also cases of glassmakers imprisoned on suspicion of intent to emigrate, and of glassmakers who had flown the coop receiving fake letters from their wives begging them to return, or being offered amnesty and travel expenses if they did.[55]

Despite all blandishments, threats, and punishments, glassmakers left Venice by the hundreds, spreading the knowledge of Venetian techniques.[56] The Venetian ambassador to England warned the Doge: "Various subjects of your Serenity, some outlaws who have taken refuge in this kingdom, where many natives of Murano may now be met, work at making looking-glasses or flint glass or teach how to make them."[57] In 1671, an English merchant warned his Venetian supplier of the consequences if a prior dispute were not satisfied: "You will discourage me for the future to send to you (or any other person at Venice) for any more Looking Glasses since we have so [many] good [ones] made here in England."[58]

But the real threat to Venetian hegemony was not from renegade Murano glassmakers who spread its technique, but from innovators in France who learned to make glass in a manner that made large mirrors possible, even affordable.

The French Development of Glass Casting

Bernard Perrot, an Orléans glassmaker, poured molten glass onto an iron plate covered with sand. The glass was rolled flat. After it cooled, it was ground and polished with iron disks, abrasive sands, and felt disks. He communicated his idea to the French Academy of Sciences on April 2, 1687. The French authorities declared, "he has invented a method, hitherto unknown, of casting glass into panels, the way one does with metal."[59] (He is reported to have practiced this method as early as 1673.[60])

However, a few years later, he was forced to shut down his glassworks, with a 500-pound annual pension as a consolation prize. The reason was to limit competition with the royal glassmakers.[61]

The first royal company had been founded in 1665, at the Faubourg Saint-Antoine, but it made mirrors from blown glass. By 1688, there was a competing company, at the Faubourg Saint-Germain, making mirrors from cast glass; its privilege was initially limited to mirrors at least forty by sixty inches. In 1693 it moved to Saint-Gobain. Finally, in 1695, Louis XIV dissolved both companies, and a new company, the Manufacture Royale des Glaces de France, was formed and given a monopoly on "glass for mirrors of all heights, lengths and widths."[62]

Foreign countries hired away French glassmakers, thereby slowly disseminating cast glass technology over the course of the eighteenth century. After 1763, the Manufacture Royale no longer produced any blown glass.[63]

Cast glass had to be further processed before it could be used to make mirrors. The glass surface facing the casting table was rough and had to be ground and polished.[64] "By today's standards ... the surface quality and flatness [of table cast glass] were poor.... Consequently it was necessary to cast at a much greater thickness than that of the polished plate. and low overall yield and high labor requirements made costs high."[65]

Plate Glass Manufacture in Modern Times

Most mirrors are rear-surface mirrors—the metal coating is behind the glass and thereby protected from scratches and oxidation. For such mirrors, it is important that the glass be clear, flat and free of distortion.

The old cylinder process was rejuvenated by the development (1903) of methods of mechanically blowing the cylinders; they could be "up to 13.4m long and 1m in diameter." But the cylinders had to be "split and flattened which was costly and harmful to the surface." Fourcault (1914) drew glass up "vertically in a ribbon from a bath of molten glass." These sheet glass processes did not need grinding and polishing, but they introduced distortions into the glass.[66]

"Just after the First World War the Bicheroux process was introduced." The glass was "rolled into a sheet between mechanical rollers.... This made a smoother sheet." Next came "a process which successfully combined a continuous melting furnace with the continuous rolling of a ribbon of glass." Attention then turned to

Manufacture of cast glass for mirrors. A. The cubical cuvette, containing molten glass, is skimmed to remove impurities. It will then be hoisted by the crane into a position above the casting table. The crane is controlled by a hand crank. Note the roller at one end of the table. (From Plate volume 21, Manufacture des Glaces, Plate XXIII, "vignette," ARTFL Encyclopedia Project (University of Chicago), Autumn 2022 digital edition of Diderot, Denis, and d'Alembert, Jean le Rond, *Encyclopédie, ou dictionnaire raisonné des sciences, des arts et des métiers* [Le Breton, 1771].) B. This shows "the table accompanied by the various tools & instruments used for the operations the vignette [C] represents." (From Plate XXIV, bottom.) C. This shows "the pour and roll operation." Around the table we see the pourers (1, 2), the rollers (3, 4), the *teneurs de main* who lay the "hands" (guides) on either side of the roller track to keep the glass from spreading off the table, (5, 6), the *grapineurs* who remove "tears" (semivitrified matter) from the glass stream (7, 8), the *grapineurs* who will "detach the 'hands' after the glass has been poured" (9, 10), the hand crank operator (11), the *tiseur* who wipes the table with his broom (12), and the workers with a cart to transport the bowl back to the kiln (13). (From Plate XXIV, top. Quoted text translated by DeepL, and some text based on the discussion of glass manufacture in Hunt [1867], 2: 471–73.)

automating the grinding and polishing, and a "twin grinder" ("which could grind the ribbon of glass on both sides simultaneously") came into service in 1935.[67]

What truly revolutionized the plate glass industry was the float glass process, conceived in 1952 by Alastair Pilkington (1920–1995). "In the float process, a continuous ribbon of glass moves out of the melting furnace and floats along the surface of an enclosed bath of molten tin." This is carried out under a nitrogen–hydrogen atmosphere, to prevent oxidation of the tin, and "at a high enough temperature for a long enough time for the irregularities to melt out and for the surfaces to become flat and parallel. Because the surface of the molten tin is dead flat, the glass also becomes flat." By the time the ribbon reaches the rollers, it has already cooled and hardened enough so that the rollers do not mark the bottom surface, so there is no need for grinding and polishing.[68]

Pilkington reports that it took "seven years and four million pounds to make any saleable glass," despite the company having given "the project the highest possible priority." After succeeding in making 6.5-mm float glass (the equilibrium thickness), Pilkington developed ways to make the ribbon thinner or thicker.[69]

Foiling with Tin, Lead, and Mercury

Now we should look at the development of the metallizing process. In the twelfth or thirteenth centuries, Nuremberg craftsmen introduced molten lead or tin into still hot, blown glass bulbs, which were rotated until covered with a film of metal; when cool, they were cut to size.[70]

In 1280 CE, the monk John Peckham, of Oxford, England, allegedly described such a mirror.[71] A generation later, Dante, in the *Inferno*, Canto XXIII, had Virgil say:

> Were I a pane of leaded glass, I could not
> summon your outward look more instantly into myself,
> than I do your inner thought.[72]

If the glass were made by the Normandy method, the mirror was convex. But glass could also be made in flat sheets by the Lorrainer method, and these could be coated with lead or a lead–antimony mixture.[73]

Some sources report that tin–mercury amalgams were used as mirror coatings as early as 1317.[74] The Venetians were the first to apply such amalgams to *cristallo*. According to a conservationist, "mirrors made of glass backed with a reflective coating of tin amalgam first came into general use in the sixteenth century. Production ceased around 1900."

The Neapolitan scholar Giambattista della Porta claimed to have seen how looking glasses were made in Venice. In his *Magia Naturalis* (1589), he explained that the artisan would make a tin foil "that is level and thin, as perfectly as he can." He then wiped this with "quicksilver [mercury] by the means of a Hare's foot, that it may appear all as silver." A "fair white paper" was placed over this. The glass was cleaned and polished, so the foil would stick to it. The glass was laid over the foil, and pressed down with one hand, as the other pulled away the paper. Finally, he laid "a weight upon it for some hours."[75]

In the *Dictionarium Polygraphicum* (1735), we find these similar directions for "foiling":

A thin blotting paper is spread on a table, and sprinkled with fine chalk; and then a fine lamina or leaf of tin, called foil, is laid over the paper; upon this mercury is poured, which is equally to be distributed over the leaf with a hare's foot or cotton. Over the leaf is laid a clean paper, and over that the glass plate.

The glass plate is pressed down with the right hand, and the paper is drawn gently out with the left; which being done, the plate is covered with a thicker paper, and loaden with a greater weight, that the superfluous mercury may be driven out, and the tin adhere more closely to the glass.

When it is dried, the weight is removed, and the looking-glass is complete.[76]

A nineteenth-century source describes some additional refinements. The table, made of stone, has a raised edge all around, and a gutter connecting to a spout. It rests on a central axle so it may be inclined by a hand screw to an angle of as much as 12 to 13 degrees. It is initially horizontal. Glass rules are placed at the edges of the tin foil to prevent the mercury from spreading too far, and the mercury is poured onto the foil "sufficient to form everywhere a layer about the thickness of a crown piece." The paper and glass are laid over it, as previously described, but care is taken "so that neither air, nor any coat of oxide on the mercury, can remain beneath the plate." The plate is weighed down and the table given a "gentle slope" so loose mercury runs off into the gutter and spout. Five minutes later, the glass is covered with flannel and weighed down more heavily. Over the next 24 hours, the inclination of the table is gradually increased. The sheets of tin foil are larger than the glass plate, and the excess is pared off. The glass is then removed from the stone table and laid on "a wooden one sloped like a reading desk." Over the course of eighteen days to a month, depending on the size of the mirror, the slope of the glass is gradually increased, until it is finally vertical.[77]

The foiling of a concave or convex glass required special procedures. To make a concave mirror, one first made a convex plaster mold of the inside of the glass and marked it so it could be placed back exactly where it was. The mold was removed, and tin foil was spread over its convex surface. The glass was placed in a box, cushioned by a sandbag, and its concavity was filled with mercury. The mold was dipped into the mercury in such a way as to slowly displace the excess, which ran into the box and from there, through a spout, into a leather collection bag. After half an hour, the collection bag was removed, and the "whole is cautiously inverted." The box is lifted off, and the mold is placed in a second box, which has a central convex support "nearly equal in diameter to the mould." The "whole is left in this position for two or three days." The tin foil is pared off the mold with a knife, and "the glass is lifted up with its interior coating of tin-amalgam."[78]

To make a convex mirror, one first makes a concave plaster mold. The mold is lined with tin foil and then filled with mercury. The convex glass is immersed in the mercury, and a sandbag laid on top of it, expelling excess mercury. "The whole is inverted" and placed on a support, and weights are placed on the mold. For making large convex mirrors, the mold was a "circular piece of new linen cloth of close texture," of "twice the diameter of the mirror," suspended on a frame.[79]

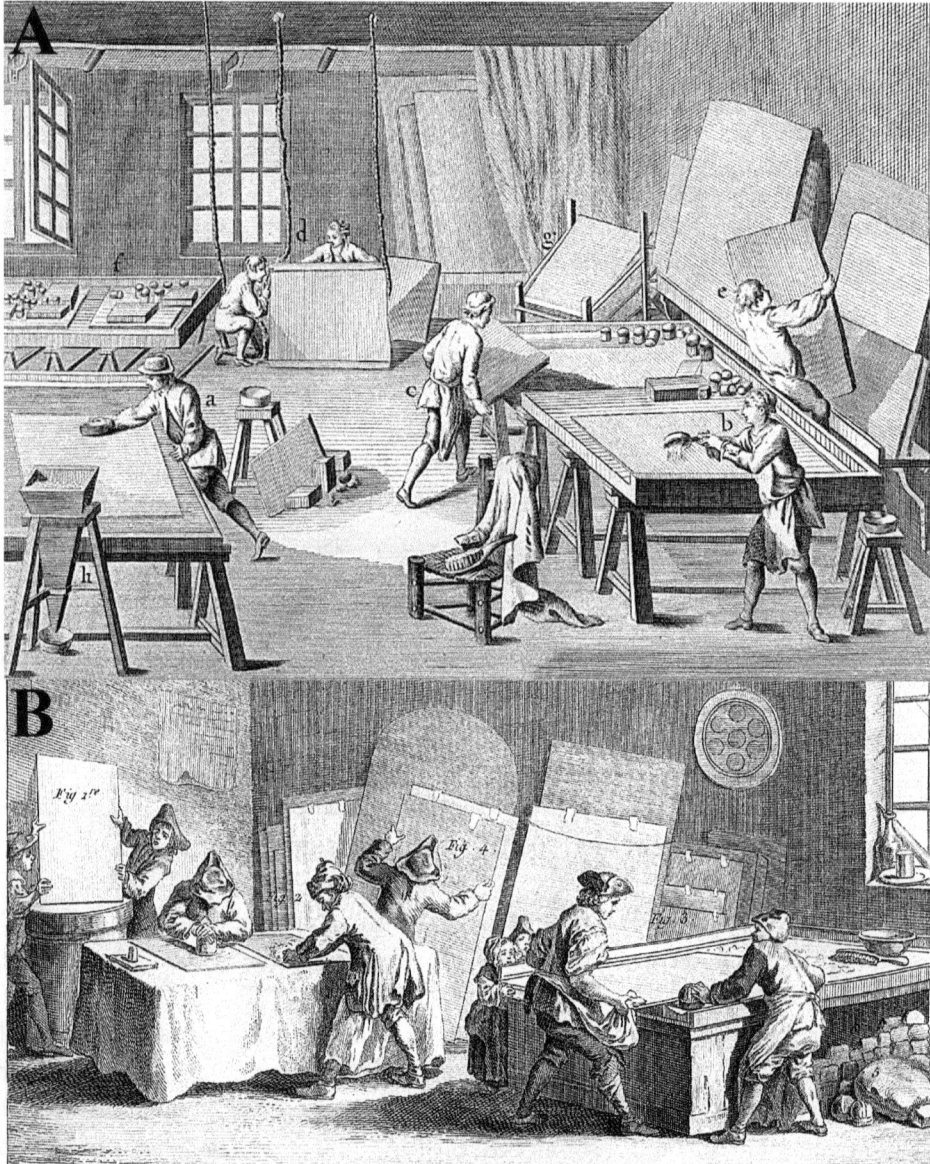

Applying the tin to the mirror. A. "Miroitier Metteur au Teint," Volume 25, Plate I from Diderot (1771), supra. This shows mirror-makers at work. "One at a is degreasing the tin [*teint*]; one at b is pouring the quicksilver onto the sheet of tin; one at c is placing the glass [*glaces*] on the same sheet of tin; others at d are placing the glasses on the drip tray; another at e is arranging the glasses to be tinned [*au teint*] at the bottom of the workshop. At f is a table where several loaded [weighted down?] mirrors have just been tinned. Opposite at g is a draining rack where the mirrors are placed. On the front, at h, is a hopper for separating quicksilver from waste."B. "Miroitier Metteur au Teint," Volume 25, "Vignette." "Fig. 1. Workers spreading glass on a stone [sic]. 2. Workers cleaning a sheet of tin. 3. Mirror-makers who tin a mirror. 4. Glass placed against the wall to drain. 5. Stones and balls for weighting down glass. (Captions translated by DeepL and edited by author. Both images are from *The Encyclopedia of Diderot & d'Alembert*, Collaborative Translation Project [https://quod.lib. umich.edu/d/did, CC BY-NC-ND 3.0].)

While nowadays the use of mercury would be of grave concern (conservators warn amateurs against resilvering old mirrors for fear of mercury poisoning), contemporaneous commentators had other worries. Sir Henry Bessemer remarked that the mercury/tin-backed mirrors had "a bluish or leaden hue that was most unfavourable to the fair sex, and spoiled the best complexion."[80] (Tin oxides, a degradation product of the amalgam, are bluish.)[81]

Early Looking-Glass Economics

How much did historical looking glasses cost? This is not as easy to judge as one would like, as the sources are not always clear whether the quoted price is for the silvered glass alone or covers the frame as well. (A solid silver frame, or one of heavily gilded wood, might cost more than the "glass" itself!) Also, mirror cost would certainly vary, not necessarily proportionately, with size.

One looking glass, two feet by four feet, and framed in hand-wrought silver, was valued in the late 1600s at three times the worth of a painting by Raphael.[82] In 1680, the cost of a 3-foot by 4-foot mirror was $40,000 in today's currency.[83]

By the eighteenth century, the cost had fallen dramatically. According to the *Plate Glass Book* (1757), for a large sheet of mirror glass, measuring 60" × 42.5", the charges were rough plate, 37 pounds, 10 shillings, 0 pence; excise tax, 18/15/0; grinding, 7/12/8; polishing, 7/12/8; silvering, 7/12/8; diamond cutting, 2/14/0; for a total of 81 pounds, 17 shillings, and 0 pence.[84] Adding in a profit for the retailer, it is clear that such a mirror would cost more than one hundred pounds. While this was still a sizeable sum, it did not require a royal budget.

In 1735, an overmantel looking-glass, of unstated size, was sold by John Belchier to Mrs. Elizabeth Purefoy for 3 pounds and 16 shillings.[85] In 1769, Thomas Chippendale sold two plates, 74 by 44 inches, at 69 pounds 10 shillings each. He also sold four 74 × 26's at 35 pounds each, and four 74 × 13's at 15 pounds 5 shillings each. This was all sold to the Earl of Mansfield, and a volume discount may have been made.[86] In 1773, he earned 160 pounds for the sale of a "fine" looking glass, 91 × 57½ inches. At the other end of the scale, in 1772 he sold servants' dressing glasses for as little as 4 shillings apiece.

The cost of glass and frame was broken down by Samuel Norman. In 1759, he supplied the Duke of Bedford with an ornate, gilt oval frame for 97 pounds 10 shillings, and with a plate for 65 pounds 3 shillings. The size was 49 × 38 inches. The same duke, in 1760, paid William Norman 183 pounds 4 shillings for a 76-inch by 44-inch plate.[87] Another bill of sale informs us that in 1769 William Twaddel supplied 18 plates of looking glass, 27½ × 17½ inches, for a total of 19 pounds 7 shillings.[88]

What does all this mean in modern terms? In Nell Gwyn's time (the late seventeenth century), a personal maid was paid 12–15 pounds a year, plus room and board (and perhaps cast-off clothes). Nell Gwyn's own "income" was 7,000 pounds in 1675; her biographer says that one pound then was equivalent to $20 in 1952.[89]

The improved economics naturally fostered the dissemination of mirrors. "It has been shown that between 1675 and 1725 ownership of mirrors (based on the inventory samples analyzed) rose from 58 percent to 80 percent in London, whilst in provincial towns the rise was from 36 percent to 74 percent."[90]

"Diamond Dust" Mirrors

Fiction set in the old American West will often describe a saloon or hotel as having a "diamond dust" mirror. Such mirrors adorned the dining room of the 1875 Hotel de Paris in Georgetown, Colorado,[91] rooms in Denver's Windsor Hotel in the 1880s,[92] and the Board of Trade Saloon in late-nineteenth-century Leadville.[93] They could also be found in private homes; James Clay installed one in the drawing room at Ashland in 1856.[94]

Sorry to disappoint, but these "diamond dust" mirrors do not contain diamond dust. Rather, they are mercury amalgam mirrors. "The amalgam is formed as the mercury reacts with the tin to form a layer of crystals in a liquid mercury-rich matrix. As the amalgam ages, the *crystals grow*, and mercury from the matrix evaporates and/or is released in liquid form, damaging the reflective surface and speeding the corrosion of the tin into structureless tin oxide ... a *sparkling effect* can result as the mercury evaporates and drains, and the spaces between the metallic crystals enlarge" [emphasis added].[95]

If diamond dust had been used as a mirror backing, you would not see an image, any more than you would on snow. A particulate surface scatters light as a result of variations in multiple reflection, depending on just how each ray strikes each crystal.

Silvering

As a youth, Justus von Liebig (1803–1873) experimented with explosive compositions, and this led to his expulsion from high school. His schoolmaster labeled Justus as "hopelessly useless." Nonetheless, by age 21 he was appointed as assistant professor of chemistry at the University of Giessen.[96]

In 1835, Liebig discovered that if ammoniacal solution of silver nitrate were heated with an aldehyde, it would be reduced to elemental silver, which could be detected when it was deposited on glass. At the time, he saw this "as a simple test for the presence of aldehydes in organic materials." It was not until 1856 that Liebig considered using a variation of this process in mirror making.[97]

There are many variations on the Liebig process, but in all of them, a solution of a silver salt is used as a source of silver ions. A reducing agent converts the silver ions to neutral silver atoms, and the latter are deposited on the glass.

In 1844, Thomas Drayton was awarded a U.S. patent on "silvering looking glasses"; it taught the use of oil of cassia or oil of cloves at room temperature as the reducing agent.[98] Unfortunately, "small specks" soon appeared "which eventually destroyed the mirror."[99]

In 1861, Cimeg patented[100] another glass-silvering process in which the ammoniacal silver nitrate solution was boiled, and the condensing steam was reduced with

Rochelle salt, the sodium–potassium salt of tartaric acid. This deposit method is slow; it may take an hour to form a thick film. For this reason, this "hot" or "Rochelle salt" process became favored for making "one-way" mirrors, which have a thin and therefore semi-reflective coating.[101]

In 1848, Drayton received a British patent which recited the use of "saccharine matter, as grape sugar" [glucose].[102] The dominant method for use for silvering telescope and other front surface mirrors came to be the Brashear process (1895), which used a mixture of nitric acid and "tablet or granulated sugar" [sucrose].[103]

An improvement on the basic method is to sensitize the glass so it more readily accepts the metal. This is usually done by "tinning": treating the glass with a dilute stannous chloride solution.[104]

Originally, silvering solutions were poured onto the glass. However, they can be sprayed on, instead. Typically, two jets are used, one supplying the ammoniacal silver nitrate and the other a fast-acting reducing agent such as hydroxylamine sulfate.[105]

Colored mirrors can be obtained just by silvering colored glass. However, reduction techniques were adapted for depositing gold (reduced gold chloride) and copper (reduced copper sulfate) on glass.[106]

Silvering eventually replaced foiling with a tin–mercury amalgam, but it was a long, drawn-out process. Mercury–tin mirrors continued to be manufactured until the early twentieth century.[107] James observes that "very high" wages were paid to Bavarian workers in that trade, as recompense for the health risks they endured, and says that they regarded "with hatred and disfavor all those who undertook to ameliorate their condition."[108]

Alternatives to Silver

In 1864, the French chemist Edouard Dodé patented a process of platinizing glass. Platinum chloride was suspended in lavender oil, mixed with litharge [lead oxide] and lead borate, and painted on the glass. "At a red heat the litharge and borate of lead are fused and cause the adhesion of the platinum to the softened glass." The process took "four to eight hours," as compared to "fifteen days" for the tinfoil process.[109]

Hans Christian Oersted discovered aluminum in 1825. He obtained it by heating dry aluminum chloride with potassium metal. In 1886, Charles Martin Hall of Oberlin, Ohio, and Paul Héroult of France, who were both twenty-two years old at the time, independently discovered the modern electrochemical process for obtaining aluminum from the aluminum oxide of bauxite.

Aluminizing of glass is typically achieved by sputtering. Grove observed metal deposits sputtered from a glow discharge in 1852. Sputtering occurs when a solid target is bombarded with high energy ions. Atoms are ejected, creating a vapor phase. The ejected atoms may be collected and deposited as a thin film on another substrate. Film deposition by arc vaporization in vacuum was first reported in 1877 by Wright. He used it to form mirrors.[110]

The thickness of the metal coating is typically less than 100 nanometers; a human hair is 2,000 times thicker (200 microns). Thicker coatings have greater surface roughness and therefore lose light by scattering.

Any metal, not just aluminum, can be applied to glass by vacuum deposition. The cost of the metal is much less important than with solid metal mirrors since so little is needed for the ultrathin coating. However, aluminum is still preferred because it is relatively inexpensive, corrosion-resistant, hard, and strongly adhering.[111]

For visible light reflection, the usual choice is aluminum or silver (more reflective but more expensive and readily tarnished). For reflection of near ultraviolet light, the clear choice is aluminum.

Gold-coated mirrors are the most commonly used infrared mirrors. Gold transmits a substantial amount of visible light, so it can be used on the window of a transparent furnace so one can observe the interior while being shielded from most of the heat. Gold does not adhere well to glass, and the solution is to apply a chrome layer first.[112]

Rhodium is a fairly good reflector of visible and infrared light, although not as good as aluminum. Rhodium's advantages are that it is highly corrosion resistant (unlike aluminum and silver) and exceptionally hard (an important consideration for a front surface mirror in an abrasive environment).

Other Coated Mirror Substrates

If a reflective metal coating can be applied to glass, it can be applied to other lightweight materials, too. This substrate must, of course, be capable of providing an optically smooth surface and, if the mirror is to be convex or concave, of taking and holding the desired shape. If the coating is on the rear surface of the substrate, the latter must also be transparent.

Gartner, a Dresden mechanic, produced "burning mirrors" (concave mirrors) made by gilding wood. The gilded wood was not a mere frame; it provided the actual reflective surface. A 1755 writer criticized his approach: The wood readily caught fire and the gold leaf easily deteriorated.[113]

Peter Hoesen, also of Dresden, made a more successful use of a wood substrate. "Hoesen carved a parabolic shape from a skeleton made of cross members of very durable wood, and lined the inner concave surface with strips of brass sheet metal measuring 5 ft. 5 in. By 2 ft 8 in." Using a Hoesen mirror of five feet diameter, "copper ore melted in one second, lead melted in the blink of an eye, asbestos changed to a yellowish-green glass after only three seconds, and slate became a black, glassy material in 12 seconds." Hoesen's reflectors were as large as 10 feet in diameter and yet were so cleverly mounted that they could be reoriented with one hand.[114]

In modern times, the two most important alternative substrates have been quartz and plastic. Quartz, like glass, is made of silicon dioxide, but it is crystalline in form. Quartz is important mostly because of its use as a "blank" in making primary mirrors for very large reflecting telescopes; its superior heat stability is important.

The Society of Plastic Engineers says that a plastic is "a natural or synthetic substance which is pliable or moldable at some stage during its forming or manufacturing."[115] Plastics are usually polymers, that is, chemical molecules made of many copies of a repeating unit (the monomer) connected together; in a linear polymer, they are like strings of beads. The repeating unit can be the same throughout the molecule (a homopolymer), or there can be a small number of different monomers involved (a heteropolymer). Naturally occurring polymers include DNA and proteins.

The first plastic—Parkesine—was intended to be a rubber substitute. It was exhibited by Alexander Parkes at the 1862 Great International Exhibition in London. It was not a commercial success. Celluloid was created in 1866 by John Wesley Hyatt. Allegedly, he spilled a bottle of collodion and was impressed by the toughness and flexibility of the dried deposit. Celluloid's first use was as a replacement for elephant ivory in billiard balls and piano keyboards. In 1907, Leo Baekeland synthesized Bakelite resin, a plastic which, once hardened, couldn't be melted down and reformed.[116]

Two of the most important plastic backings for mirrors are acrylic and polyester. "Acrylic" is a nickname for polymethyl methacrylate, introduced by ICI in 1936. Under the trademarks Plexiglas (Rohm & Haas) and Lucite, it was used to build aircraft canopies during World War II. Acrylic mirrors are usually heat-formed into the desired shape and then silvered. Acrylic weighs about half as much as glass and is more flexible. For a given thickness, it is fourteen times more shatter-resistant. However, it is less rigid than glass, hence less suitable for making large mirrors.[117]

In 1941, Rex Whinfield and James Dickson created polyethylene terephthalate, better known, after the war, as polyester, Dacron, and terylene.[118] Mylar polyester films were developed by Du Pont in 1952. They are flexible, and metallized Mylar films can be stretched over a lightweight frame to create a very large reflective surface of any desired shape.

An alternative to the polyester film is cloth. National Products, Inc., of Louisville, Kentucky, marketed "Flex-sheet" mirrors, in which small squares, rectangles, triangles, and so on of metallized glass are mounted on a cotton cloth.[119]

Metallized Glass Versus All Metal Mirrors

Why did glass mirrors, which could break, displace polished metal mirrors, which couldn't? One possible explanation is that glass sheets could be made in larger sizes than could metal sheets. Unfortunately, that cannot be the answer. At the second millennium BCE tomb of Rekhmire, vizier to Pharaoh Thutmose III, there is a tomb painting showing the casting of a pair of large bronze temple doors.[120] Those doors were certainly larger than the eighteenth-century glass mirrors at Versailles, let alone the glass mirrors of the Renaissance. On Cyprus, "ox hide" ingots—copper slabs weighing up to thirty kilograms—have been found.[121] If one were formed into a sheet 1 centimeter thick, its dimensions could be 50 centimeters (20 inches) by 80 centimeters (32 inches). That would make a respectable size mirror. More impressively, during the Shang dynasty (1400–1100 BCE) of China, craftsmen created a bronze cauldron which weighs 875 kilograms.[122] It is, of course, a royal artifact, but nonetheless demonstrates that the ancients could craft very large pieces of bronze.

In Roman times, even silver was formed into massive pieces. Pliny speaks of banqueting couches and carriages plated entirely with silver, and of chargers (plates) of silver weighing as much as five hundred pounds.[123]

Another theory is that polished metal mirrors, especially those of great size, were too expensive. Again, this is not likely. Considering how extravagantly the pharaohs provided for their own afterlife, would they really begrudge the expense? As for others, while gold and silver were costly, copper was much less so, and bronze was used in many commonplace items. Moreover, glass mirrors were extraordinarily expensive for several centuries after they were first made.

The likeliest explanation relates to the mechanical and chemical properties of the mirrors. For example, there may have been some concern over the resistance of the mirror to twisting and bending. Metal mirrors must be quite thick to maintain the necessary rigidity; without it, the image is distorted. However, relatively thin glass has good bending and torsional (twisting) stability.[124]

The stiffness of a material is measured by Young's modulus. When the material is formed into a flat panel, its structural efficiency (contribution to stiffness of a unit mass of the material) is the cube root of the Young's modulus divided by the density.[125] The structural efficiency of crown glass is 1.6, while the best of the metals used in ancient mirrors gave 0.53 (copper and bronze).[126]

Density is also important in its own right. Metal mirrors were much heavier than glass mirrors of equal dimensions; copper is almost four times denser than glass.

There were other problems with metal mirrors. Copper and gold mirrors did not reproduce colors faithfully. Silver mirrors did but were subject to tarnish. Iron mirrors rust. Glass was clear and did not tarnish or rust.

Metal Plating

In many cases, metal objects, ranging from jewelry to bicycles, are plated with other metals. Usually, the bulk metal is cheap and strong, and the plate metal is decorative, or scratch, acid, and oxidant resistant. Generally speaking, the plating offers bright specular reflections which add to the object's aesthetic appeal. Hence, even though the plating need not be of a mirror per se, it seems appropriate to touch upon plating methods in this work.

Over the years, various plating methods have been developed. If the metal plate were sufficiently malleable, like gold, it could be hammered on to the underlying metal. In French plating, the latter was heated first. Several other methods used heat to facilitate the joinder of the metals. One could cover the base metal (say, copper) with another (say, tin) having a lower melting point. The composite would be heated sufficiently for the latter metal to melt and diffuse ("weep") into the underlying base metal. In a variation called "close plating," the tin was used as an intermediate layer to effect adhesion between the bulk metal and another, such as silver.[127]

In "fire gilding," known since Roman times, a paste-like gold amalgam (a solution of gold in mercury) was prepared and painted onto a metal. Or mercury was applied first, and then gold leaf was brushed on top (the "vermeil" method).[128] (The use of mercury in gilding is described by Pliny, but there is some dispute as to the particulars of the use of heat.) Either way, the mercury was then evaporated away (much to the disadvantage of the artisans) by heating the metal.[129] It took 100 kilograms of gold and eight years (1835–1843) to "fire gild" the 21.83-meter-diameter copper dome of St. Isaacs in St. Petersburg.[130] "It is reported that some sixty craftsmen died from the resultant mercury poisoning."[131] Fire gilding yields the best gold coatings, but it is rarely used nowadays because of the toxicity of the mercury vapors.[132]

In an alternative method, "water gilding," the base metal was immersed in a solution of gold chloride. Or the metal was polished with the ashes of a rag impregnated with a gold chloride solution ("friction gilding"). Unfortunately, only a very small amount of the gold was transferred out of the solution.[133]

Finally, and most importantly, one metal could be electrodeposited onto another. Electrodeposition is the use of an electric current to cause a solution of a metallic salt to deposit the metal on a surface. It was made possible by the voltaic pile (a crude battery) developed by Allisandro Volta in 1799. In 1806, Luigi V. Brugnatelli gilded two large silver medals by this technique.[134] However, as a result of a run-in with Napoleon Bonaparte, Brugnatelli kept a low profile, and the technique languished in obscurity for many years.[135]

The electrodeposition of nickel, lead, iron, copper, tin, zinc, silver, antimony, bismuth and manganese was reported by Golding Bird in 1837, and that of platinum and cobalt by the Count of Ruolz-Montchal in 1841.[136] (Applying the technique to yet other metals, as suitable salts became available, was obvious.) Inventors then turned to finding which salt of a particular metal was the best choice as the source of the deposit. Thus, R. Bottger in 1843 advocated use of nickel ammonium sulfate in nickel plating.

Gold, silver, and copper were deposited onto base metals for decorative purposes. In England, in the 1840s, the Elkingtons and their employees discovered how to obtain an even coating and popularized silver electroplating of jewelry, candlesticks, and flatware.[137]

The purpose of precious metal plating was to preserve attractiveness, while maintaining strength and minimizing costs. Other metals were plated as protection for the underlying substrate; nickel, chromium and rhodium are particularly important in this context.

Zinc is very reactive and hence is never found as a native metal. The first use of zinc was unintentional; the Egyptians made copper artifacts from copper ore contaminated by zinc, and hence inadvertently produced a copper–zinc alloy, brass. Today, zinc is most likely to be associated with mirrors through its use in "galvanized iron or steel"; a galvanized metal is one coated with zinc, usually electrolytically. Galvanization was patented in 1837.

A "little round zinc mirror" is mentioned in the early-twentieth-century novel *The Groceryman and Peck's Bad Boy*.[138] In it, the Bad Boy inspects his eyebrows, which were singed off by a mishap involving fireworks. Two dozen zinc mirrors, costing 5 shillings, were among the trade goods which Australian Graham Officer brought with him to the Solomon Islands in 1901.[139]

Nickel is named after "Old Nick"; medieval miners mistook nicollite (NiAs) ore for copper and were dismayed by its arsenic-mediated effect on their health.[140] It was identified as an element by Cronstedt in 1751 and obtained in pure form in 1804 by Richter. Nickel's great advantage was that it did not rust. It was also hard, strong, and difficult to melt.

The commercialization of nickel plating in the 1870s was the work of Isaac Adams, Jr. (1836–1911) and Edward Weston (1850–1936). Weston received over 300 patents, including many for refinements in the art of electroplating. The bicycle was one of the commercial articles which benefited from nickel plating; the brass handlebars, spokes, and rims were so treated. For example, the nickel-plated 1883 Columbia Expert was the first bike to be ridden across the North American continent.[141]

The merits of nickel were such that "nickel-plated" became a metaphor, meaning of top quality. Thus, the New York, Chicago & St. Louis Railroad Company was nicknamed the "Nickel Plate Road." In Oz, the Tin Woodman, after becoming Emperor of the Winkies, was literally nickel plated, an imperial "perk."[142]

Chromium, which is the element responsible for the color of emeralds, was discovered in 1789. It can be given a mirror surface by polishing (its reflectivity for visible light is about 70 percent). Chromium is extremely hard and corrosion resistant. If the outermost layer of chrome plate is oxidized, the resulting chromium oxide is itself transparent and protects the rest of the plate from further oxidation. When used as decoration, the chrome coat is usually very thin (0.13–1.3 microns; the diameter of human hair is 17–181 microns[143]). When used industrially to protect a bulk metal, the coating is thicker (a few microns up to 500 microns). Decorative chrome is usually applied over nickel.[144]

In 1927, electroplated chromium (over nickel) was commercially used for the first time, on an auto bumper.[145] In 1929, Cadillac became the first car maker to make chrome plating the standard finish for bright parts.[146] The 1934 Elgin Blackhawk bicycle, manufactured by Columbia, had chrome "everywhere."[147]

Rhodium was discovered in 1803 by Wollaston. It reflects 75–80 percent of visible light and is extremely hard and corrosion-resistant. In the 1930s, coatings of rhodium were used to protect silver from tarnishing.[148] Typically, rhodium plating is extremely thin (0.5 to 2.5 microns thick), which is not surprising since rhodium costs more than gold.[149] Rhodium-plated mirrors are used by dentists because their reflective surfaces are not damaged by autoclaving.[150] Rhodium-coated silicon mirrors are even used in high-intensity X-ray focusing systems.[151]

Exotic Metal Mirrors

Today, solid metal mirrors are used mainly when handling very-high-intensity beams. The metal then has the advantages of high thermal conductivity (so it can dissipate heat quickly) and high melting point (so the remaining heat does not deform it). Copper is favored for this work. Mirrors used for this purpose sometimes have channels cut into them to allow the circulation of coolant.[152]

Beryllium Mirrors

Beryllium mirrors are something of a special case. Beryllium, an alkaline earth element, was discovered in 1798, but the pure metal was not isolated until 1828. Its principal ore is beryl. Beryllium combines high stiffness with low density; its structural efficiency as a flat panel is higher than glass! It is also very hard and corrosion resistant. Of course, it is also much more expensive than glass. Beryllium can also irritate the skin, or cause pulmonary disease if beryllium dust is inhaled, which complicates the fabrication process.

Liquid Metal Mirrors

Metal mirrors that are liquid at room temperature exist, but their principal use has been in telescopes, so they are discussed in detail in Chapter 3. However, we briefly note here that they may be made of any of the following materials: (1) mercury, (2) gallium or a gallium alloy, or (3) a fluid supporting a reflective metallic film. In 2021, it was reported that a gallium-based mirror could be switched electrically between a state in which it reflects specularly (a light ray is reflected in just one direction) and one in which it is reflected diffusely (the light ray is scattered in multiple directions). The electric current, applied in one direction, causes "a reversible chemical reaction, which oxidizes the liquid metal in a process that changes the liquid's volume in such a way that many small scratches on the surface are created, which causes light to scatter." If the current is reversed, "the liquid metal's surface tension removes the scratches, returning it to a clean reflective mirror state."[153]

As we show in Chapter 3, if the mirror is of the third type, and the fluid is a ferrofluid (a colloidal suspension of ferromagnetic or ferrimagnetic nanoparticles), its surface can be deformed at will by applying magnetic fields. Thus, we have a mirror which can change its shape.

The Quest for the "Perfect" Mirror

A metal mirror may reflect up to 90–95 percent of the light that hits it, depending on the metal, its thickness, and the wavelengths of the incident light. Unless the mirror is extremely thin, the rest of the light is absorbed. For most purposes, that is good enough. However, if your device uses multiple mirrors, the 5–10 percent losses per reflection compound, like interest on an unpaid credit card balance, resulting in poor performance. With five successive 90 percent reflections, you end with just a little over 50 percent of what you started with.

Metal mirrors are particularly problematic for laser use; when light is reflected back and forth hundreds of times between two mirrors, those mirrors better be extremely reflective.

There is a solution—actually, two solutions, each with its own limitations. First, if light inside a prism strikes a face at a sufficiently oblique angle, it will undergo "total internal reflection." This is explained in the introduction to the next part of this book. But "total internal reflection" won't help with laser design.

Second, it is possible to make an extremely reflective mirror out of carefully chosen and arranged layers of transparent materials, without any metal at all. To understand why that is so, we need to explain the wave theory of light and its implications. We do that in Chapter 7.

PART II

Mirrors at Work and Play

CHAPTER 3

Extended Vision

The Science of Mirror Use

There is some basic science that it is fruitful to set out here, before we explore how mirrors are used.

Specular Versus Diffuse Reflection

Specular (mirror-like) reflection is possible if light strikes a smooth surface. When light strikes a microscopically rough surface, there is diffuse reflection; the light is scattered in random directions, depending on the orientation of each "facet" of the surface relative to the incident light.

Reflection and Refraction

When a light ray strikes a surface, it may be reflected, transmitted, or absorbed … or some combination thereof. The path of the reflected or transmitted light depends on the angle at which it strikes the surface—the angle of incidence. This is defined so that if the incident light is perpendicular ("normal") to the surface, the angle of incidence is zero. The angle of reflection is similarly defined.

It has been known since antiquity that the angle of reflection equals the angle of incidence, and that the incident ray, the reflected ray, and the "normal" (the line perpendicular to the surface at the point of incidence) must lie in the same plane.

We may also define an angle of *deflection*, between the incident and reflected rays. It follows, from the "law of reflection" just described, that this is twice the angle of incidence. This is exploited in numerous optical devices.

If the angle of incidence is not zero, the transmitted ray is bent ("refracted"). The ratio of the sine of the angle of incidence to the sine of the angle of refraction is a constant for a surface between two given media (e.g., air and glass)—this is called "Snell's Law." It was ultimately determined that each medium could be assigned what was called a refractive index, and the ratio of sines was then equal to the ratio of the refractive index of the second medium to the refractive index of the first. The higher the refractive index, the slower is the speed of light in the medium, and high refractive index materials are sometimes said to be "optically dense" or "optically thick."

The refractive index of light in a given medium is color (wavelength) dependent, and in most materials, blue light is refracted more than red light—this is why a simple triangular prism, which refracts light twice, breaks ("disperses") white light into its component colors.

If the incident light ray is in the medium with the lower refractive index (e.g., in air, attempting to enter glass), then the reflection is said to be external, and if it is in the medium with the higher refractive index (e.g., in glass, attempting to enter air), it is said to be internal.

In the latter case, if the angle of incidence is greater than the "critical angle," all of the light is reflected—this is called total internal reflection.

Why does total internal reflection occur? It follows from Snell's law. In the internal reflection scenario, the ratio of the second index to the first, by definition, is less than one. So there will be some angle of incidence, less than 90 degrees, at which the angle of refraction is 90 degrees (along the interface, so, no light transmission), making the sine of the angle of refraction be unity, and the sine of the angle of incidence equal the ratio of the indexes.

As to why Snell's law is correct, that is a consequence of the wavelike properties of light, but for the purposes of this book, we don't need to explain how one follows from the other.

Total internal reflection is exploited by "reflective prisms" and in fiber optics. In fiber optics, a flexible fiber (originally glass or quartz) conducts light inside itself by virtue of multiple total internal reflections. However, with bare fibers, the light would escape if the fibers touched, if the fiber were scratched, or if a fiber came in contact with a material having a higher refractive index. Fiber optics was made more practical by cladding the fibers in a protective sheath of a clear material with a refractive index lower than that of the fiber proper.[1]

Mirror Shapes

A mirror may be flat, concave, or convex. Its effect on light is dictated by the angle of incidence of the light and the law of reflection.

A flat mirror forms an image that is upright and virtual. A virtual image can be seen in the mirror, and appears to lie behind it, but it cannot be projected on a screen or wall.

A convex mirror causes the light rays to diverge; it forms an image that is upright, smaller than the object, and virtual. Because it provides a wide field of view, it is used in stores to spot shoplifters and at the mouths of driveways to show oncoming traffic.

A concave mirror causes the light rays to converge on its focal point and then diverge. For a spherical concave mirror, the focal point is halfway between the mirror and its center of curvature. Assume the reflected object is on the optical axis (line of symmetry). If it is closer than the focal point, its image will be upright and virtual; if it is further away, its image will be inverted and real. At the focal point, there is no image at all. If the object is closer than the center of curvature, its image will be magnified, and if further away, it will be reduced.

A concave parabolic (properly, paraboloid) mirror does not have a constant radius of curvature. Incident light rays, parallel to the axis of symmetry, are reflected to the focal point. For this reason, paraboloid mirrors may be used in solar collectors and solar furnaces (see Chapter 4). Or, if a light source is placed at the focal point, a collimated (axis parallel) reflected beam is generated. This may be exploited in flashlights, headlights, searchlights, and the like. They are also used in some telescopes.

With an ellipsoidal mirror, if you put a light source at one focal point, the light will be reflected to the other focal point. A hyperboloidal mirror will reflect light from one focal point so that the rays appear to come from the other focal point. Both are used in telescopes.[2]

Retroreflection

A mirror will reflect light straight back to the source if the incident light is at a zero angle of incidence. Sometimes, we want to achieve a similar "retroreflection" even though the angle of incidence isn't zero.

A limited form of retroreflection is achieved by placing two mirrors at a 90-degree angle to each other. If the incident ray is perpendicular to the "hinge" connecting the mirrors, it will be doubly reflected, and the second reflected ray will be parallel to but opposite the incident ray.

A "interior" corner cube reflector—three mirrors, each at a 90-degree angle to the other two—can retroreflect any ray striking one of the three mirrors. A panel consisting of 100 corner cube reflectors was left on the moon by Apollo 11 in 1969. By targeting it with a laser and measuring the time for the light to return to Earth, the Earth–Moon distance could be calculated.[3]

Retroreflection may also be achieved with a transparent sphere. It is the result of a combination of an incoming refraction, an internal reflection, and an outgoing refraction.

Retroreflective beads have been used in "the production of reflecting motion picture screens, reflecting road signs and road markers and reflecting advertising signs." Rudolf and Paul Potters began manufacturing such beads in 1914.[4]

For retroreflection to work, you want a bead that is as close to spherical as possible. In 1943, Rudolf Potters received a patent on an improved method. Potters acknowledged that "it has long been known that a molten material such as lead or glass when dropped from a height will take on a spherical shape. This fact has been made use of in the manufacture of lead shot and in the manufacture of large sizes of glass beads, such as in the making of marbles." However, this conventional method had problems producing small-diameter beads; it was difficult to make the droplets small enough, and "in many instances such extremely small droplets are blown out by air currents to such an extent that the particles adhere together."[5]

Potters's patent claim recited dropping irregularly shaped glass particles of roughly the desired size into a flame. A regulated draft keeps the particles suspended in the flame until their surfaces are melted into a spherical shape and ultimately carries them upward, out of the heat zone.

Retroreflection is maximized for a transparent sphere in air if its refractive index is 2. However, a lower refractive index material may be used because it's cheaper (recycled window glass, 1.5) or because you want to spread out the retroreflection (for highways, the incident light is coming from the headlights, but you want the retroreflected light to reach the driver's eyes). Rain reduces the effectiveness of the beads, but not of corner cube reflectors.[6]

For some applications, half of the sphere is coated with aluminum. This increases reflectivity, but only if one can ensure that the sphere is on the surface of its substrate, with the aluminum side down.

Reflective Prisms

Reflective prisms are often used as substitutes for conventional mirrors, and we cover them, as mirror equivalents, in this book. An optical prism is a highly transparent object "with polished plane faces where the light is reflected or refracted," and in which at least two of the faces "are not parallel."[7] It may have three, four, five or more such faces. Prisms may be used to disperse or reflect light; we're interested here in reflective prisms. The efficiency of reflection may be increased by "silvering" the reflected surface, or by choice of a geometry resulting in total internal reflection.

The advantage of a prism over a mirror system is that you don't have to worry about the reflective surfaces moving out of alignment. However, prisms are heavier and, in larger sizes, more expensive than their mirror equivalents.[8] In addition, with glass prisms, there is significant absorption of ultraviolet and infrared light.[9]

Reflectance and Reflectivity

The reflectance of a surface is the fraction of the light energy that is reflected, relative to the total incident light energy. Normally, only specular reflection is considered, but it is possible to also include diffuse reflection. The reflectivity is the reflectance when the material is thick enough that a further increase in thickness has no effect. It is wavelength specific, and the figure on page 200 shows the variation in reflectivity at normal incidence for several metals.[10]

Literature values for the same material may vary, depending on how the surface was prepared, how smooth and clean it was, how the surface was altered by chemical reactions with the environment, the spectral distribution of the light source, and the measurement conditions (angle of incidence, detector sensitivity, etc.).

Beam Splitters (Transflectors)

Sometimes we want to split a beam of light into two separate beams. When light strikes a transparent material, it is partially reflected and partially transmitted. The relative intensity of reflection and transmission depends on the angle of incidence and whether the light is polarized. (But we ignore polarization until Chapter 7.)

Probably the most common use of a plain glass as a mirror is in a diagonal mirror (angle of incidence 45 degrees). The total reflection would be about 10 percent of the light.[11]

We can increase the reflectivity by applying a uniform but thin coat of a metal like silver or aluminum to the front surface. (If we used a thick coat, we would get all reflection and no transmission, and therefore no beam splitting.) Depending on the thickness, the mirror might reflect anywhere from one-third to two-thirds of the incident light. However, "a mirror which has a surface coating of silver sufficiently thick to transmit twice as efficiently as it reflects will absorb about 30 per cent of the light that is incident upon it." In 1874, du Hauron proposed that instead of using a mirror with a uniform but thin coat, one use a mirror "composed of alternate areas that were completely transmitting and completely reflecting." The ratio of these areas would determine the balance between reflection and transmission. The key insight is that where it is completely reflecting, because the silver coat is thick, the absorption is small, "not more than 5 per cent." The silvering may be deposited in bands, or in a checkerboard pattern.[12]

A quite different approach to beam splitting was presented by Brewster's "Swiss cheese mirror." This was a metal mirror in which holes had been drilled, so some light would be transmitted.[13]

Ghost Reflections from Glass

A pane of glass of course has two surfaces, front and rear, and reflections will occur at both surfaces. Let's assume the two surfaces are perfectly flat and parallel. If the angle of incidence is nonzero, then as a result of a combination of an incoming refraction, rear surface reflection, and outgoing reflection, the rear reflected ray will emerge from the glass at the same angle but at a slightly different point from the front reflected ray. So we get two images, not one; their offset depends on the incidence angle and the thickness of the glass.

There are a number of ways of addressing this. If the angle of incidence is fixed for our application, we can make the mirror slightly wedge-shaped in cross section, just enough to compensate for the refraction. Or we can color the glass so it absorbs the light passing through the front surface. We can also make the transparent material very thin, so the offset created by refraction is very small. (Early pellicle mirrors were about 0.005 inches thick and "made of cellulose, celluloid, or cellophane that has been stretched taut in a frame."[14]) Finally, we can apply an antireflective coating to the rear surface.

Wave Optics

Light also has wavelike properties. These are a bit more complicated to explain, and I defer the discussion of those properties, and the inventions that depend on them, to Chapter 7.

Periscopes (and Polemoscopes)

The periscope, as its name implies (*peri*, around + *scope*, to see) allows you to look around a corner. It is fundamentally a device for seeing without being seen.

The simplest "true" periscope is just two mirrors, at opposite ends of a tube, mounted so that they are parallel to each other and at a 45-degree angle to the axis of the tube.

Nonmilitary Periscopes

A passage in the *Huai Nan Wan Bi Shu* ("The Ten Thousand Infallible Arts of the Prince of Huai Nan"), written during the Han dynasty (second century BCE), declares, "suspend a large mirror high up, put a basin of water underneath it, and you can see the people around you."[15] An illustration shows the mirror is actually offset from the basin and inclined partly toward the target.[16]

In the Western world, Lucretius (99–55 BCE) may have been aware of the periscope arrangement. In *De Rerum Naturae*, he wrote:

Things which are hiding behind a mirror, or in some tortuous recess, however out of the way.
Are all brought out by the repeated reflection;
It is the play of the mirror which brings them to light.[17]

In the *Catoptrica* sometimes attributed to the first-century (CE) engineer Hero of Alexandria (doubters say "pseudo-Hero"), we find, as part of a discussion of the utility of mirrors: "For who will not deem it very useful that we should be able to observe, on occasion, while remaining inside our own house, how many people there are on the street and what they are doing?"[18]

Hero proposed to accomplish this with a single mirror. For that to work, the mirror had to be at or above the level of the top of a high or tall window and facing downward. What you saw would depend on where you stood, and you would be effectively looking obliquely downward at a passerby (see figure).

However, Hero also provided a diagram which shows how an observer can look into a mirror and see not himself but a statue. This diagram shows two mirrors, with the lower mirror inclined to face upward and to the left, and the upper mirror inclined to face downward and to the right.[19] Hero's periscope was a building fixture, not a tube-like portable device, used to present an "apparition" of a statue hidden from the direct sight of a person. However, if the positions of the person and the statue were reversed, the two mirrors would have been acting just like a periscope.

Another periscope innovator was certainly Giambattista della Porta. In *Magia Naturalis*, he describes the use of one or more mirrors as "Glasses" to see around an obstacle.

"In plain Glasses those things that are done afar off, and in other places."
So may a man secretly see, and without suspicion, what is done afar off, and in other places, which otherwise cannot be done. But you must be careful in setting your glasses. Let there be a place appointed in a house or elsewhere, where you may see anything, and set a glass right over against your window, or hose, that may be toward your face, and let it be set straight up if need

were, or fastened to the wall, moving it here and there. Inclining it till it reflect right against the place. Which you shall attain by looking on it, and coming toward it. And if it be difficult, you cannot mistake, if you use a Quadrant or some such instrument. And let it be se [*sic*, "set"] perpendicular upon a line, that cuts the angle of reflection, and incidence of the lines, and you shall clearly see what is done in that place. So it will happen also in diverse places.[20]

Della Porta recognized that you might need a second mirror to see over an obstacle.

Hence it is, that if one glass will not do it well, you may do the same by more glasses. Or if the visible object be lost by too great a distance, or taken away by walls or mountains coming between, moreover, you shall fit another glass just against the former, upon a right line, which may divide the right angle, or else it will not be done, and you shall see the place you desire.[21]

By "divide the right angle," I believe della Porta

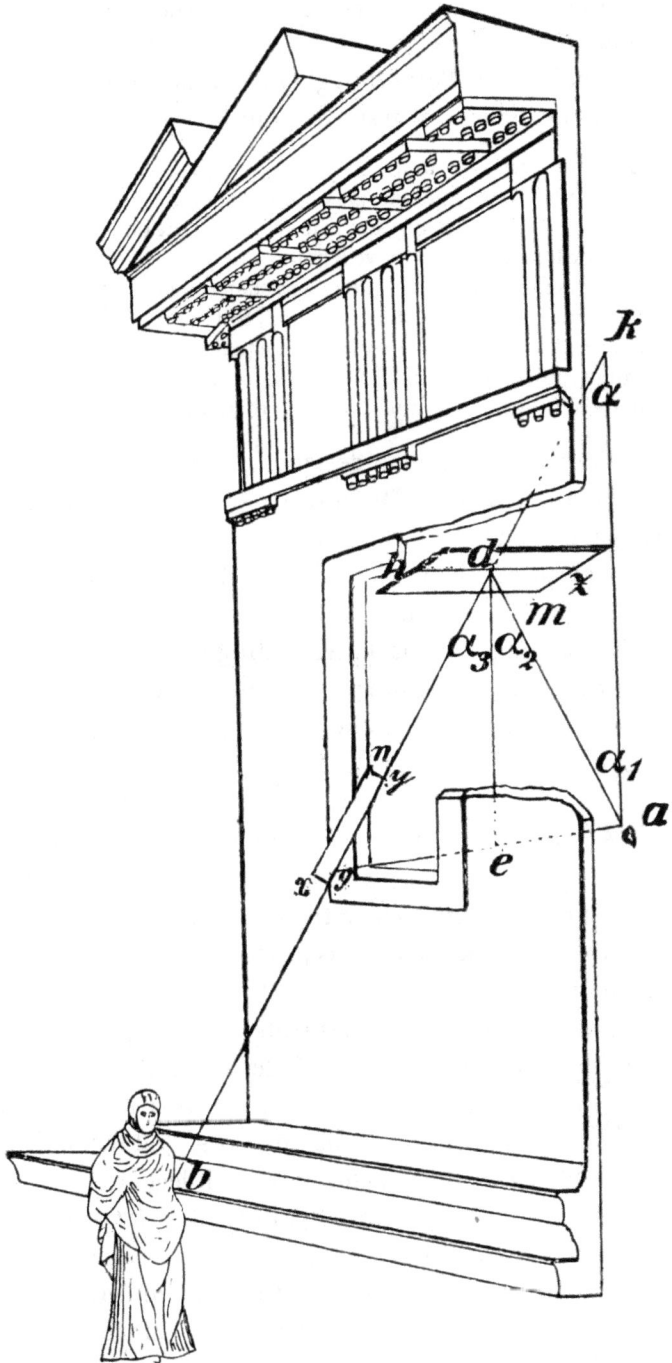

Hero's setup for seeing the street. Fig. 89 (354), in Catoptrica XVI, from Nix, L., and W. Schmidt, *Heronis Alexandrini Opera Quae Supersunt Omnia, Vol. II, Fasc. I, Mechanic et Catoptrica* (G. Teubneri, Leipzig, 1900). The original Greek text of *Catoptrica* is lost. What we have is a thirteenth-century Latin translation believed to have been made by William of Moerbeke and then ascribed to Ptolemy with the title *De Speculis*; see Cohen and Drabkin 261 note 3.

meant placement at a 45-degree angle to the line of sight from the first mirror to the second.

Finally, della Porta thought of a periscope with an adjustable mirror, by which you could alter what you spied upon:

> But if otherwise you desire to see any high place, or that stands upright, and your eye cannot discern it. Fit two Looking-glasses together longways, as I said, and fasten one upon the top of a post or wall, that it may stand above it, and the object may stand right against it. The other to a cord, that you may move it handsomely when you please, and that it may make with the first sometimes a blunt, sometimes a sharp angle, as need requires, until the line of the thing seen may be refracted [sic, "reflected"] by the middle of the second glass to your sight, and the angles of reflection and incidence be equal. And if you seek to see high things, raise it. If low things, pull it down, till it beat back upon your sight, and shall you behold it. If you hold one of them in your hand, and look upon that, it will be more easily done.[22]

Emmanuel Maignan designed a periscope for Cardinal Spada; constructed in 1645, it allowed Spada "to observe, from within a room with closed windows who was passing on the street below."[23]

Nowadays, the principal nonmilitary use of the periscope is as a toy. However, adults use them, too. Nowhere in the world is the periscope more popular than in Ærøskøbing, Denmark. According to Rick Steves, periscopes are everywhere: "Notice the 'snooping mirrors' on the houses. With these little neighborhood periscopes, antique locals may be following your every move."[24]

Military Periscopes

A periscope certainly comes in handy when raising your head could mean getting shot at. Hence, the periscope found use in land battles. Burke Davis asserts that the first use of a periscope in trench warfare was in the American Civil War.[25] However, the trench periscope was proposed centuries earlier by the Polish astronomer Johannes Hevelius (1611–1687).[26] Hevelius, an astronomer, had extraordinary eyesight; his naked-eye measurements of stellar positions proved to be as accurate as those that Edmond Halley (of Halley's Comet fame) made with telescopes. I do not know for sure why he was interested in what came to be known as "ditch mirrors," "rampart mirrors," and "trench mirrors," but he did live through the Thirty Years War (1618–1648), which included several sieges.

Hevelius's *Selenographia sive lunae descriptio* described a device he called a *polemoscopium* in 1647; he claimed to have invented it in 1637. The term "polemoscope" is from Greek polemos "war" + Latin scopium "to see," and Hevelius suggested using it during a siege to see over a rampart what the enemy is doing. However, a single mirror would not permit that unless you were willing to hold the polemoscope above your head and look upward through the tube.

A careful study of Hevelius's drawings and text suggests that his *polemoscopium* was a modular device that could employ one or two mirrors, as well as lenses and sectional extensions. In his Fig. B, we find a right-angled piece with a diagonal flat mirror inside, two linear tubes, and a linear piece with a diagonal mirror and a facing

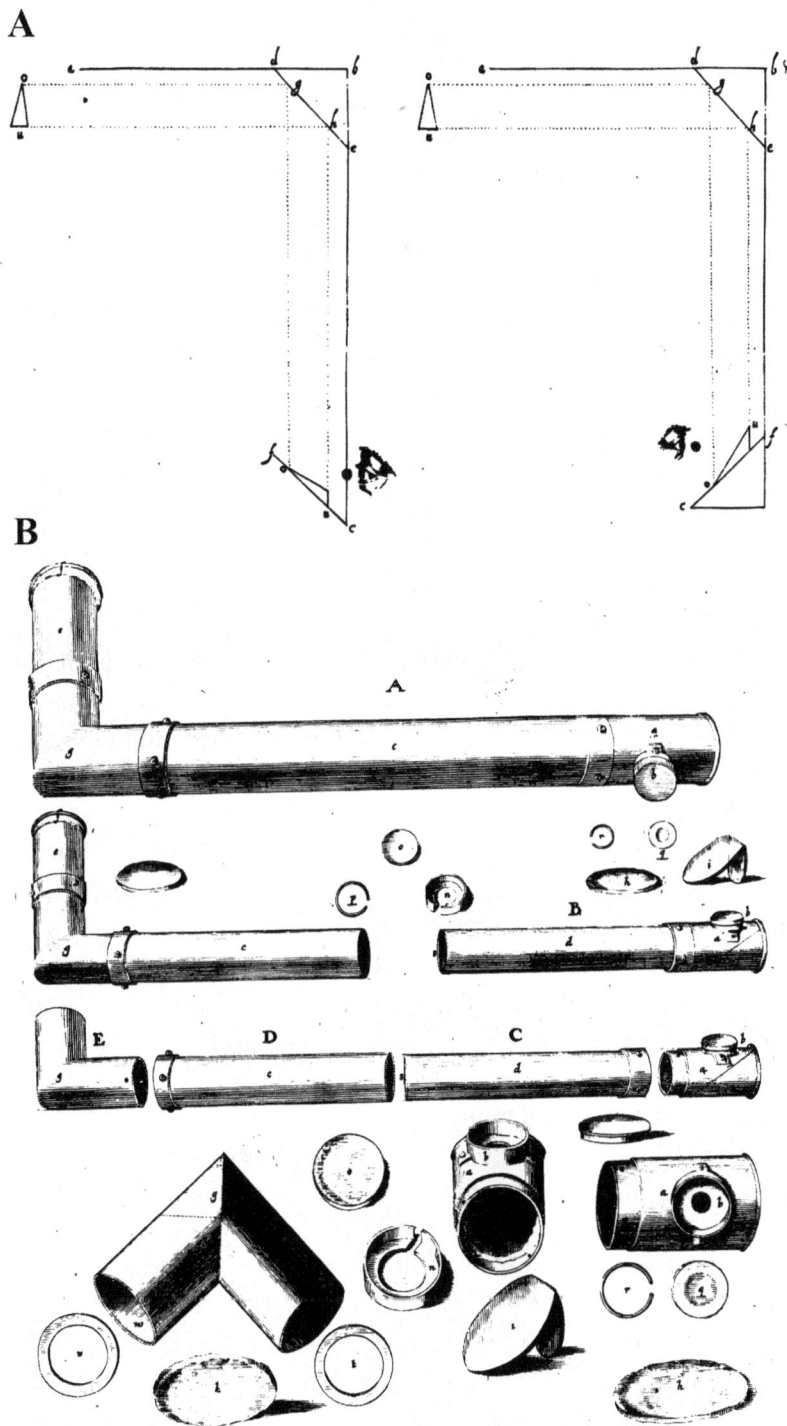

A. Operation of two-mirror polemoscopium. Fig. C, page 28a, Hevelius, Johannes, *Selenographia sive lunae descriptio...* (1647). Note that the left, facing forward configuration presents an erect image and the right, facing backward configuration, an inverted image. B. "Polemoscopium" with accessories. Fig. C, page 26a, supra.

side opening at one end. My interpretation of Fig. B is that one could use as many of the linear tubes as needed to provide the required length, and one could choose whether to employ the "eye end" mirror or leave it off. Hevelius's Fig. D shows a clear two-mirror operation with two different arrangements: object in front of and above the observer's eye, and object behind observer and above his eye.

In modern usage, a periscope has two mirrors; a polemoscope has just one. The latter allows the user to watch what is happening to his side while seemingly looking forward. Polemoscopes were used at the theater (so one could covertly watch a fellow guest while seeming to examine the action on stage), and in that context were also called "side opera glasses." They have also been used by private detectives and by photographers.

In 1902, William Youlten obtained a patent on a "hyposcope," a periscope adapted to be attached to a rifle. In one embodiment, the vertical tube was attached to the rifle butt, with the objective tube sighted just above the barrel, and the eyepiece tube well below. Since that meant that the butt would not serve its normal purpose, there was a separate "skeleton butt" that would be braced against the shooter's shoulder.[27] The following year, he received a quite favorable review from the Small Arms Committee at the Royal Small Arms Factory in Woolwich. It suggested "that a small number be purchased and issued to command for trials," but at the time it was envisioned that it would be an "addition to the equipment of permanent fortifications."[28]

In 1914, Youlten received two British patents on improved versions of his hyposcope, and "during the First World War, this device was advertised for private purchase" as an attachment to the new "short magazine Lee-Enfield" (SMLE) rifle.[29] A somewhat similar device was patented by J. Chandler in 1915, but its mirrors were not held within tubes but rather by brackets.[30]

Necessity is the mother of invention, and several soldiers in the trenches improvised their own "periscope rifles."[31] It is perhaps worth noting that depending on the fine details of the construction, one could be superior to another in terms of ease of construction, mounting and demounting, stability, adjustability, and so forth. At Gallipoli, on May 19, 1915, Lance Corporal William Beech of the New South Wales 2nd Battalion jury-rigged a periscope mount for his rifle and told Major Blamey that it was "an arrangement so that you can hit without being hit." Impressed, Blamey brought Beech to headquarters at Anzac Cove; his system was tested and found to work; and Beech set up a crude workshop for manufacturing it in large numbers. The mounts were made from packing boxes, and glass for the periscopes was scavenged from the naval vessels. The contraption was accurate to "200 or 300 yards."[32]

By the end of World War I, the French, Germans, and Americans (and probably other combatant nations) had manufactured periscope rifles.[33] They do not appear to have been used much in World War II, but in 2023–2024, several soldiers in Ukraine improvised their own "periscope rifles."[34]

Surface Navy Periscopes

The American Civil War did witness the first use of the periscope by the surface navy. This innovation was credited to Chief Engineer Thomas Doughty of the Union ironclad USS *Osage*:

> During the campaign of the Red River, while he was serving aboard the monitor Osage, Confederate cavalry, from the banks of the river, kept up a steady series of surprise attacks upon the Union vessels which had no way of seeing over the banks. This led Doughty to seek

A. U.S. Army periscope conversion of the Springfield rifle. "The periscope is so mounted that the shooter can stand below the lip of the trench parapet, protected from the overhead fire of shrapnel, which is not true of all periscopes. Also it is so hung that the recoil of the rifle swings the lens away from the eye, instead of pushing the optic all over the face as is the case with some periscopes. An extension enables the trigger to be pulled from the level of the shooter." (Crossman, Edward C., "Indirect Fire from Springfields," *Popular Science*, March 1918, 388.) B. Schematic of World War I periscope. The 2:1 gearing is "to compensate for the difference of azimuth between the top rotating prism and the lower prism which is fixed." (Talbot, Frederick A., *Submarines: Their Mechanism and Operation* [William Heinemann, 1915], 89.)

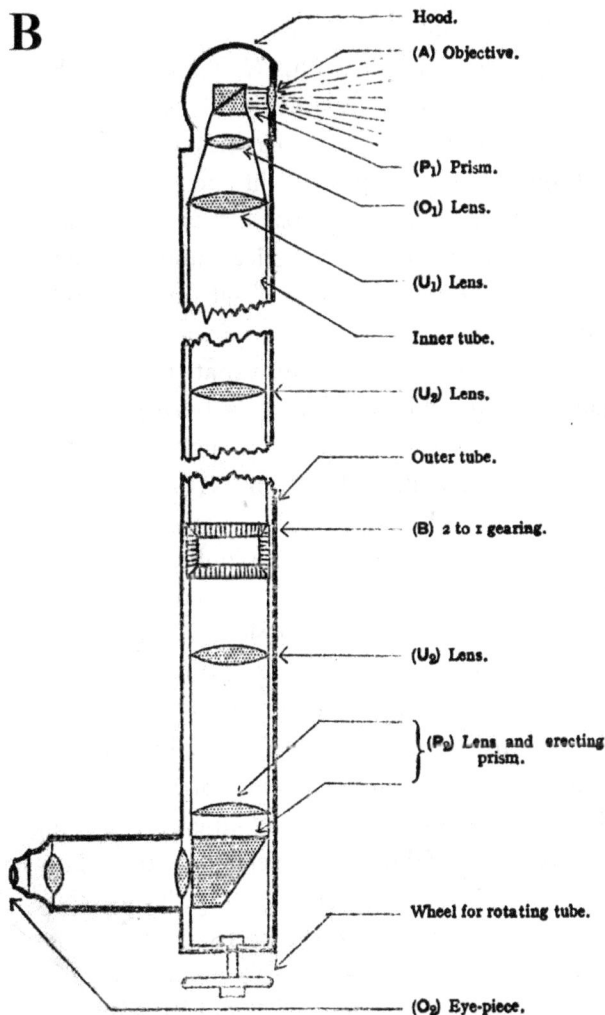

Hood.
(A) Objective.
(P₁) Prism.
(O₁) Lens.
(U₁) Lens.
Inner tube.
(U₂) Lens.
Outer tube.
(B) 2 to 1 gearing.
(U₃) Lens.
(P₂) Lens and erecting prism.
Wheel for rotating tube.
(O₂) Eye-piece.

some new method of watching the shores. He took a piece of lead pipe, fitted it with mirrors at either end, and ran it up through the turret.[35]

On April 12, 1864, the Union flotilla was sailing down the Red River. "At Blair's Landing, dismounted cavalry supported by artillery, engaged the Union fleet. The 430-ton wooden side-wheeler *U.S.S. Lexington* … silenced the shore battery but the Confederate cavalry poured a hail of musket fire into the rest of the squadron." The Union ironclads sought to engage the cavalry. Lieutenant Commander Thomas O. Selfridge, Jr., commander of the *Osage*, reported:

> On first sounding to general quarters, … [I] went inside the turret to direct its fire, but the restricted vision from the peep holes rendered it impossible to see what was going on in the threatened quarter, whenever the turret was trained in the loading position. In this extremity I thought of the periscope, and hastily took up station there, well protected by the turret, yet able to survey the whole scene and to direct an accurate fire.[36]

Submarine Periscopes

An early submarine innovator, New Jersey school teacher John Holland, did not put a periscope on board his *Holland II* (the "Fenian Ram"—so called because his work was funded at the time by the Fenian Brotherhood, Irish revolutionaries who hoped to use Holland's submarines to sink British shipping). This was an unfortunate oversight on his part; "[w]ith no periscope to see the way, the *Holland II* collided with a Weehawken ferry boat and sank in June 1882."[37] Even "Holland's successful HOLLAND VI, the first practical submarine delivered to the United States Navy in April 1900 lacked a periscope"—it surfaced so the target could be seen through its turret viewports.[38]

Some naval writers claim that the periscope was invented in 1902 by Simon Lake.[39] Inspired by Jules Verne's *Twenty Thousand Leagues Under the Sea*, Simon Lake built his first experimental submarine, "the Argonaut, Jr.," in 1894. Its successor, the "Argonaut," sailed from Norfolk to New York in 1898, drawing plaudits from his muse, Jules Verne.

While Simon Lake was undoubtedly a submarine innovator, the first proposal for the use of a periscope in a submarine was apparently by Marié-Davy, a Frenchman, in 1854. This was a simple periscope with two mirrors.[40] There is also reference to the use of periscopes by Claude Goubet.[41] Lake's contribution was a rotating, retractable periscope, featured in his USS *Seal*.[42]

The use of totally reflecting prisms in place of mirrors was proposed in 1872.[43] The 180-degree rotation of the upper prism caused the image to be seen upside down.[44] This was remedied by an erecting prism geared to the upper prism so it "turns with half the angular velocity of the top" prism.[45]

The ability to magnify the image is desirable, but incorporating a single "astronomical" telescope into the periscope would result in an inverted image, and an "erect" terrestrial telescope is "long and inconvenient."[46] On World War II submarine periscopes, the light reflected by the upper prism passed first into an inverted (reducing) telescope, then into a normal (enlarging) telescope, before reaching the lower prism.[47] The second telescope rights the inverted image from the first one.

TANK PERISCOPES

In 1936, Rudolph Gundlach patented a rotary periscope that "allowed a tank commander to obtain a 360-degree field of view without moving his seat." It was first used in the Polish 7-TP light tank. The "patent was sold [licensed?] to Vickers-Armstrong for use in British tanks." The technology was later used in the American Sherman tank and the Soviet T-34 and T-70 tanks.[48]

INVERTED PERISCOPES

Periscopes may be constructed to look down instead of up. Underwater periscopes have been used for studying and catching fish. Artificial lighting is generally needed for their use.[49]

Using inverted periscopes, "U-2 pilots peered down on their targets from seventy thousand feet."[50] And as Alan Shepard's spacecraft hurtled toward the Pacific Ocean, he used his periscope to see the sights below:

> The periscope, located two feet in front of him, had two settings, low and high magnification. On low at the 100-mile altitude, there theoretically should have been a field view of about 1900 miles in diameter, and on high, a segment 80 miles in diameter. Shepard was able to distinguish clearly the continental land masses from the cloud masses. He first reported seeing the outlines of the west coast of Florida and the Gulf of Mexico. He saw Lake Okeechobee, in the central part of Florida, but could not see any city. Andros Island and the Bahamas also appeared in the scope.[51]

Telescopes

The first telescopes employed lenses. They were called refracting telescopes because convex lenses focus light by refracting (bending) it. Unfortunately, a simple lens will refract light of different wavelengths (colors) to different degrees, so each color has its own focus distance. Focus on an object, and it will have a reddish or bluish halo. This problem is called chromatic aberration. Isaac Newton discussed it in his *Opticks*,[52] and commented, "'tis a wonder that telescopes represent objects so distinct as they do."

If the lens has a spherical cross-section, then paraxial rays (rays parallel to the axis of the lens) passing through the lens near its center will not meet at the same point as those passing through the periphery of the lens. This problem is called spherical aberration, and results in a fuzzier image.

To minimize these optical defects, the early astronomers used relatively weak lenses. In consequence, to achieve high magnification, the focal length of the objective had to be lengthened. That meant lengthening the telescope. By the early 1670s, Johannes Hevelius had built a 140-foot telescope. This was called an "aerial telescope" because it was suspended in the air. Pity the Renaissance astronomer who was observing on a gusty night! Newton comments that "very long tubes are cumbersome, and scarce to be readily managed, and by reason of their length are very apt to bend, and shake by bending, so as to cause a continual trembling in the objects."[53]

The first reflecting telescope was built by Niccolò Zucchi (1586–1670), a professor of mathematics at the Jesuit College of Rome, in 1616. He used it to observe the belts of Jupiter in 1630. His book *Optica philosophia experimentalis et ratione a fundamentis constituta* (1652–1656) may have influenced the later, better known reflector designs of Gregory and Newton. Zucchi's telescope used a bronze concave mirror instead of a lens. He viewed the mirror image through a lens, possibly handheld. Some authorities say that he had to put his head in front of the mirror in order to make observations—which would have been something of a nuisance. Others say that the mirror was tilted to avoid obstruction by the observer. With a tilted mirror, the light would be reflected obliquely, and the observer could stand to one side of the telescope.[54]

In 1630, French astronomer Marin Mersenne (1588–1648) proposed using a second, concave mirror to reflect the light down through a hole in the center of the larger primary mirror. (He may also have suggested that these mirrors be paraboloid in shape.) Unfortunately, René Descartes persuaded him not to proceed, apparently because of the difficulty of securing high-quality concave mirrors of sufficient size. But Mersenne explained his design in his *l'Harmonie Universelle* (1636).[55]

A similar design was proposed by James Gregory (1638–1675), a Scottish mathematician, in his treatise *Optica Promota* (1663). The telescope was to have used both a concave paraboloidal mirror and a concave ellipsoidal mirror.[56] The image formed would have been right-side-up, so the Gregorian telescope could have been used in the daytime to observe terrestrial sights, not just at night to see the heavens. One of the disadvantages of the Gregorian reflector design was that it featured an eyehole in the primary mirror, which reduced its light-gathering power. Another was that an ellipsoidal surface is hard to grind. Worse, "if an optical system contains two sequential reflectors, regardless of their shapes, the combined effect is to magnify any geometrical imperfections in either surface."[57] Gregory commissioned craftsmen to build a working telescope according to his plans, but without success.[58]

The credit for actual realization of the reflecting telescope goes to one of the intellectual giants of world history. Sir Isaac Newton (1642–1727), declaring that "the improvement of telescopes of given lengths by refractions is desperate," adopted a radically different approach, employing reflection, and "using instead of an object-glass a concave metal."[59] This speculum had a diameter of about 2 inches, and a thickness of one-third of an inch, and it was ground to the shape of a sphere with a diameter of about 25 inches. While not mentioned in Newton's *Optics*, it may safely be assumed that he placed it in the bottom of a tube and caught the reflected rays on a 45-degree secondary mirror, which in turn redirected the light to a planoconvex eyepiece. This Newtonian reflector was about six inches long and magnified 30 to 40 times. (Newton says that the primary mirror was made of copper[60] but more likely it was speculum metal.[61])

Unlike Gregory, Newton did not place his trust in craftsmen to reduce his design to practice. "I asked him where he had it made," recalled John Conduitt, "he said he made it himself, & when I asked him where he got his tools he said he made them himself & laughing added 'if I had stayed for other people to make my tools & things for me, I would have never made anything of it.'"[62]

With his new scope, Newton saw "Jupiter distinctly round and his satellites, and Venus horned."[63] Newton displayed it at a meeting of the Royal Society of London in December 1671, and shortly thereafter he was voted in as a Fellow.[64]

The great advantage of reflectors (telescopes with mirrors) over refractors (telescopes with lenses) is that they do not refract light. When light is reflected, all wavelengths are redirected at the same angle, so chromatic aberration does not occur.

The original Newtonian design had a spherical primary mirror. Like a spherical lens, a spherical mirror cannot focus parallel rays of light down to a single focal point; it suffers from spherical aberration.

In 1723, John Hadley (1682–1744) replaced Newton's spherical primary mirror with a paraboloidal one, thereby avoiding this problem. There is no doubt that Newton was aware of the advantages of a paraboloid shape over a spherical one. In analyzing refractor (lens-based) telescopes, he declared, "the imperfection of telescopes is vulgarly attributed to the spherical figures of the glasses, and, therefore, mathematicians have propounded to figure them by the conical sections";[65] those, of course, would include the parabola. But Newton calculated (erroneously) that the contribution of spherical aberration to the scattering of the rays was only $\frac{1}{5449}$th that of chromatic aberration.[66] Having solved the latter problem by replacing lenses with mirrors, Newton was no doubt of the opinion that the additional sharpness achievable with a paraboloid mirror was insufficient to justify the effort necessary to grind a mirror to that shape. Nonetheless, Newton's telescope may have inadvertently used the superior paraboloid shape, as a result of flaws in his polishing technique.[67]

Hadley's further contribution was that he devised a reliable method of monitoring the approach to a parabolic cross section. First, he ground the mirror to a spherical shape. Then he ground the mirror more deeply in the center than at the periphery. Without his assay method, this would have been entirely hit-or-miss. But he "placed a tiny illuminated pinhole at the mirror's center of curvature and examined the reflected cone of light in the vicinity of the image. From the appearance of this cone, Hadley could infer the state of the mirror's surface and was thus able to pass, by successive polishings, from a spherical to a paraboloidal figure."[68] Like Hadley's telescope, modern Newtonian reflectors use a paraboloidal primary mirror.

Another problem with the Newtonian reflector—indeed, with early reflectors in general—was loss of light. The metal reflected only about 60 percent of the light,[69] most likely because of tarnishing. Newton recognized both the problem, and a possible solution: "because metal is more difficult to polish than glass, and is afterwards very apt to be spoiled by tarnishing, and reflects not so much light as glass quick-silvered over does, I would propound to use instead of the metal a glass ground concave on the foreside, and as much convex on the backside, and quick-silvered over on the convex side."[70] In other words, he had conceived of a back-silvered glass concave primary mirror. Newton also suggested replacing the 45-degree secondary mirror with a prism of glass or crystal.

One of the early triumphs of the reflecting telescope was the discovery of the planet Uranus on March 13, 1781, by Friedrich William Herschel (1738–1822), an erstwhile professional musician, using a seven-inch scope of Newtonian design. (It is conventional in discussion of telescopes to identify them by the diameter of the primary mirror or lens, rather than by the length of the scope.) He initially thought it a comet, but by careful observation, determined otherwise. Americans may be thankful that the astronomical community rejected his notion of naming the newly discovered celestial object "Georgium Sidus" after Herschel's patron, King George III.[71]

Another great discovery was when the 42-inch speculum metal mirror of the third Earl of Rosse's (1800–1867) Birr Castle telescope revealed the spiral structure of M51.[72] (The Andromeda Galaxy had already been observed with the naked eye, but its spiral structure was then unknown.)

Nonetheless, problems in mirror grinding and in maintaining an untarnished surface discouraged the early adoption of reflector telescopes.

Other Telescope Designs

In the Newtonian reflector, an on-axis planar mirror moves the focal point of the primary mirror (spherical or parabolic in shape) outside the main telescope tube. The eyepiece tube is perpendicular to the main tube. In the older Gregorian design, it was parallel to the main tube and aligned with it. Of course, while Newton avoided the need for an eyehole in the primary mirror, his secondary mirror would prevent some of the incoming light from reaching his primary mirror in the first place.

The science teacher and priest Laurent Cassegrain (circa 1629–1693) described a new telescope design in 1672. A Cassegrain telescope is a wide-angle reflecting telescope with a concave mirror that receives light and focuses an image. A second, convex mirror reflects the light through a gap in the primary mirror, allowing the eyepiece or camera to be mounted at the back end of the tube.

While not pointed out by Cassegrain, the combination of a concave mirror and a convex one tends to limit the adverse effects of geometric imperfections in either surface. Despite this advantage, the Cassegrain reflector sank into obscurity for almost three hundred years, under the weight of Newton's scathing criticism of it.[73]

In what is now called the "classical" Cassegrain design, the primary mirror is paraboloidal and the secondary mirror is hyperboloidal. This avoids spherical aberration without the need for a corrective lens. It is unclear whether Cassegrain himself conceived of the hyperboloidal secondary mirror or whether it was a later development. Accurately grinding both paraboloidal and hyperboloidal mirrors would have been extraordinarily difficult in the late seventeenth century.

If the primary mirror of the Cassegrain reflector were spherical, it would suffer from spherical aberration. A correcting plate (a lens) was added (in front of the primary mirror) in 1930 by the Estonian astronomer and lens maker Bernard Schmidt (1879–1935), creating the Schmidt–Cassegrain telescope.[74] Since it uses both a mirror and a lens, it is called a *catadioptric* design. The Schmidt correction lens was flat on the front side and had a complex curve on the rear side.

A. Bouwers of Amsterdam, Holland, in February 1941 and Dmitry Maksutov of Moscow, Russia, in October 1941 independently invented an alternative correction lens which was curved on both surfaces. It is called a meniscus corrector shell, and the overall telescope design which incorporates it is called a Maksutov–Cassegrain reflector. In 1957, John Gregory realized that the secondary mirror could be dispensed with if a small central portion of the rear surface of the meniscus corrector shell were silvered to make it reflective. The result was the "Mak" reflector.[75]

A quite different approach was taken by William Herschel for the 40-foot long, 48-inch diameter telescope he built in 1789: He dispensed with the secondary mirror entirely. To avoid blocking the incoming light with his own head, he tilted the primary mirror. (As previously noted, this might have been done by Zucchi over a century earlier, but this type of telescope is nonetheless now known as a Herschelian reflector.)

The Reflective Surface

While speculum metal was used by Newton and others, it was a problematic telescope mirror material. "The alloy was brittle and difficult to cast, and still reflected only about 60 percent of the incident light even when freshly polished.... The major problem with speculum was that the copper caused the polished surface

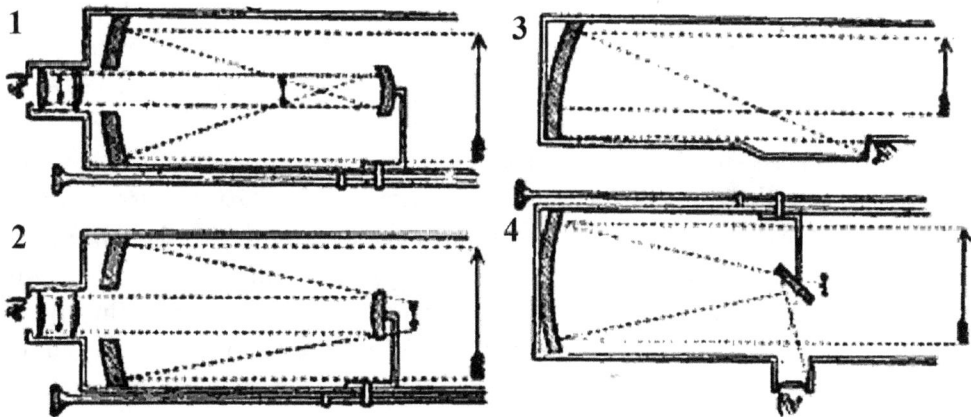

Early reflecting telescope designs: 1) Gregorian, 2) Cassegrainian, 3) Herschelian, 4) Newtonian. (From "Telescopes," *Nelson's Encyclopedia* 10: 118 [1907].)

to tarnish rapidly, necessitating frequent repolishings. These repolishings destroyed the careful configuring of the curved mirror surfaces—every time the mirror was polished it had to be remade." There was a substantial cost, and a loss of weeks of observing time.[76]

In the 1720s, James Pound wrote, "it is to be hoped that [Hadley or others] will in a short Time find out a Method, either of preserving the concave Metal from tarnishing, or ... of making a good concave Speculum of Glass quicksilver'd on the Back-part." James Short indeed attempted to "make glass mirrors back-coated with a tin-mercury amalgam, but he soon switched to speculum metal."[77]

There were two problems with Short's attempted solution. "For careful measurements the reflective coating must be on the front of the mirror, not the back, to eliminate light losses and distortions, as well as spurious reflections and aberrations, caused by the thickness of the glass." Also, "using the tin and mercury amalgam method to coat the front surfaces of scientific mirrors yielded results that were dull and irregular."[78]

The first silvered glass reflecting telescope, just four inches in aperture, was built by Karl August von Steinheil (1801–1870), a friend of Liebig's, in 1856.[79] He later reported that at a 45-degree angle a flat silvered glass mirror reflected 92 percent, as compared to 76.5 for a foiled glass mirror, and 67.18 percent for a metal mirror made of Lord Rosse's alloy.[80]

The physicist Léon Foucault (1819–1868) was another pioneer in the development of the "silvered-glass reflecting telescope." In 1850, he deposited silver on glass by the Leibig process; this "fingernail-sized" reflector was used in his comparison of the speed of light in air and water. A few years later, he was using the Drayton process to make larger silvered glass mirrors, and he presented his first small telescope (diameter 88 mm) to French colleagues in January 1857.[81]

By May 1858 he had a 36-centimeter mirror. His first mirrors were spherical, but as he ascended to larger, "faster" mirrors, he produced ones that were ellipsoidal or paraboloidal. He also developed optical tests that facilitated determination of the shape of a mirror. His "largest telescope mirror, nominally of 80-cm diameter, [was] completed in 1862."[82] It was used first in Paris and later in Marseilles. It reflected 92 percent of the light striking it.[83]

While glass was fragile, it was still easier to handle than speculum metal, which one writer has called "wilfully perverse." The story of the 1870 Melbourne Cassegrainian reflector is instructive. The Australians decided not to use the newfangled silvered glass mirror. They ordered a 48" (1.2 meter) speculum mirror from Dublin. It was only with the third attempt at casting that success was achieved. The mirror was shipped with a protective coating of shellac. When the Australians removed the shellac, they damaged the reflective surface. Rather than shipping the mirrors back to Ireland, the Australians decided to polish it themselves, with unhappy results. G. Ritchie wrote, "I consider the failure of the Melbourne reflector to be one of the greatest calamities in the history of instrumental astronomy."[84]

Finally, while silver mirrors, like speculum metal, will tarnish, the silver of a silvered glass mirror could be dissolved away and replaced with a fresh coating, leaving the mirror shape unaffected.[85]

One problem with reflectors was that they were much more sensitive than refractors to temperature effects, to the flexion of the telescope tube, and to misalignment of the optics. Nonetheless, for large telescopes, they had substantial advantages. As lenses were increased in size, they had to be made thicker, which increased their absorption of light. This was particularly a problem for astrophotography, as the film was most sensitive to violet and ultraviolet light, and flint glass strongly absorbed these radiations. The large lenses also had to be supported at the edges and hence were liable to warping. In contrast, silvered mirrors strongly reflected violet and ultraviolet light, and large mirrors could be supported all the way across the rear of the mirror "blank," rather than just at its edges.[86]

The first large reflector was the Hooker 100-inch reflector at Mount Wilson. At the time of its construction, the largest reflector in use was just a 36-incher. The new scope was completed in 1917, but its primary mirror maintained a tie with the past: "The mirror blank was cast by the St. Gobain Glass Works in France. Throughout its history, this firm had derived much of its revenue from the production of the green-colored bottles used to protect vintage French wines. Not wanting to depart from tradition, the mirror for the 100-inch was cast out of green glass!"[87] More to the point, the St. Gobain Glass Works was one of the pioneers in the manufacture of silvered glass mirrors, having made the glass blanks for Foucault.

"Impurities in the air cause the silver to tarnish and moisture loosens small flakes of the metal from the glass." The Mount Wilson observatory mirrors were "burnished … and resilvered at least twice a year.… Telescopes located near cities require resilvering at even shorter intervals."[88]

For the 100-incher, it took "five hours" for the "mirror in its cell" to be "lowered from the telescope tube to the silvering room below." The "old silver coat is completely dissolved off with strong nitric acid.… After several rinsings with tap water a strong caustic potash solution is applied," and rinsing is repeated. Then a solution of "rock candy reducing solution" is added to the water in "the concavity in the glass." A "silvering solution" ("silver nitrate and caustic potash") is poured in, while the mirror is "vigorously rocked." The solution "rapidly turns almost black … presently the silver begins to adhere to the glass giving it a bright reflecting surface." The mirror is lightly scrubbed, and more silvering solution is added. The mirror is rinsed, dried, and polished. "The silvering solutions require three pounds of silver nitrate, containing 1.8 pounds of metallic silver," but only about a quarter dollar coin's worth is actually deposited. "The thickness of an average silver coat is about $1/150,000$ inch."[89]

For telescopes, aluminized mirrors replaced silvered ones in the late 1930s.[90] The first large telescope mirror to be aluminized was the 100-incher of the Hooker telescope at Mount Wilson observatory, which received this treatment in 1935. The mirror was originally cast at the St. Gobain Glass Works in France and began astronomical service in 1917.

While aluminum is not quite as reflective as silver, it is much more durable. To recoat a mirror, it must be lifted out of its frame. The Mount Wilson mirror weighed four tons; obviously, the less it had to be played with, the better. Prior to the aluminization, it had to be resilvered every few months. The aluminum coating must also be renewed, as the mirror surface is marred by urban air pollution (Los Angeles is nearby) and chemical damage from pollen (thanks to a national forest), but these treatments are just annual.[91]

Telescope Mirror Blanks

The glass initially used in making telescope mirror blanks was crown glass. Its biggest disadvantage was a relatively high coefficient of thermal expansion ($8 \times 10^{-6}/°K$). In the 1930s, crown glass was replaced, at least for "major telescopes," by Pyrex borosilicate glass. This glass was much less sensitive to temperature changes than plate glass (coefficient just $3.2 \times 10^{-6}/°K$). The first use of Pyrex glass in a large telescope was in the Macdonald 82-inch reflector. In the 1970s, fused quartz blanks (coefficient $0.4 \times 10^{-6}/°K$) became available. These were followed by glass ceramics (coefficient $\leq 0.05 \times 10^{-6}/°K$) and "ultra low expansion fused quartz."[92]

Another issue with glass is that it is heavy; it has essentially the same density as granite. That means that big mirrors need big support structures, and it also means that the optical quality can be impaired if the mirror deforms under the influence of its own weight. Deformation was traditionally fought by making the mirror thick, but of course that also increased the weight.

The weight of a glass blank could be reduced by combining a smooth faceplate with a honeycombed back. This gives it "high rigidity for low weight," and also allows it to come into thermal equilibrium with the outside air more quickly.[93]

The glass blanks are formed into their proper shapes by casting and then ground and polished as needed.

Now let's look at the "prehistory" of spin casting. In the winter of 1850, Dr. Krecke of the Utrecht Meteorological Observatory filled a basin, suspended by a strongly twisted rope, with mercury, and observed that when it was set free to rotate, the "surface of mercury took on a parabolic shape." This led him to propose that one might "melt a metal in a mold or basin rotating around a vertical axis: the metal should then take on the parabolic shape, and on cooling, while the mold rotates regularly, the surface should retain the desired form."[94] He was not successful, probably because of problems with maintaining a perfectly vertical axis and uniform rotational speed.

Nonetheless, the principle was sound, and applicable to molten glass, too. But it was not successfully practiced for fabrication of telescope mirror blanks until relatively recently. In 1992, the Steward Laboratory "transformed 10 tons of glass into a 6.5-meter mirror blank" by melting the glass in a furnace rotating "at 7.4 revolutions per minute." The same technique was later used to make the 8.4-meter mirror blanks for the Large Binocular Telescope.[95]

Note that the mirror blank must be spun not just during the initial melting, but also while it slowly (over several months) cools down.[96]

Segmented Mirrors and Adaptive Optics

According to Joel Davis, for conventional, single-mirror telescopes, "doubling the aperture increases the cost of building the telescope six times."[97] The Keck telescope (Mauna Kea, Hawaii) has twice the aperture of the venerable Palomar scope, but it does not employ a single mirror. Rather, it is a mosaic of 36 hexagonal mirrors, a "giant insect eye" that looks up at the sky. Each segment—1.8 meters in diameter, 74 millimeters thick, and 880 pounds—is far easier to make, transport, and, if need be, replace, than the great Palomar speculum. Each segment is independently oriented by hydraulic means to face the correct direction, giving the segmented mirror an effective diameter of 10 meters; its total weight is only 14.4 tons. In contrast, the smaller (6 meter) but monolithic Zlenchukskaya mirror weighs 41 tons.[98]

The Hubble Space Telescope (HST, 1990–) uses a 94-inch Ritchey–Chretien reflector design, a variation on the Cassegrain system. The secondary mirror is much smaller than the primary. To increase the effective field of view, the design uses mirrors which are hyperboloids, not the usual paraboloid. They are difficult to grind accurately, and indeed astronauts had to apply corrective optics to the HST. The first telescope of this kind was a 20-inch aperture scope built in 1930. Another, with a 40-inch primary mirror, was constructed later that decade for the U.S. Naval Observatory in Washington, D.C., and ultimately moved to Flagstaff, Arizona.[99]

The optical system of the James Webb Space Telescope is "a 3-mirror anastigmat design, consisting of primary, secondary, and tertiary mirrors." It is designed for infrared astronomy; hence, the reflective surface of its primary mirror is a thin layer of gold, which strongly reflects infrared. To minimize heating, it has an open design (no tube!) and is protected by "tennis court-sized sunshields." Since it is therefore extremely cold (–364°F), the mirror blank is made of beryllium, which retains strength at low temperatures. Beryllium is a light metal, and to make the mirror even lighter, the backs of the blanks are machined in a "honeycomb structure." The concave, 6.64-meter primary mirror of is composed of eighteen hexagonal segments; there is a hexagonal gap in the center. The convex secondary mirror is 0.738 meters in diameter. The tertiary mirror is concave and of similar size.[100]

The Largest Telescopes

At present, the largest reflecting telescope in the world is arguably the "large binocular telescope" at Mount Graham International Observatory in Arizona; as the name implies, it has two mirrors (each 331 inches, 8.4 meters; "collecting area equivalent to a single 11.8m"[101]).

Primary mirror of James Webb Space Telescope, with segments labeled. In the original color photograph, the golden color of the mirror segments is obvious. A reflection is visible in segments C4 and B4. The secondary mirror (SM) is in "stowed" (folded) position. The AOS (aft optics system) is in the center of the primary mirror and includes the tertiary mirror (TM) and the fine steering mirror (FSM). (https://webbtelescope.org/news/webb-science-writers-guide/telescope-overview and https://jwst-docs.stsci.edu/files/97976981/97976983/1/159607 3035908/OTE_entrance_pupil_annotated_Oct2016.png.)

The largest unsegmented monocular telescope is the 323-inch (8.2-meter) Subaru at Mauna Kea Observatory in Hawaii.[102]

The largest segmented mirror telescope, according to Wikipedia, is the Gran Telescopio Canarias, 409 inches (10.4 meters), in the Canary Islands, Spain. However, the South African Large Telescope (SALT) is 11 meters, and when the Giant Magellan Telescope (Chile) is completed, it will be 25.4 meters.[103]

X-Ray Telescopes

The universe speaks to us across the electromagnetic spectrum, but the ability of earthbound telescopes to perceive what it is saying is hampered by the atmosphere. While it is fortunate that the atmosphere protects us from X-rays and gamma rays (if they reached Earth freely, life would not exist here), there is much that we can learn from X-ray emissions.

Because of its proximity to us, our sun is the brightest X-ray source in our sky. The first X-ray picture of the sun was taken in 1963, from a rocket ship. However, there are many celestial X-ray sources of great interest. X-rays are produced by supernovae (exploding stars), supermassive black holes (the X-rays are emitted by gases falling into the black holes), and X-ray binaries (a binary system consisting of a neutron star and a normal star; the X-rays are emitted by gases falling from the normal star into the neutron star).

To form an X-ray image, a telescope must focus the X-rays. When a normal telescope mirror reflects visible light, most of the light is striking the mirror surface perpendicularly, or nearly so. This is not practical with X-rays; even metals barely reflect X-rays at normal incidence. Fortunately, reflectivity increases as the angle of incidence changes. (Annoyingly, in the X-ray field, the angle of incidence is often expressed as the angle from the horizontal rather than the vertical.) Because a metal's index of refraction for X-rays is less than 1.0, there is a critical angle which creates a condition of total external reflection, analogous to total internal reflection of visible light inside diamonds or prisms. For gold and nickel, the most commonly used X-ray reflectors, the desired angle for 1-keV photons is about 1 degree from horizontal, that is, just grazing the surface.[104] Thus, X-ray telescopes make the X-ray photons careen off these mirrors like pebbles skipping across water.[105]

A simple paraboloidal mirror, like that used in visual astronomy, cannot be used; the "images of off-axis objects will be severely blurred."[106] In order to form the image, two mirrors are needed. The first telescope used to image an X-ray source other than our sun featured the Kirkpatrick–Baez design; the X-rays reflect off two parabolic troughs, with axes perpendicular to each other. The Wolter design is more popular; it features the combination of parabolic and hyperbolic surfaces.

The problem with these X-ray telescope configurations is that they have a narrow field of view. In contrast, "lobster eye" telescope optics mimic the compound eyes of the lobster, which permit a wider field of view.

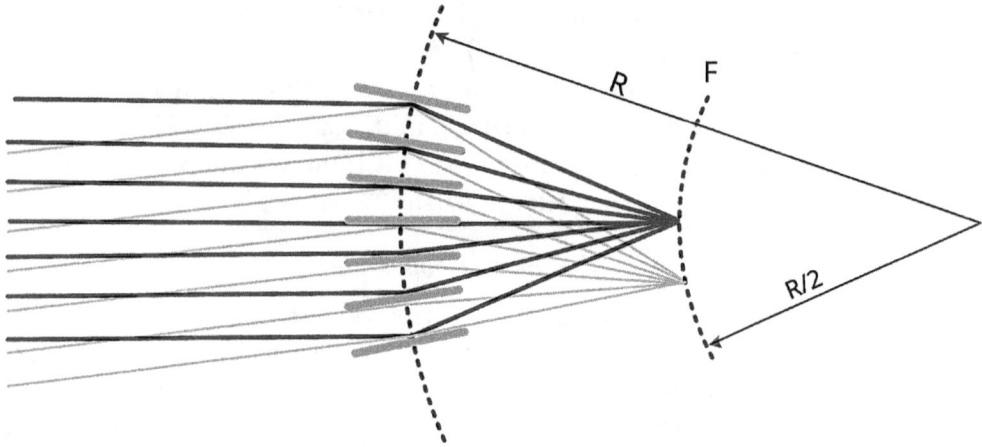

Schematic of lobster-eye optics. "X-ray radiation from various sources reflected from the walls of cells located on a curved surface of radius R is focused at different points of the focal surface F at a focal distance R/2." Note that these are grazing reflections. (Wikimedia user "Vsatinet" CC BY-SA 4.0.)

"Instead of a standard lens, a lobster's eye consists of a large number of square pores evenly distributed across a sphere, with each pore pointed towards a common centre. The light reflects off the very smooth walls of each pore and is focused onto the curved retinal surface…. The ratio of the width of the pores to their length requires the light to be reflected at very shallow angles—as is required to focus X-rays…. The lobster eye geometry for X-ray imaging was [proposed] by Angel in 1979…. The size of the FoV [field of view] of a lobster eye telescope depends only on the angular extent of the spherical optic and detector."[107]

The telescope's full optic is a spherical dome and consists of an array of individual micropore optics (MPOs), each corresponding to a spherical patch. Each MPO takes the form of a lead glass presenting an array of micropores pointing toward a "common centre of curvature." The pores, all of square cross section, might be 20 or 40 microns wide. Even though the angle of incidence is about 1 degree from horizontal, the reflectivity of the bare glass is only ~20–40 percent for X-ray photons of 1.49 keV energy. Hence, "to improve the reflection … the pores are coated with iridium." The pore length is typically over fifty times the pore width.[108]

There are many variations on this theme, but to explore them would be to go beyond the scope of this text. We turn instead to giving a specific example, the Chinese Academy of Science's Einstein Probe, which was launched in January 2024. It carries both a wide-field "lobster-eye" telescope and a traditional "Wolter mirror" telescope. The wide-field telescope consists of "twelve identical modules," each providing "36 MPOs in a 6 by 6 array." This has "a total FoV [field of view] of more than 3600 square degrees."[109] The wide field scope can survey X-ray sources which are then studied more closely by the Wolter-type telescope. The latter has a field of view of just 1 degree of the diameter.[110]

Liquid Mirror Telescopes

Mercury is a metal that is liquid on Earth's surface between –39 and 357°C (Wikipedia). Like other metals, it is strongly reflective; its visible light reflectance is about 78 percent.[111] There are three obvious problems with mercury mirrors: They are liquid at room temperature; they are heavy for their size (because of mercury's high density); and they give off poisonous vapors.

As it turns out, the liquidity of mercury can be something of an advantage. As a result of the combined effects of gravitational and centrifugal forces, the surface of a liquid spinning about a vertical axis assumes a paraboloidal shape, which, we know, is an excellent shape for a telescope mirror.

In 1850, Capocci wrote, "if a circular box filled with mercury were given a suitable rotational movement, and that this movement were well executed and uniform, he would end up arranging the surface of the liquid in such a way as to gather the reflected rays perfectly at a point, which would be at a greater or lesser distance on the perpendicular raised to its center. At this point, we would place an eyepiece which, being at a great distance from the mirror, could give high magnifications…. Even if such an instrument could only be used to see a single star (of the Lyra, for example) or a nebula as it passes through the zenith, but with very high magnification, it would be well worth having a telescope of such a new kind."[112]

"The first published account of a working LMT [Liquid Mirror Telescope] was provided by Skey (1872)." With his 0.35-meter telescope, "he demonstrated the ability to alter the focal length" at will; it is inversely proportional to the square of the rotational velocity. In 1908–1909, Robert Wood sought to identify and minimize the various influences that could degrade the image obtained by an LMT. For example, if the mirror weren't leveled, there were "tidal waves in the mercury pool," and "fluctuation in the angular velocity of the basin" would cause ripples and focal length changes.[113] While he was able to reduce the ripples, and he made some astronomical observations, he ultimately abandoned work on the LMT, and there was then a long hiatus in LMT development.

Further optical improvements were achieved in the 1980s by rotating the basin with a synchronous, quartz oscillator-driven motor, with the basin suspended on air bearings.[114] Paul Hickson built a 106-inch (2.7-meter) LMT in 1993. In 2000, he completed, at a cost of about a million dollars, a 236-inch (6-meter) LMT. "A conventional telescope with a regular solid glass mirror of the same size would require an outlay of about $100 million. A large part of the savings comes from not making, polishing, testing and mounting a standard mirror." It was the thirteenth largest telescope in the world at the time of its completion.[115] In 2022, the 3.6-meter International Liquid Mirror Telescope began operation in the Himalayas.[116]

While mercury's liquid state at room temperature makes it possible to spin it into a paraboloidal shape, and good ventilation somewhat alleviates the toxicity issue, it has the serious disadvantage of high density. A nontoxic gallium–indium

(76:24) mixture melts at 16°C, has a density less than half that of mercury, and is more reflective. It wets surfaces easily, so one can make a very thin layer of it. Ga-In was used to make just "proof of concept" mirrors of one meter diameter.[117] Galinstan, which also contains tin, melts at –19°C (Wikipedia).

Another option is a ferroliquid, essentially an ordinary liquid "to which one adds ferromagnetic particles coated with a chemical agent that prevents coalescence." The liquid can then be "shaped by an externally applied magnetic field," rather than by spinning, and you are not limited to a parabolic shape. Commercially available ferroliquids are dark and would need to be "coated with reflective layers."[118]

I previously mentioned that the paraboloid figure was achieved by spinning the liquid about a vertical axis. That implies that the LMT must be a "zenith telescope"— one that points straight up. The advantage is that it doesn't need an elaborate mount so that it can elevated and traversed. The disadvantage of that is that it can only study whatever passes overhead each night. As a corollary, it can't physically track a celestial object, and faint objects require long exposures. However, if film is replaced by a CCD (charge-coupled device) sensor, then the object's drift within the telescope's field of view can be corrected for by shifting the electronic image during the exposure.[119]

For some astronomical work, this is tolerable. For example, a zenith telescope may be used for cosmological surveys in a small patch of sky, or for photometric variability studies. (A spinning liquid mirror may also be used as a laser light collector in LIDAR studies of the atmosphere overhead.)[120]

Still, it would be desirable to be able to use an LMT to view celestial objects that are lower in the sky. One approach is to leave the mirror pointing vertically, view it obliquely (as with a Herschelian reflector), and use one or two additional mirrors to correct for the aberrations introduced by the oblique view. The "aberrations can, in principle, be corrected in small patches to zenith distances as high as 45 degrees."[121]

A complication is that the required correction would vary depending on the zenith distance. In 1994, Shuter and Whitehead offered a proof of concept that a spun-up colloidal suspension of iron particles in mercury could be reshaped by magnets into a spherical surface. The required aberration correction would then be "constant, independent of zenith angle."[122]

Alternatively, if one had a highly reflective liquid of high viscosity (one that flows slowly), and spun it quickly, it might be used to make a liquid mirror telescope that could be tilted to some degree. The liquid would not have time to flow much to one side before it was tilted in the opposite direction. A limited experimental study (1 degree tilt) suggested that "the maximum tilt angle increases linearly with viscosity" and for a 10-degree tilt, a liquid with the "viscosity of glycerin at 10°C" was needed for "reasonable surface quality." Since the highly viscous liquids aren't themselves highly reflective, one must also deposit a reflective metal film upon them. A stable metal liquid-like film (MELLF) can be formed by suspending silver nanoparticles at the interface between water and a dense organic solvent.[123]

There have been proposals to put rotating liquid mirror telescopes on the moon. The Moon has only one-sixth the surface gravity of Earth, so a much larger mirror could be supported. Also, it rotates at only about one-thirtieth the speed of Earth, so there is less of a problem of the target drifting out of the field of view. Mercury would freeze, "but they could employ low-temperature ionic liquids that do not evaporate. The liquid would have to be made reflective by depositing a metallic film on its surface while the mirror is spinning. Because the Moon has no atmosphere, an air bearing would not be practical. Instead, the mirror would probably float on a magnetic field stabilized by superconducting elements."[124]

Liquid mirrors may have uses outside of astronomy. A prototype "3-D scanner that uses a liquid mirror as its objective" has been constructed.[125]

Microscopes

Many early microscopes could be handheld; a transparent specimen could be examined by placing it in front of the lens (possibly attaching the specimen to an integral stage) and holding it up to the light.

The study of transparent specimens was improved by the introduction of a substage mirror, which illuminated them from below. This was pioneered in Edmund Culpeper's Tripod Microscope (1725). In essence, he put a hole in the object stage (the platform holding the specimen) and placed a pivoting concave mirror beneath it. Since you no longer had to point the objective lens at the sun or a candle, this enabled development of heavier, more elaborate tabletop models.[126]

Opaque specimens required that light be reflected off the specimen, rather than transmitted through it. However, when high magnifications were used, opaque specimens posed a problem: The short focal length required meant that the objective lens obstructed light from reaching the specimen.

A "Lieberkuhn" reflector (named after the anatomist Dr. Johann Lieberkuhn, 1711–1756, who popularized it) made it much easier to examine opaque objects. The mirror was mounted on the objective lens, with the concave surface facing away from the microscopist. With a handheld microscope, the specimen was held in front of the lens, and light was concentrated by the mirror upon it. The resulting illumination was very even.[127] Each objective lens had to have its own speculum so that their focal lengths matched.[128] The design had been proposed a century earlier by René Descartes in *Dioptrique* (1637).[129]

A substage illuminator could be used in conjunction with a Lieberkuhn reflector for examining opaque specimens. While the light source is below the specimen, the light would pass upward, around the specimen, but then be concentrated down upon it.[130]

"Beck's vertical illuminator" (1885)[131] provided "top lighting" by interposing a beam splitter between the eyepiece and the objective lens. The light source is in a side tube and is focused by a condenser lens on the splitter. This reflects light down through the objective lens onto the specimen. This is the method of "top lighting" typically used in modern microscopy.[132]

Instead of a Lieberkuhn reflector, the geologist Henry Sorby (1826–1908) used both parabolic and flat mirrors, set off from the optical axis, to provide oblique illumination of mineral and metal specimens.[133]

The term "solar microscope" was sometimes applied to the combination of a microscope with a camera obscura (see Chapter 5); it sat at a window and projected the eyepiece image onto an interior wall. The sun was the strongest available light source and thus permitted the brightest projected image. "The first solar microscopes were intended to be mounted so that the sun could shine directly into the instrument, but the inconvenience of this soon became apparent." In 1742 John Cuff directed sunlight "into the microscope by means of a movable mirror operated by a system of levers and cords from inside the room." The mirror had two axes of rotation.[134]

The solar microscope could be combined with the heliostat (see Chapter 4), which would cause the mirror to automatically track the sun. This was suggested by Morawetz and Volkmar in 1866.[135]

As in the case of telescopes, one of the problems of using lenses for magnification was chromatic aberration. Since the problem was greatly exacerbated if multiple lenses were used, single lens were the norm until this problem was overcome.

A. Microscope with Lieberkuhn reflector (c) attached to objective lens (b). Here the sample is held in forceps (f), but it could be mounted on glass. If the object is transparent, a "dark well" (3) is placed underneath it, so it is only illuminated from above. (Quekett, John Thomas, *Quekett's Practical Treatise on the Use of the Microscope* [Hippolyte Bailliere, 1848], Fig. 66.) B. Substage mirror of microscope constructed by Claude-Siméon Passemant circa 1750, possibly made for Louis XV. The mirror pivots on two axes independently. Courtesy The Met, Accession 1986.1a-d. (Purchase, Mr. and Mrs. Charles Wrightsman Gift and Gift of Mr. and Mrs. Charles Wrightsman, by exchange, 1986. Background of original replaced with white for visual harmony with A and C. The Met.) C. Barker's catoptric microscope (1736). This was essentially "a Gregorian telescope of very short focus ... 9 to 24 inches." Note the specula AB and CD, and the two lenses inside the eyepiece GHIK. (Mayall, John, *Cantor Lectures on the Microscope* [W. Trounce, 1886], 39.)

Reflecting microscopes, like reflecting telescopes, were free of chromatic aberration as the angle of reflection is indifferent to the wavelength of the incident light.

In 1672, Sir Isaac Newton wrote: "I have sometimes thought to make a Microscope, which in like manner [to a telescope] should have, instead of an Object-glass, a Reflecting piece of metal.... For those Instruments seem as capable of improvement as *Telescopes*, and perhaps more, because but one reflective piece of metal is requisite in them."[136] There is no indication that Newton actually built such an instrument. However, the construction of a "catoptric microscope" was reported to the Royal Society by Robert Barker in 1736.[137]

The principal advantage of the reflecting microscope was the absence of chromatic aberration. This advantage was lost when the achromatic lens was developed in 1758. Since the lenses in microscopes are small, the mirror's advantages of scale—so persuasive in the case of telescopes—did not come into play.[138] However, a catoptric microscope was manufactured by Jecker in the nineteenth century.[139]

The World Seen in 3D: The Stereoscope

The stereoscope was invented by the physicist Sir Charles Wheatstone (1802–1875) in 1838. Wheatstone was one of the principal developers of the telegraph. He also invented the "enchanted lyre" (a lyre that seemed to play itself; the vibrations were passed down a brass wire to the lyre from a musical instrument in the room above) in 1821, and the concertina in 1828.[140]

We have depth perception because we have two eyes, set so they can both focus on an object. This is called "binocular vision." When we make a drawing or a photograph of a scene, it is flat, and we must infer the distance to objects from how small and fuzzy they appear, whether they are blocked by other objects, and so forth.

The stereoscope (*stereo* form + *scope* to see) recreates binocular vision. Human eyes are set about 2.5 inches apart, so each eye sees a scene from a slightly different angle. This permits us to perceive depth. In Wheatstone's "reflecting mirror stereoscope," two drawings, crafted to be combined into a three-dimensional (3D) picture, were held at a set distance from each other. Each drawing was of the same scene, but, like the view from one eye or the other, from a slightly different vantage point. Between them were two mirrors, set at an angle. The "V" formed by these two mirrors pointed toward the viewer, with each mirror presenting the reflected image of one of the two drawings. The eyes merged the images formed in these mirrors together. The fusion of the images therefore recreated the original sense of depth.[141] Calvert comments:

> Wheatstone's stereoscope used two mirrors to separate the optical paths. Each eye saw only the image in its mirror, and focused upon it. The angle between the mirrors could be changed to allow for convergence of the optic axes, and the distances of the pictures could be adjusted. This stereoscope was better suited to laboratory investigations than to the enjoyment of stereograms in the sitting room, largely because of the possible adjustments, which made it hard to set up, and its unwieldy size. Also, the stereograms had to be produced as mirror images, so they would be viewed correctly.[142]

Nonetheless, Sir John Herschel called this device "one of the most curious and beautiful for its simplicity in the entire range of experimental optics."[143]

In the later, more popular, Brewster stereoscope design, the prints were mounted side by side, and prisms were used to diverge the optical paths to match the convergence of the eyes. The prisms could also be associated with lenses, to provide magnification.

While the Brewster stereoscope proved more salable, precision mirror stereoscopes are still on the market, especially for analysis of aerial photos. There are also some inexpensive modern mirror stereoscopes:

> The "View Magic" is a viewer designed in America which requires the two images to be mounted vertically rather than horizontally. The viewer uses mirrors to align the images correctly so that they appear to be horizontally mounted. Mounting the images in this way removes one of the limitations on stereo photography i.e. the size of the print. Human vision is designed to have a wide angle of view horizontally i.e. letterbox format but traditional stereoscopic photographs tend to be square. The reason for this is that the traditional horizontal mounting of the images means the centre of both images must be approximately 60mm apart i.e. the same as the distance between the eyes. The images can therefore only be 60mm wide and although height is not restricted in theory, most viewers used a fixed mount for the composite image i.e. the viewer could not be moved relative to the image being viewed which restricts the viewing height. These restrictions do not apply to the "View Magic" device as it is a hand held viewer and can therefore be used with any width prints. The only imitation is the height which can be a maximum of 4" as the mirrors are set for this print height and cannot be adjusted.[144]

The View-Master was introduced at the 1939 New York World's Fair. The stereo pairs were mounted in "reels," which were cardboard disks holding the tiny color slides. Originally, most "reels" were educational, for example, views of Boulder Dam (reels 4–8, 1946–1947).[145] There was even, in 1962, a 25-reel atlas of human anatomy.[146] "Stereo-Stories" were introduced in 1948; "Fairy Tales" was a six-reel set.[147]

The View-Master was originally manufactured by Sawyer's. In 1951, it acquired its competitor Tru-Vue, which enjoyed licenses to produce stereo stories about Disney characters. This of course led to a shift toward more entertainment material. The shift became more pronounced after 1966, when Sawyer's was acquired by General Aniline & Film (GAF). In 1981 the View-Master line was spun off to a new company, View-Master International. In 2009, as a result of more corporate restructurings, the line came into Mattel and is now marketed as a Fisher-Price product. Mattel stopped production of the scenic reels.[148]

Photographic Cameras: Image Formation

In the first successful photographic camera, that popularized by Louis Daguerre (1799–1851), a lens projected the image onto a plate, and chemicals were used to fix the image into a permanent form. Unfortunately, the lens of the day was extremely "slow" (poor in light-gathering power; the effective aperture was about f/14, whereas a modern fast lens might be f/1.4, a hundred-fold difference), and hence long exposure times were needed. This made portraiture difficult.[149]

The first U.S. photographic patent (1840) was granted to Alexander Wolcott (1804–1844) for his "method of taking likenesses by means of a concave reflector and plates so prepared that luminous or other rays will act thereon." His preferred camera was box-shaped, fifteen inches long, holding at one end a concave mirror seven inches diameter, with a twelve-inch focal length, and at the other end was an equal size opening, equipped with a plate holder. The plates, unfortunately, were thus in-between the subject and the camera. They were therefore small, just 2.5 × 2 inches, so light could get around the back side of the plate, strike the mirror, and be reflected onto the sensitive side.[150]

Wolcott opened the world's first portrait studio in 1840. Richard Beard licensed rights to Wolcott's camera and became England's first portrait photographer; the portrait would "cost the sitter between one and four guineas"; exposure would be "from three seconds to as much as five minutes, depending upon the weather." That was still better than the 20 to 30 minutes required with Daguerre's lens.[151] Another advantage of the "reflector camera" was that it produced images that were the correct way around, rather than the normal laterally reversed picture from the Daguerre camera.

The Wolcott design dominated photography for only a few years; faster lenses made the mirror approach obsolete.

Catadioptric interchangeable lenses are available for single-lens reflex (SLR) cameras. As the name implies, they use a combination of mirrors and lenses. Their advantage is that they are relatively light and compact for their focal length. Their principal disadvantage is that because the mirror obstructs the optical axis, they have a fixed, "slow" aperture and the images have weird, donut-shaped bokeh (background blurs).

Mirrors are also used in an adapter to permit an ordinary single-lens camera to create a stereoscopic pair. "Some cameras, typically 35mm with standard 50mm lens, can be fitted with a beam splitting device which produces two images on a single frame. This device basically uses mirrors to bring the two images side by side so they fit into one standard 35mm frame."[152]

I previously mentioned the polemoscope, a device for covertly watching someone or something off to the side.

The Deceptive Angle Graphic (1901), a "detective camera," had "a pair of 'dummy' stereo lenses" on its apparent front, but on its apparent side there were apertures behind which the taking and finder lens were placed (concealed inside the box). Thus, it was a twin-lens reflex camera. Both the ground-glass screen and the "film" (plate or roll) compartment were in the apparent rear, so there must have been two diagonal mirrors, one for each lens.[153]

In 1909 a photographer described "photographing sideways" by means of a diagonal mirror mounted in a tube made in front of the lens. The problem at the time was that this mirror had to be front-silvered glass, and was thus easily tarnished or damaged, since the available permanent metal mirrors (or glass prisms) were heavy and expensive.[154]

The 1985 Spiratone catalog listed the Circo Mirrotach, which was "designed to look like part of the lens for 'discreet' right angle shots." Toward that end, it had clear glass in front. "It was attached to the front of the actual lens (a telephoto of 100mm focal length or greater) with an adapter ring, and it contained a front-surface mirror." Spiratone urged that "you get natural expression, avoid self-conscious reaction of subject."[155] A similar attachment is presently sold by Fotodiox.[156]

Camera shake limits the shutter speeds that a photographer may safely use when shooting without a tripod. This problem can be ameliorated by electronic or optical image stabilization. The latter typically takes one of two forms, those in which lens elements are used, and those in which the sensor is moved. A 2024 Samsung published patent application proposes image stabilization by rotating a "reflecting module" with respect to two axes perpendicular to the optical axis. This module, comprising a mirror or prism, directs the incoming light onto the optical axis direction and thus toward the lens module. The inventors argue that the reflecting module would weigh less than the lens module and thus that achieving image stabilization by moving it rather than the lens module significantly reduces power consumption.[157]

Photographic Cameras: Viewfinders

Mirrors (or prisms) were also used in camera viewfinder systems. The first cameras did not have viewfinders; you focused the image on a ground-glass screen, then removed the screen and inserted a photographic plate.

This was a cumbersome process, improved upon in the twin-lens reflex (TLR) camera. The TLR had two identical lenses, one focusing on the plate (or later, film) and the other on a ground glass viewfinder screen. The viewfinder was on the top of the camera, and the image produced by the viewfinder lens was reflected upward by a fixed diagonal (45°) mirror, hence the term "reflex." The camera would be held at waist level, making it easier to support. The "Twin Lens Carlton," advertised in the 1894–1895 *Photography Annual*, was an early example.[158] However, the best known TLRs are those manufactured by Rolleiflex, beginning in 1928.[159]

The twin-lens reflex camera had several disadvantages: parallax error (the difference in field of view between the two lenses); difficulty in tracking movement (the mirror caused a lateral reversal); and (for most models) a lack of interchangeable lenses.

The TLRs were eclipsed in the twentieth century by single-lens reflex (SLR) cameras. The basic concept was that the mirror would be interposed between the shutter and the lens during composition, but when the photographer released the shutter, the mirror would swing out of the way.

The first SLR patent was issued to Thomas Sutton in 1861.[160] According to his provisional specification, at the upper back edge of the open end "is hinged a moveable plane reflector, which may be made of polished metal, silvered glass, or other suitable reflecting material." Sutton's "focusing screen is fitted into an open panel in the top of the camera, so that when the face of the reflector is turned towards the lens at an angle of 45° with its axis, an erect image is thrown upon the screen."

"Sutton proposed to size the reflector so that it would shield the sensitive plate from the light which passes through the lens." And when the reflector was turned up, "the image formed by the lens will be thrown on the sensitive plate." Thus, Sutton used the mirror as a shutter. "The reflector may be turned up and down on its hinge instantaneously by any suitable contrivance with an external handle."

In 1884, the artist Calvin Rae Smith received a patent for an SLR, likewise with a swinging mirror-cum-shutter. However, there were some differences. The mirror was attached to a cylindrical wedge with an "exposure opening," and it swung on a spindle that was spring-loaded so that when the catch was released, the mirror would swing up to the shooting position. The speed of the movement, and thus the length of the exposure, depended on the tension of the spring. A second catch was used to secure the swing for a "time exposure."[161]

What was more important is that Smith (more precisely, his brother Everdell William Smith) actually commercialized his "Monocular Duplex." Initially it used plates, but it was later modified to take roll film.[162]

Whether you use the mirror as a shutter (Sutton) or couple the mirror to the shutter so they move together (Smith), the interdependence limits the choice of shutter speeds. In 1901, the pioneer bird photographer Francis Hobart Herrick wrote, "the reflecting camera does the work of the two lenses with a single lens and bellows, and in the recent designs is provided with a focal plane shutter.... When the object is focused, a lever is pressed which raises the mirror and automatically releases the shutter." His was manufactured by the Reflex Camera Co.[163] Another of this type was the Primus 7a.[164]

The focal plane shutter featured two curtains (the "roller blind") moving horizontally or vertically, and immediately in front of the film. The greater the gap between the curtains, the longer the exposure was. In the late nineteenth century it was the highest speed shutter, allowing one to freeze motion—if the lens and film were fast enough to capture the requisite amount of light.

The very first reflex camera with a focal plane shutter was possibly the Holland (1891). It had a maximum speed of $\frac{1}{1000}$th of a second, but the film available was only ISO 1 to 3[165]; for comparison, when I became a serious photographer in the late 1970s, I typically purchased film in the ISO 25–400 range.

In 1909, there were two types of reflex cameras. "In the first, the mirror is raised by a spring upon the shutter release being operated, the mechanism releasing the shutter when the mirror has reached the up position, where it remains until put down. In cameras of the second group, the mirror is raised not by a spring but by the operator's pressure on the release key … and then, on further pressure releases the shutter. On removal of the finger from the release key, the mirror falls."[166]

Bear in mind that as soon as the mirror was raised, the viewfinder would go black, and it would stay that way until the mirror was returned. This was particularly a problem with the spring lift type. On the other hand, with the gravity return type, the mirror could fall back before the shutter had closed, cutting short the exposure.

Rather than have a separate lever for returning the mirror on the spring lift type, one could arrange for the (manual) film advance mechanism to also pull the mirror back down. But that would still mean enduring a blackout period of a second or more.

A much later development was the true "instant return" mirror—one actuated by the shutter mechanism after the exposure was completed, rather than by the shutter release. "This feature was first introduced on the Hungarian Gamma Duflex, whose first prototypes were made in 1946 and series production started in 1947.… However the first camera produced in quantity with an instant-return mirror was the Asahiflex II[B], released in 1954. The introduction of reliable instant-return mirror mechanisms and the subsequent elimination of the 'mirror blackout' was an important step in the acceptance of SLRs by a wide public."[167]

In the Asahiflex IIB, as the film was advanced, the shutter was cocked and a heavy spring was tensioned. Pressing the shutter release button freed this spring, which not only swung the mirror up (ultimately tripping "a latch that releases the shutter"), but also tensioned a second, light spring. "When the shutter closes, it trips a latch that releases the energy stored in the light spring. This energy is used to pull the mirror back down."[168]

An alternative to the movable mirror was a fixed but "semi-transparent pellicle mirror." On the Canon Pellix (1965), this transferred "30 percent light to the viewfinder and the rest (70%) to the film during exposure." The advantage was that it preserved through-the-lens viewing up to and through the exposure. The later Sony SLT α33 and α35 (2010) were digital cameras that combined a "built-in electronic viewfinder" with a "translucent" reflex mirror.[169]

Like the TLR, the early SLR was typically held at waist level. An early twentieth-century photographer suggested that this was a problem "not once in twenty times," and that when one had to shoot over a fence, one could "hold the camera in an inverted position at arm's length directly above the head, and view the picture by looking up into the hood." Alternatively, one could place an "accessory mirror in the hood or placed at an angle at the mouth of the hood.… The hood of the Graflex is permanently fitted with a mirror for this purpose." The accessory mirror was preferably front-silvered, to avoid ghost reflections.[170]

A camera with an eye-level viewfinder (vertical focusing screen) is easier to aim at the subject, since the lens axis is closer to the photographer's line of sight and the subject is in the photographer's peripheral vision. In 1933, Kurt Staudinger received a German patent on the use of a combination of mirrors to provide a "vertical, upright and laterally correct image."[171] Later, the Wrayflex I (1951), an SLR with an eye-level viewfinder, used two mirrors to erect the image, but left the image reversed. And there was a Wrayflex II prototype (circa 1957) that added two more mirrors in order to correct the lateral reversal.[172]

Another way of erecting an inverted image is to use a pentaprism. This, as its name implies, is a prism with five reflecting sides. The incoming light is reflected twice, emerging at right angles to the former. The image is inverted. Since the camera lens itself both inverts and reverses the image, the result is that the eye sees an upright but reversed image. A roof pentaprism replaces one side of the normal pentaprism with two surfaces at 90°, like a steep roof (so the prism is actually six-sided). The incoming light, reflecting off both of these surfaces, undergoes a compensatory lateral reversal, so the eye sees a normal image.

Both Wrayflex (in the United Kingdom, May 21, 1947) and Corsi Telemacho (in Italy, June 3, 1947) filed patent applications proposing the use of a pentaprism in a camera. The Telemacho design became the eye-level viewfinder of the Rectaflex SLR, which went on sale in 1948, whereas the Wrayflex II, with a pentaprism, didn't enter the market until 1959. The patent drawings show an ordinary, not a roof, pentaprism. Just to muddy the waters further, the Zeiss Contax S (1949) claimed to offer an "always upright and non-reversed" image, implying the use of a roof pentaprism.[173]

In the days before digital cameras with articulated viewfinders, there were also rotatable right-angle viewfinders that attached to the camera's eyepiece. This was useful for copy stand work, in which the lens was pointed down, or when the camera had to be held above one's head or by one's knees.

High-Speed Motion Picture Photography

Standard motion picture film cameras run the film behind the lens at a frame rate of 24 frames per second. To capture fast motion, you need a higher frame rate. However, the standard camera does not run the film continuously; it is advanced, held in place while the shutter is open, then advanced again.

The faster the advance, the greater is the stop-start stress on the film, and at "about 500 pps [pictures per second] the film needs to be moved continuously. This means that the image must also be moved in synchronism with the film."[174] This is called "optical compensation."[175]

"The usual way to achieve this is to replace the normal shutter with a rotating parallel-sided glass block (a square prism). The rotation of this prism ... moves the image in synchronism with the film movement, and at the same time acts as a shutter.... These cameras provide sufficient speed to capture most natural phenomena such as insect flight, a chameleon's tongue, or an exploding seedcase, and show them in ultraslow motion with excellent resolution."[176]

Once frame rates exceed 10,000 fps, film transport becomes problematic.

However, a fixed length of film could be placed inside a drum and an admittedly short-duration recording made as a result of the relative motion between the image reflected by a rotating mirror and the film.[177]

A *rotating* drum camera was invented by Cearcy David Miller in 1936; his interest was in studying "combustion and knock in the cylinders of internal-combustion engines." The camera was built in 1938 and was able to operate at "up to 40,000 photographs per second." His mirrors were right triangle prisms with the symmetric faces "coated with an aluminum film for back-surface reflection." These faces acted like "a pair of mirrors," and an unstated number of the prisms were "arranged in a continuous row around the inner surface of a rotating drum," with the "film at the outermost edge of the drum." "The optical system also included three lenses."[178]

In 1940, Miller filed a patent application describing a somewhat different approach. In his preferred embodiment, this featured a rotating hexagonal mirror, with concave faces. As it rotated, it would reflect the image to, successively, each of an array of relay lenses, disposed in a circular arc. These, in turn, would focus the image to a particular position on the surface of a stationary drum. Miller asserted that this design could achieve a frame rate of "up to 1,000,000 per second."[179]

This was one of several high-speed camera designs used by the Manhattan Project. In 1952, after the technology was declassified, Berlyn Brixner described the development of a high-speed camera of the stationary drum type that "operated at up to 3,500,000 frames per second." Its rotating mirror was "small, thin, two-faced."[180]

It is extremely important that these high-speed cameras be started at just the right time. For example, a frame camera used by Brixner at Los Alamos provided only "up to 96 consecutive pictures from a single event."[181]

Modern rotating mirror cameras use CMOS (complementary metal oxide semiconductor) sensors rather than film but work on the same principle; the Cordin Model 510 is capable of 25 million frames per second. It uses a helium-driven turbine and a beryllium mirror, but the captured sequence is only 128 frames long.[182]

The cameras already described are "frame" cameras. For studying the detonation wave front generated by the trigger for the atomic bomb, they weren't fast enough. Julian Mack developed (or supervised the development of) a "streak" camera that married the rotating mirror technology to that of the racetrack finish-line camera. Instead of having a shutter, it had a slit opening. On the racetrack version, the slit was at the finish line and the film was transported behind it, in the direction opposite the racehorse.[183] In the streak camera, the light passed through the slit and struck a face of a rotating mirror, which reflected it to a particular point on a stationary film. The image is distorted but it could "photograph events happening in one hundred-millionth of a second."[184]

This was not a new idea; in 1881 Arthur von Goettingen and Arnold von Gernet used a rotating mirror to produce a streak record of flame-front propagation on a photochronograph.[185] The latter is a rotating drum covered with photographic paper.[186]

The developers of the first atomic bombs intended to use conventional explosives to create a shock wave that would compress the fissile material, bringing it to critical mass and thus initiating the nuclear reaction. With uranium-235, this took "about one millisecond," but with plutonium-240, it was much faster: "approximately ten nanoseconds."[187] The streak camera revealed that this "implosion process" was not behaving as the physicists expected, and they had to revise their calculations.[188]

Photocopiers

In a photocopier, the document may be scanned by a rotating mirror and the image transferred, through a copying lens and a fixed mirror, to a rotating drum.[189] The drum doesn't carry film but rather is positively charged and photoconductive. Light reflected from the light areas of the document strikes the drum and causes electrons to be emitted, locally neutralizing the positive charge. Black toner is negatively charged and adheres where the drum remains positively charged. The copy paper itself is positively charged and the toner transfers to it.[190]

Chromoscopes and Color Projectors

In 1802, Thomas Young proposed that human color vision arises as a result of the stimulation of three different kinds of "fibers" (visual receptors), and in 1822, Hermann von Helmholtz hypothesized that they responded to red, green, and blue (or violet) light, respectively.[191]

In 1855, the physicist James Clerk Maxwell wrote:

> Let a plate of red glass be placed before the camera, and an impression taken. The positive of this will be transparent wherever the red light has been abundant in the landscape, and opaque where it has been wanting. Let it be put in a magic lantern, along with the red glass, and a red picture will be thrown on the screen. Let this operation be repeated with a green and a violet glass, and by means of three magic lanterns, let the three images be superimposed on the screen ... by properly adjusting the intensity of the lights, &c., a complete copy of the landscape ... will be thrown on the screen.[192]

The "additive primary colors" are nowadays considered to be red, green and blue (not violet). Add three such lights together and you get white.

It is obviously problematic to superimpose the projections from three different slide projectors. If the distances to the screen vary, they will be of different size, and even if the distances are the same, there will be distortion (keystoning).

A chromoscope is a viewer, allowing a single person to simultaneously view the three component transparencies in such a way as to form the complete color image. A chromoscope may be converted into a color projector (so a group may view the color image) by providing a sufficiently strong light source, replacing the eyepiece with a projection lens, and directing the lens at a screen or wall.

In Louis Ducos du Hauron's 1862 chromoscope, ordinary mirrors were used to

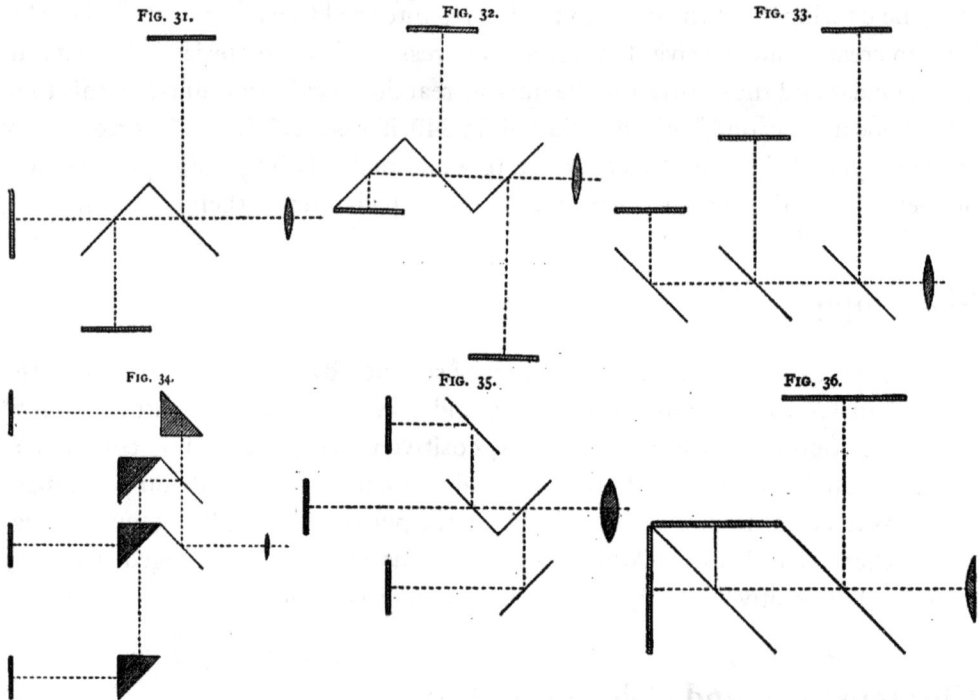

FIG. 31.   FIG. 32.   FIG. 33.

FIG. 34.   FIG. 35.   FIG. 36.

A potpourri of chromoscope mirror (or prism) arrangements, from Figs. 31–36 of Lecture III, in Shepherd, E. Sanger, *Cantor Lectures on the Photography of Colour* (Society for the Encouragement of Arts, Manufacturers & Commerce, 1900). The Fig. 31 configuration was devised by du Hauron in 1862.

shine light through the transparencies, into a box, and the component images were combined by three unsilvered diagonal mirrors, reflecting the light onto the optical axis of an eyepiece lens. The middle mirror reflected light from one side of the box, and the other two mirrors, from the other. The optical path lengths from the transparency to the eyepiece varied from one to the other, so the superposition would have been imperfect. In 1869, du Hauron adopted a form in which all three transparencies were on one side of the chromoscope (and thus could have been exposed on a single plate), but the optical paths still varied.[193]

    The first chromoscope to equalize the optical paths appears to have been Charles Cros's "chromometer" (1878?).[194] But it still had problems. "The optical path is too long, making the picture very small. Also, the glass mirrors, which both transmit and reflect, create annoying double images."[195]

    Frederic Ives's 1894 patent disclosed several solutions to the double image problem. His preferred solution was to make the "transparent mirrors of colored glass of such character" that the glass absorbs the color that the glass is intended to reflect. Thus, if the mirror is intended to reflect the red light, the glass is cyan. The reflection off the front surface is unaffected, but the light passing into the glass is absorbed before it can be reflected off the rear surface, or after reflection and before it reemerges.[196]

    Ives also came up with a more compact chromoscope by moving one of the diagonal mirrors outside the box. The reduction in light path length (from the transparencies to the eyepiece) enlarged the image obtained. This arrangement was used in Ives's commercialized stereoscopic chromoscope, the Kromskop (1895), with the refinement that a diagonal clear glass was inserted into the "red" light path in such a way that the "red" and "green" light would both pass through two pieces of glass on the way to the eyepiece.[197] There was also a projection version of the Kromskop.[198]

In 1907, the stereoscopic Kromskop sold for $50, the nonstereoscopic Junior Kromskop for $25, and their gas-burning "night illuminators" for $12 and $10, respectively. There was also the Lantern (projection) Kromskop for $65. The latter was to be "used in conjunction with any ordinary lime light or electric lantern," which was included with the $12 "special stand." "With the lime light good results are obtained up to four feet square, and with the arc light up to six feet square or more."[199] Per Measur-

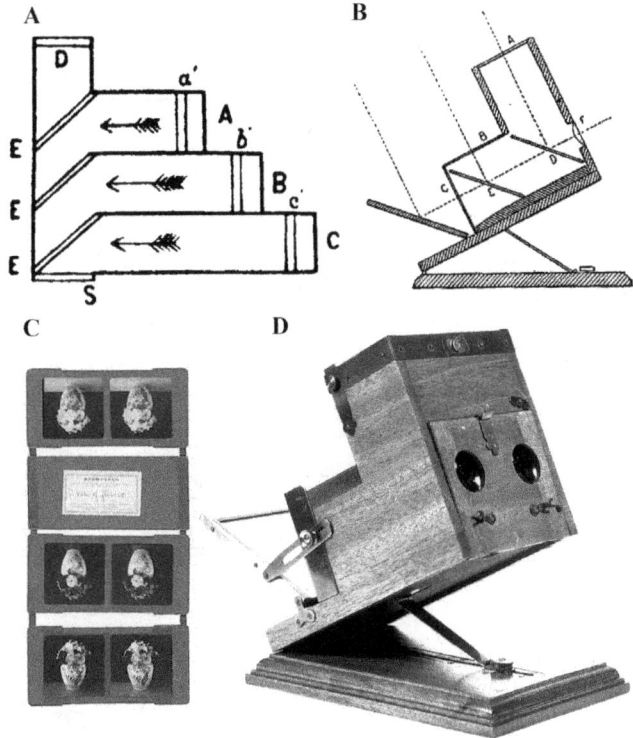

Color separation viewers and separation sets. A. Cros's chromometer, from Cros (1879). Light shines through transparencies across openings A, B, C, then through liquid color filters at a', b', c', and is reflected off or transmitted through the transparent plate glasses E, combining into a colored image seen by spectator S. B. Sectional view of Ives's Kromskop, from Ives (1898), 3. "A, B, and C are red, blue and green glasses, against which the corresponding images of the color record are placed when the instrument is in use. D and E are transparent reflectors of colored glass.... Beyond C is a reflector for illuminating the images at C—those at A and B being illuminated by direct light from above." Cf. Fig. 5 of Ives's 1894 patent, which discloses that this last reflector is opaque. C. Ives Kromogram set of glass stereo slides (circa 1895), depicting "vase of flowers." The three kromogram panels are connected by strips or cords; the geometry of the chromoscope required a greater separation between the top panel (red) and the middle (blue) one. Note also the inversion of the bottom panel (green) relative to the other two. (Item ST 23773.5, Museums Victoria Collection.) D. An Ives stereoscopic Kromskop (circa 1895) photographed by David Demant, in the collection of Museums Victoria. (Item ST 23773.1, Museums Victoria Collection, CC BY 4.0.)

ingWorth, one dollar in 1907 has the same purchasing power as $34.61 (based on CPI) or $79.51 (based on the "Consumer Bundle") in 2024.

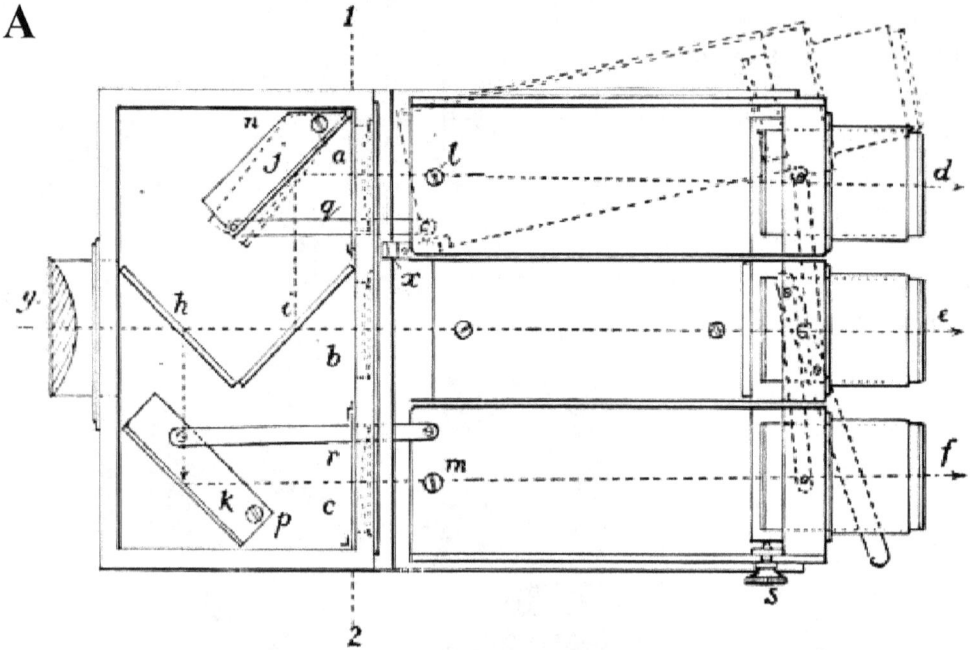

Lantern Kromskop (color projector). A. Schematic. B. Perspective view. (From Figs. 10 and 11, respectively, from Ives, Frederic, *Krōmskōp Color Photography* [Photochromoscope Syndicate Ltd., 1898].)

The Kromskop was used to view Kromograms, which were three-panel color separation negative holders. "Whilst the quality of colour in the Kromograms was praised, the system was prohibitively expensive and ultimately too bulky and complex to be a commercial success."[200] Also, the stereo Kromograms were shot by photographers with a poor understanding of how to compose a good stereo photograph.[201] Kromograms sold for $1.00 (stereo) or 75 cents (mono) in 1907.[202]

Color Separation Cameras

While colored lights are additive, colored inks are subtractive; they absorb the colors they do not reflect. The subtractive primary colors corresponding to the additive primary colors red, green and blue are cyan, magenta, and yellow, respectively. Add them equally to a white paper and you get black.

"Ives created what may be considered the first three-color halftone print" in 1881.[203] The standard four-color process uses cyan, magenta, yellow, and black inks (the latter is used because it is cheaper than combining three colored inks to obtain black).

Color printing typically involves the separation of a color image into its component subtractive primary colors. Each component image is printed in the corresponding ink, and since they are superimposed, the colors add to reproduce the desired color image.

The simplest color separation method was to photograph the color image three times, each time through a filter transmitting only light of one of the additive primary colors (red, green, and blue). Inverting the resulting component image yields the proper component cyan, magenta, or yellow image for printing. (The method of obtaining the black ink component is outside the scope of this book.)

"One-Shots"

Color separation cameras made it possible to obtain color photographs before color film was acceptable (cheap, fast, good color rendition, easy to get developed), and "one-shot" cameras simultaneously captured all three component images. The color separation cameras differ from the chromoscopes and color projectors of the last section in that the light path is reversed, so a full color image is separated into three component images, rather than three component images combined into one. The separation was effectuated by mirrors (plain glass or partially silvered) or possibly by prisms.

In 1874, du Hauron obtained a French patent which disclosed equalized light paths,[204] but I have not been able to locate a diagram showing its geometry. Two of the three mirrors were of plain glass, so there would have been reflections from both their front and rear surfaces.

In order to make Kromograms, Ives needed a suitable camera. "In his original camera, patented in 1892, the images are arranged in trefoil upon the glass plate."[205] However, in the commercial version, the images were "so arranged as to fill an oblong plate, the centers being on the line."[206] This "one plate, one exposure" camera

A

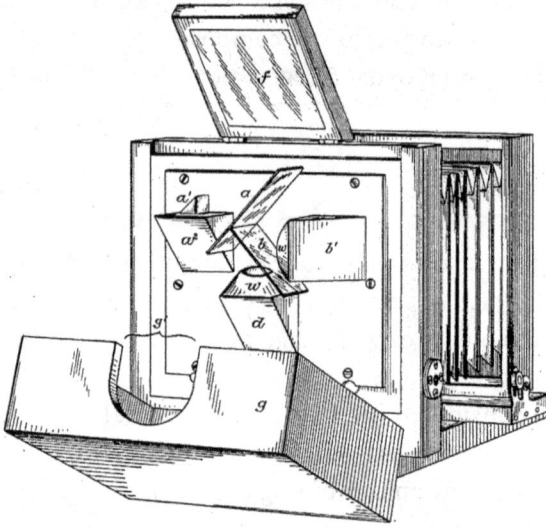

B

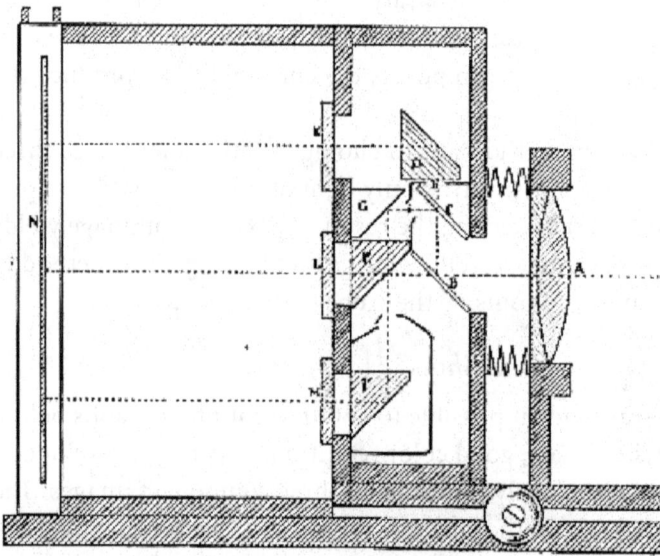

sold for $75 in 1907.[207] According to MeasuringWorth, that had the purchasing power of $2,596 (by CPI) or $5,963 (by "Consumer Bundle") in 2024.

"About three quarters of all the light goes to the lower image, which is made to represent the fundamental color to which the plate is least sensitive." Nonetheless, at least with 1898 emulsions, it was "necessary to give exposures of one minute and upwards on well lighted landscapes, and proportionately longer in the studio." This negated the nominal advantage of a "one-shot" color camera, namely, to permit shooting a moving subject.

There are many

A. Perspective view of the 1892 Ives camera. Light from the scene is reflected off mirror f and interacts with the beam splitting transparent mirrors a and b. The beam reflected off a is directed to the plate by prisms a1 and a2; and that off b by b1. The light transmitted through both mirrors is directed to the plate by prism d. (From Fig. 1, USP 475,084; patented May 17, 1892.) B. Schematic of the Kromskop camera. The light passes through lens A and then is split into three parts by transparent reflectors B and C ("of colored or thinly silvered glass") and directed to the oblong plate N at the back of the camera through the "silvered prisms" D, E, F, after passing through the color screens K, L, M. (From Fig. 13, Ives, Frederic, *Krōmskōp Color Photography* [Photochromoscope Syndicate Ltd., 1898].) Reflector B could be swung out of the way in order to have a brighter image while the photograph was being composed and focused (33). While Ives received several additional patents relating to simultaneously forming the component images on a single plate (or three plates arranged in a row), they appear to all disclose somewhat different optical systems. See USPs 655,712 (1900); 660,442 (1900); 668,989 (1901).

more patents on "one-shot" color separation cameras than those I have discussed here.

In 1923, Snodgrass wrote, "at the present time there is no three color camera available which is priced within the reach of the average worker."[208] In 1939, National Photocolor "Lerochrome One-Shot Cameras" were priced between $180 and 750, depending on film size.[209] According to MeasuringWorth, $180 in 1939 dollars corresponds to $4,086 (per CPI) or $9,116 (per "Consumer Bundle") in 2024 dollars; the CPI figure is about 10 percent more than a 2024 Nikon D6 35mm DSLR body. The same year, Hiscox suggested, without providing any design specifics, that "if the camera fan is handy with tools he can build" his own one-shot color camera.[210]

After World War II, "Truman, Attlee and Stalin met in Berlin," and a National Photocolor "One Shot" was used to photograph them. The "three color-separation negatives were radioed to the U.S. for rush printing."[211]

The principal disadvantage of "one-shot" color cameras using mirrors (or prisms) to split a single entrance beam three ways was light loss: "First of all there is a loss of at least 25 per cent of the light caused by absorption by the mirrors which make up the light splitter. Then there is a further loss of at least 66 per cent of the

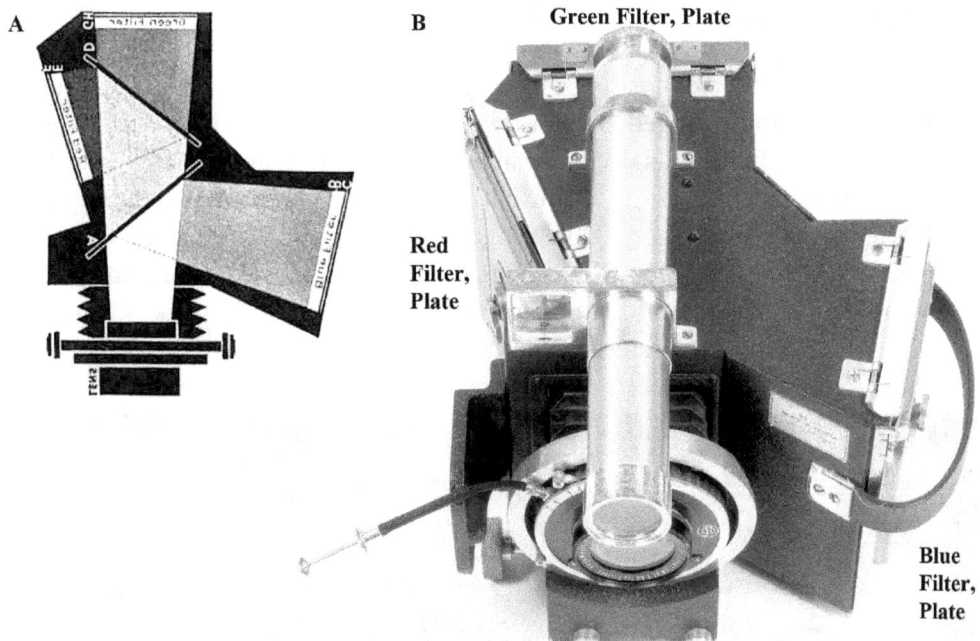

A. Schematic of the Devin Tri-Color Camera (Devin Colorgraph Co.). (From Lerner, Harry, "Color Photography and Engineering," *Technical Data Digest*, 10[1]: 64–69 [1944], at 65.) Note the odd shape. The schematic has been laterally reversed to match B. It used pellicle mirrors; see "Devin Tricolor Camera," http://camera-wiki.org/wiki/Devin_Tricolor_Camera, which also provides a photograph. The manual is available at https://www.butkus.org/chinon/ devin/devin.htm. B. Photograph of 6.5 × 9 cm Devin Tri-Color camera (manufactured circa 1935). (Courtesy of Scott Bilotta, the photographer and copyright owner. Additional photos are available on his webpage, http://www.vintagephoto.tv/devin_img.shtml.)

remaining light due to the fact that the lens beam is divided into three equal parts, and each sub-beam is used to make an exposure through a filter." Finally, the filters themselves absorb some of the light they are supposed to transmit; by way of example, of the filters used in 1947, the red filter transmitted almost 80 percent of the red light, but the green and blue filters transmitted less than 40 percent of their nominal color.[212] Of course, the greater the light loss, the longer the necessary exposure was.

As you might imagine, some inventors thought it simpler to dispense with mirrors and just use three lenses, each with a different filter. But no matter how those lenses were arranged, they would not see quite the same scene, and thus the resulting negatives could not be properly registered (aligned) for printing.[213]

Repeating back cameras were also developed. These took three exposures in quick succession, all through the same lens, but through different color filters. If used to photographic a static scene, they avoided the registration problem. However, moving subjects would generate "color fringing."[214]

Color separation cameras and chromoscopes became obsolete as a result of the advances in color film technology.

Montage Camera

The classic method of creating a photomontage was by "multiple printing or … cutting out and binding together color transparencies into a thick, unwieldy 'sandwich.'" Jon Abbot built a "dual-view" camera with a cube-shaped body. It had "two sets of bellows, each containing a conventional lens … mounted at right angles to each other on adjoining 'faces.'" The ground glass screen and film holder were on a third side. A "half-silvered pellicle mirror is mounted inside of the camera body at an angle of 45 degrees to the focal plane." The main lens'd image is transmitted through the mirror; the side lens'd image is reflected; both are thus directed to

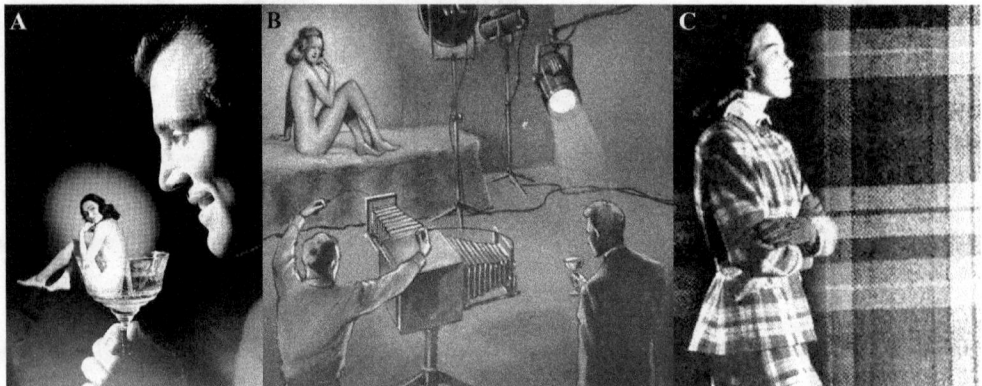

A. Abbot's nude-in-champagne glass photograph. B. Abbot's studio setup. C. Photo of Abbot's camera. D. Fashion model with fabric pattern. (From Lipton, Norman C., "Illusion," *Popular Photography*, August: 34–5, 114–15 [1949].)

the focal plane, although the reflected image undergoes a lateral reversal. In 1949, National Photocolor sold the "NPC Montage Camera," designed by Abbot, for "about $1000."[215]

The camera was intended for studio photography, and the two shots didn't need to be taken simultaneously. Indeed, they might have received different lighting and exposures. They could combine distance and close-up shots. In one Abbot photograph, a nude seemed to be sitting with her knees folded over the lip of a partially filled champagne glass held in the hand of a man looking down at her. (A sketch shows how this was set up.) "However, a more common use was posing a fashion model with an enlarged pattern of the fabric she was wearing."[216]

Ophthalmoscope

Every year, perhaps, we visit our eye doctor for a checkup. At some stage in the process, the ophthalmologist shines a light into our eye, studying it intently. We take this examination for granted, because we have never thought out how it is done.

Look into the pupil of someone's eye, and it looks black. Try to shine a light in while you look at the pupil face-on, and your head blocks light from entering into the other's eye in the direction needed to reflect it back in your direction. Oblique illumination does not work; the human retina is only poorly reflective and oblique light rays would be multiply reflected before they could emerge through the pupil; the emergent light would be too feeble. How then, do eye doctors actually see into your eye?

In 1847, Charles Babbage (1791–1871), best known as the inventor of a mechanical computer, showed the ophthalmologist Thomas Wharton-Jones a primitive ophthalmoscope. It "consisted of a bit of plain [*recte*, plane] mirror, with the silvering scraped off at two or three small spots in the middle, fixed within a tube at such an angle that the rays of light falling on it through an opening in the side of the tube, were reflected into the eye to be observed, and to which the one end of the tube was directed. The observer looked through the clear spots of the mirror from the other end."[217] Unfortunately, Wharton-Jones was near-sighted, and without placing a "weak diverging lens" behind the mirror, all he would have been able to see was a "red splodge." Therefore, he told Babbage that the device was worthless.[218] Babbage's design was not published until 1854.[219]

The *augenspiegel* ("eye mirror") proposed by physicist and physician Hermann von Helmholtz (1821–1894) in 1851 used an unperforated mirror. Helmholtz recognized that if a plain glass were placed between the physician's eye and that of the patient and tilted so that the reflected ray from a candle lay on the same line, the reflected ray would enter the patient's pupil and be reflected straight back by the retina. Some of that light in turn would be transmitted through the glass and reach the physician's eye. "By this arrangement," Helmholtz said, "the pupil of the eye … appears to shine with a red light."[220]

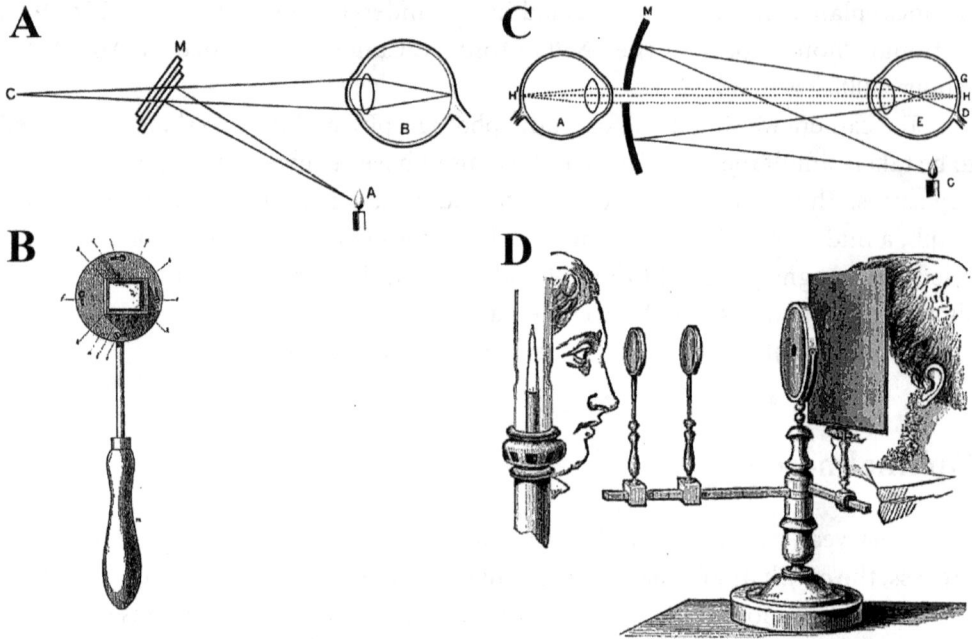

Early ophthalmoscopes. A. Schematic of Helmholz ophthalmoscope. (Hartridge, Gustavus, *The Ophthalmoscope* [J&A Churchill, 1907], Fig. 27.) B. Frontal view of Helmholtz ophthalmoscope. (Helmholtz, supra, Fig. 2.) C. Schematic of Ruete ophthalmoscope. (Fig. 28, supra.) D. Perspective view of Ruete ophthalmoscope. (Fig. 4197, Buck, Albert H., ed., *A Reference Handbook of the Medical Sciences*, Vol. 6 [W. Wood, 1916].)

However, "should it be desired … to recognize distinctly the structure of the retina…, then one must endeavor to make the illumination as strong as possible. That can be done in two ways, namely, by a proper choice of the angle under which the incident light is reflected from the mirroring plane, and by an increase in the number of the reflecting plates."[221]

Helmholtz, as a physicist, was aware that Fresnel had developed equations for quantifying the intensity of reflected light as a function of the angle of incidence (and refractive index), and he extended them to the situation of multiple parallel plates. However, in terms of maximizing the light ultimately reaching the physician's eye, he had to consider that the brighter the light shining into the patient's eye, the more the pupil would contract. He concluded, semiempirically, that the best angles of reflection were about 70 degrees if one plate was used, 60 degrees for three plates, and 55 degrees for four.[222] However, the angle that was best for minimizing reflection from the cornea was about 56 degrees,[223] for reasons having to do with the polarization of light (see Chapter 7).

In practice, for the physician to obtain a good focus on the patient's retina, a concave or convex lens is interposed in the optical path, depending on the specifics of the physician's and the patient's eyesight. In 1852, Egbert Rekoss added "2 rotatable discs, each containing a series of lenses," to Helmholtz's ophthalmoscope.[224]

Babbage's perforated mirror ophthalmoscope was reinvented by Franz Donders.

However, the flat mirror that both employed was an "inefficient gatherer of light." In G. Theodor Ruete's 1852 ophthalmoscope, the light source was placed beside the patient's head, level with but on the contrary side to the eye to be examined, and its light was reflected into the latter by a concave mirror, which could be turned laterally as needed. The standard model had a mirror "three Paris inches in diameter, and 10" focal length." This mirror had a central perforation, through which the physician could see the reflection from the retina, and screens were provided so neither the physician nor the patient saw the direct light of the light source. Between the patient's eye and the mirror there were two lenses, mounted pivotally on pedestals which could be moved along a horizontal rod, closer to or further away from the patient. It was a table unit, whereas the Helmholtz ophthalmoscope was handheld.[225]

The ophthalmoscope "permitted the clinical correlation of signs and symptoms with findings in the retina, vitreous, and optic nerve."[226] The subsequent evolution of the ophthalmoscope included the provision of more intense light sources, binocular views, and image capture attachments.

A 2000 article noted that "medical students are expected to become competent at fundus examination with a direct ophthalmoscope," but most don't have their instrument because of its high cost—even a "less expensive" modern ophthalmoscope then cost £80–130. Roger Armour described a homemade ophthalmoscope, with a single "reflecting card," made for just 75 pence (mostly the cost of the pen torch), and useful for practicing the observation of the retina.[227]

Seeing Inside the Body: The Endoscope

The ancient Chinese Emperors only claimed to have mirrors that allowed them to see inside the body; scientists developed the real thing, the endoscope (*endo*, inside + *scope* to see).

Before the endoscope, there was the simple speculum, used by Hippocrates II (460–375? BCE) "for investigation of the rectum." And Abu Al-Qasim (d. 1013 CE) "held a glass mirror in front of the vulva and thus reflected light into the vaginal vault."[228]

The endoscope combined illumination and inspection in a single device. The first such instrument was the *Lichtleiter* ("light conductor") of Philipp Bozzini (1773–1809), a Frankfurt physician. According to his 1806 article, it was a large, tripartite instrument, "fashioned in the shape of a vase," and "measuring thirteen inches in height, roughly three inches in width, and approximately two inches in depth." The middle part contained the light source, a "wax candle ... held in place by springs."[229]

According to Engel, a "concave mirror" was "situated behind the candle," to project "the light through an opening in the front while a fenestra on the opposite side was fitted with an eyepiece."[230] Sircus says that Bozzini "peered through a small hole in the center of the mirror."[231] Ramai reports that "Bozzini used double aluminum tubes to separate lighting and imaging. The tubes were angled with mirrors to relay images back to his eye while concurrently illuminating distal parts of internal anatomy." (These tubes and mirrors would have been in the upper part.)[232] Ellison asserts that "a combination of flat, concave, and convex" mirrors were placed inside the tubes "in such a manner that the image was transmitted to the eye while the image of candlelight was reflected to the distal tip of the instrument and into the interior body."[233] It is difficult to reconcile these disparate descriptions, but there is no doubt that his endoscope used mirrors. Bozzini developed attachments for inspection of the "vagina, urethra, the female bladder, the rectum and the upper air passages."[234]

It is known that his device was successful for examination of the rectum and vagina prior to surgery. Nonetheless, Bozzini ran afoul of both the "not invented here" syndrome (he was a refugee from French-occupied Mainz) and the rivalry between the Vienna-Josephs Academy (which controlled military medicine) and the Vienna Medical Faculty.[235] The device also had its limitations; "the light conductor was much too weak and the field of vision too small."[236]

The next major advance was the portable endoscope of Antonin Jean Desormeaux (1815–1894). In 1853, he generated a brighter light than that of a candle by burning a mixture of alcohol and turpentine.[237] "His endoscope was a system of mirrors and lenses…. One of the disadvantages was the enormous heat generated by the light source, which led to burns."[238] After 1885, a new light source, an incandescent electric filament, entered endoscopic practice. This could be water-cooled.[239]

Endoscopes were rigid instruments until the 1930s. They could be difficult to insert into a patient; the first gastroscope was demonstrated in 1868 with the assistance of "a professional sword-swallower." And there was a danger of perforation.

A semiflexible endoscope was developed by Rudolf Schindler and George Wolf in 1932; this had a "lower third" consisting of "a flexible bronze spiral covered in rubber," and containing "an inner tube filled with short-focus lenses which could be bent in any direction to an angle of 34° without visual distortion." Later, means for controlling the movement of the tip were devised.[240]

The last invention of interest to us was the fully flexible endoscope employing fiber optics. This was first used on a patient in 1956.[241]

One-Way Mirrors

There really is no such thing as a "one-way" mirror, that is, a mirror which can only transmit light in one direction. If there were, you could build a box in which all the walls were one-way mirrors, oriented so that the light could get in, not out. Light is energy, so you would be collecting energy at no cost.[242] A "one-way" mirror is really a "two-way" (transparent) mirror that is lit on only one side.

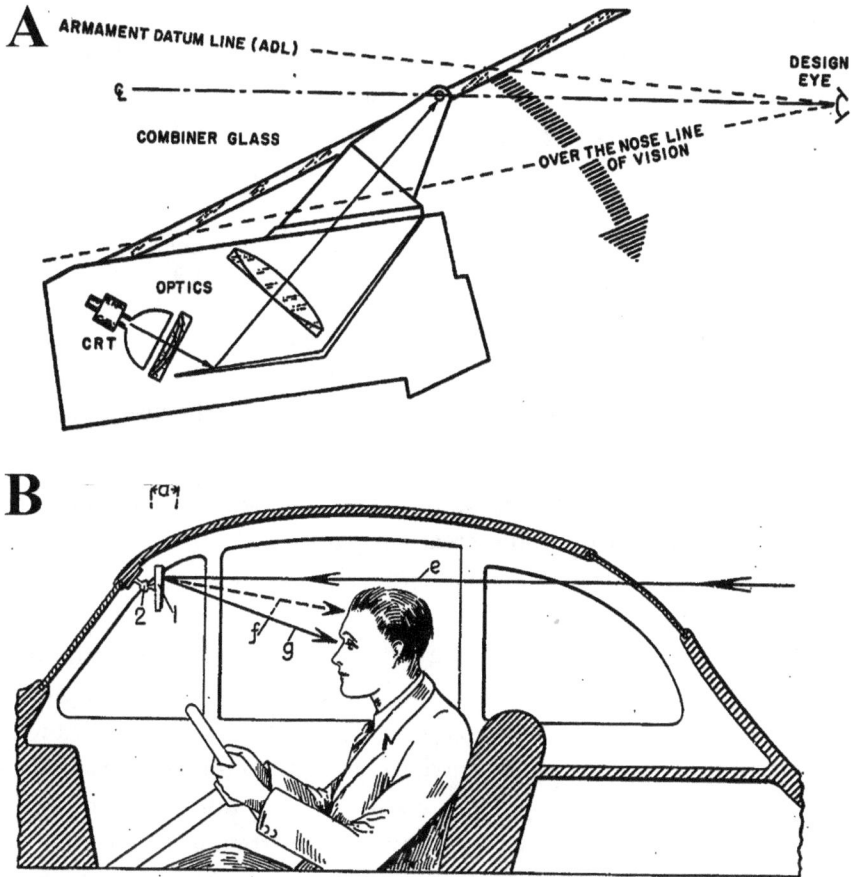

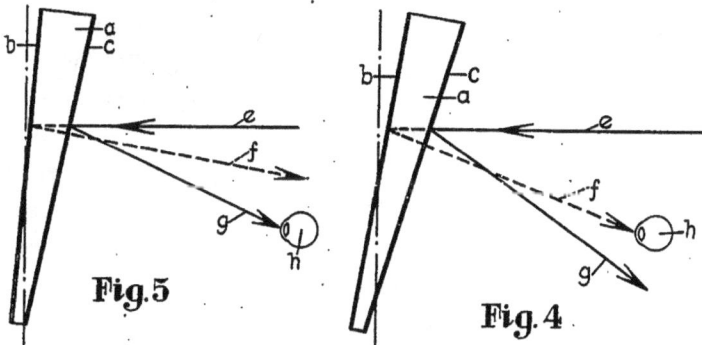

A. Schematic of the AN/AVQ-7 head-up display. Flight and targeting data are processed and displayed as symbols on the CRT. The CRT image is reflected, and the reflected rays pass through a lens and then are reflected by the "combiner glass" (the partially reflective mirror) toward the pilot's eyes. (Fig. 10-2 in Reeves, Johnny R., *Aviation Fire Control Technician 3&2, Part 2*, NAVEDTRA 10387-D [GPO, 1984].) B. Rearview mirror of Colbert. Patent Fig. 3 shows the mirror in use, in night driving position. Patent Fig. 4 shows the mirror in day driving position. The rear view is reflected to the driver's eye by the back surface. Patent Fig. 5 shows the mirror in night driving position. The view is reflected to the driver's eye by the front surface.

A transparent glass also reflects light from both sides. If lit on both sides, then you can see through it in both directions. But now suppose you light one side and not the other. Someone on the lit side would just see a reflection, as the reflected light would be much stronger than the light transmitted from the dark room. An observer on the dark side would see only a dim reflection of the dark side and hence would see through to the lit side.

The "one-way mirror" principle explains several phenomena so commonplace that we take them for granted:

- If we look at an office building at night, dark offices have windows that look like mirrors, while lit offices can be seen into.
- We see our reflected image in a window more clearly when we look out it at night.

It is not clear when it was first recognized that the combination of an unsilvered glass and differential lighting would create the "one-way" mirror effect, but it was certainly known by the late nineteenth century, when it was exploited in the "Pepper's Ghost" illusion (see Chapter 6).

Modern "one-way" mirrors increase reflectivity by means of a thin, partially reflective, aluminum film or coating, but the first reflective coating used was silver.

A U.S. patent on a "transparent mirror" was awarded to Richard Wilson in 1903. It contemplated placing a coating of amyl acetate, guncotton, and fusel oil over "a silvering such as is usually employed for making ordinary mirrors." Wilson claimed that this "rendered the same transparent without destroying its reflective properties."[243] While the coating might have protected the silver, it would not have made it more transparent.

Another patent was awarded to Emil Bloch in 1903, but it claimed a specific reflective treatment (Rochelle salts and tin chloride). His argument was that the normal thin silver coating was "easily destroyed and is of very short life when used for an advertising display device," whereas his coating was cheaper and more durable. He mounted electric lights behind the mirror and placed an advertising poster between the mirror and the lights. When the lights were off, the mirror seemed ordinary. When the lights were turned on, the poster was illumined from behind.[244]

"One-way mirrors" are typically used in malls, casinos, department stores, and police interrogation rooms (so a witness can see a suspect but not vice versa).

They are also used to superimpose information on a visual field—for example, in the head-up display in a fighter plane. Thanks to a partially reflective mirror, a ghostly radar display can float before the pilot's eyes.[245]

Rearview Mirrors

Iron mirrors have been found in Iron Age "chariot burials" in England, and Friend speculated that these mirrors were mounted on the chariot "to enable the driver to see what is going on behind" him.[246] But it is doubtful that this was the case. The mirrors were "mainly found at female graves" and had ornamental handles.[247]

Hero of Alexandria wrote that it was possible, by means of mirrors, "to see those who are coming behind."[248]

The invention of the rearview mirror is often attributed to Ray Harroun, although accounts differ as to exactly why he employed it. The Indianapolis 500 racing track, a 2.5-mile oval, was completed in 1909. The first race at the Speedway was held May 30, 1911. One source says that in those days, mechanics usually rode along with drivers (a commentary on the mechanical reliability of early racing cars?) and watched out for traffic. However, driver Ray Harroun chose to leave his mechanic behind and drive with a rearview mirror. He won the race, perhaps because a rearview mirror is a lot lighter than a human observer.[249] Another source states that in the race he was driving his Marmon Wasp, and that it was a one-seater with a configuration that "made it difficult for him to turn his head and look behind him."[250]

Harroun did not claim to be the inventor; he said that "in 1904 he saw a driver of a horse-drawn buggy using a rearview mirror."[251]

In 1909, Dorothy Leavitt suggested that women drivers carry a "fairly large" mirror "with a handle" when on the road, and "occasionally hold [it] up to see what is behind you."[252]

In 1914, Chester Weed acknowledged that "it is now usual for an automobilist to apply to his car, either to the wind shield or the fender, or to some other accessible place, a reflecting mirror." And he noted that these mirrors were of two types, either a convex (reducing) or flat (true) mirror. The reducing mirror gave a "full rear view" but was "deceptive" as to the distance, whereas the true mirror gave "only a comparatively small view of the road in the rear." He patented "a reflecting mirror attachment for automobiles, comprising a true mirror and a reducing mirror, and means for supporting it."[253] The mirrors could either be separate but attached to a common rod, or one could be set within the other.

An ad in *Hardware World* (1920) advertised Berger & Company's "Cop Spotter" rearview mirror, made of plate glass and mounted in a "nickel-plated frame." This was an aftermarket accessory that could be "attached and adjusted on any open or closed car in a few minutes." The price was four dollars.[254] The ad said "patent pending," but Elmer Berger failed to actually obtain a patent.

A problem with a silvered glass mirror was glare at night. The day/night rearview mirror was a rear surface mirror in which the front surface and the "silvered" rear surface were not parallel; that is, it had a wedge-shaped cross-section. In daytime use, you would see the reflection off the rear surface. When you flipped the control lever to the "night position," the rear surface would be aimed upward, and you would see just the weak reflection off the front surface.[255] Walter Bell's first patent (1929) on such a "prismoidal" mirror contemplated use of colored glass, to further reduce glare, but his second (1934) said that it was "preferably transparent."[256] William Colbert, considering the reflectivity (4.5%) of the front surface glass to be inadequate for night driving, patented applying to it a "mirror film of semi-transparent chromium of a reflectivity of approximately 10 per cent to 30 per cent."[257]

Rearview mirrors aren't limited to vehicles. A 1953 *Popular Mechanics* article referred to sunglasses equipped with two "small mirrors to provide a backward view.... The glasses can be used to advantage by cycle deliverymen, hunters, detectives and fishermen."[258]

CHAPTER 4

<<<<<<<<<<<<<<<<<<<<<<<<<<<<<<<<<<<<<<<<<<<<<<<<<<<<<<<<<<<<<<<<<<<<<<<<

Solar Reflections

Archimedes's Burning Mirrors

Archimedes of Syracuse (287–212 BCE) is famed for many mathematical and engineering feats, including allegedly setting Roman warships on fire with "burning mirrors." We can ask three questions about the Archimedes saga:

- Could the feat be performed at all?
- Could the feat have been performed in classical times?
- Did Archimedes actually do it?

The Historical Record

The historical evidence for the exploit is meager. Three early accounts of Archimedes's defense of Syracuse from the Roman attack in 213 BCE make no reference to burning mirrors whatsoever: Polybius (Greek, circa 200–118 BCE), *Universal History*; Livy (Roman, 59 BCE–17 CE), *History of Rome from its Foundation*; and Plutarch (Greek, circa 45–120 CE), *Parallel Lives* ("Marcellus").[1]

According to Gibbon, the "conflagration is hinted by Lucian (in Hippia, c. 2) and Galen, (l. iii. de Temperamentis, tom. I. p. 81, edit. Basil) in the second century."[2]

Anthemius of Trailes's treatise, *On Burning Glasses* (circa 500 CE), "mentions offhandedly that Archimedes may have used a parabolic mirror to focus the sun's rays on the invading Roman ships. He also gives a description of what Archimedes' burning mirrors might have looked like, but, as he states, his description is strictly his own creation."

While Book XV of the *Roman History* of Dio Cassius (Greek, circa 155–235 CE) described the siege, it survives only in the form of paraphrases from twelfth-century Byzantine authors John Zonaras and John Tzetzes.

John Zonaras, *Epitome ton Istorion* 9, 4, reported:

Marcellus crossed into Sicily and proceeded to besiege Syracuse. The city had submitted to him, but then had revolted again as the result of a false message sent by the treachery of certain men. Now he would have subdued it very speedily, as the result of a joint assault upon the wall by land and sea, had not Archimedes with his inventions enabled the inhabitants to resist for a very long time. For this man by his devices suspended stones and heavy-armed soldiers in the air, and these he would let down suddenly, and presently draw them up again. And he would lift up ships, even those equipped with towers, by means of other appliances which he dropped upon them; and raising them aloft, would let them drop suddenly, so that when they fell into the water they were sunk by the impact. *At last in an incredible manner he burned up the whole Roman fleet. For by tilting a kind of mirror toward the sun he concentrated the sun's beam upon it; and owing to the thickness and smoothness of the mirror he ignited the air from this beam and kindled a great flame, the whole of which he directed upon the ships that lay at anchor in the path of the fire, until he consumed them all.*[3]

John Tzetzes, *Book of Histories* (Chiliades) 2, 109–128, declared:

And when once Marcellus, the Roman general, was assaulting Syracuse by land and sea, this man [Archimedes] first by his engines drew up some merchantmen [ships], and lifting them up against the wall of Syracuse dropped them again and sent them every one to the bottom, crews and all. Again, when Marcellus removed his ships to a little distance, the old man gave all the Syracusans the power to lift stones of a wagon's size, and hurling them one at a time, to sink the ships. *When Marcellus withdrew them a bow-shot thence, the old man constructed a kind of hexagonal mirror, and at an interval proportionate to the size of the mirror, he set similar small mirrors with four edges, moving by links and by a kind of hinge, and made the glass the centre of the sun's beams—its noontide beam, whether in summer or in the dead of winter. So after that, when the beams were reflected into this, a terrible kindling of flame arose upon the ships, and he reduced them to ashes a bow-shot off.* Thus by his contrivances did the old man vanquish Marcellus.[4]

Tzetzes's "hexagonal mirror" seems inspired by Anthemius's hypothesis.[5]

One way of resolving the disagreement in the history books would be to show that, on the basis of science and engineering considerations, Archimedes could not possibly have accomplished and lived up to the legend.

Mirror Shape

Some scholars believed that the "burning mirror" tale is unbelievable simply because the paraboloid mirror was not known in Archimedes's time. The first written description we have of its properties is that of Diocles, many years later. However, it is quite possible that the focusing properties of the parabolic mirror were known to Archimedes; he wrote a text on catoptrics that is now lost to us.

Moreover, even a spherical mirror will concentrate light; it just won't focus parallel rays to a single point and hence is less efficient than a paraboloid mirror.

In any event, knowing what shape would work and being able to make it are two different things. In 1747, Georges-Louis Leclerc, Comte de Buffon (1707–1788), pointed out that it would have been impossible to accurately fabricate a metal or glass concave mirror with a diameter of 15–20 feet and a focal length of 200 feet, even with a spherical rather than a paraboloid shape.[6]

Giant Mirror Versus Mirror Array

Moreover, a paraboloid mirror is advantageous only if one wishes a mirror to focus to a single focal length. Even if the Syracusans constructed a perfect, bronze paraboloid mirror, it could be used effectively only at one distance. They would have to wait for the enemy ship to come into range and hope that it would catch fire before it lumbered so close that the mirror became ineffective once more.

It would be better to use an array of mirrors, whose elements could be adjusted as needed to provide the right focal length. (Tzetzes's reference to mirrors hinged to a central mirror suggests that he, if not Archimedes, contemplated an adjustable array, and a much larger array was tested by Buffon in 1747, described later.) Ideally, the mirror array would approximate the surface of a paraboloid. But even a spherical array of mirrors would be more useful than a single paraboloid mirror, unless the latter was able to burn the target effectively instantaneously.

If Archimedes had used an array of mirrors, one imagines that as a practical matter, the mirror-bearing defenders would be lined up along a straight wall. In the solar energy industry, some solar concentrators are reflectors that are straight in one dimension and curved in another. The ideal curve is then the parabola, and if the straight dimension is horizontal, the concentrator is called a parabolic trough. One could, of course, have instead a circular trough, and it would be less effective than the parabolic trough, which in turn would be less effective than a paraboloid reflector (which has a parabolic cross section in two dimensions).

The trough concentrates light along a line of focus, rather than a point. There would have been no advantage to construct a trough with a length greater than the length of the target ships, and if they remained bow first, then a trough length greater than the beam would be wasted.

The attentive reader may also object that the paraboloid mirror focuses light to a single point only if the sun and the target are both directly in front of the mirror. Rorres points out since the ship is at sea level and the mirror on top of a wall, this cannot be the case. To achieve focus, one must therefore use an "off-axis" paraboloid mirror and these "are quite difficult to design and construct."[7]

However, one can use a normal "right" paraboloid mirror (symmetric around its axis) and tolerate the resulting coma (inability to similarly focus rays which are oblique to the optical axis of the mirror). While coma reduces the concentration efficiency of the mirror, it does not render it useless. Moreover, if the paraboloid surface is approximated by an array of mirrors, then it is possible to adjust individual mirrors to compensate for the off-axis position of the sun and the target, so as to improve the focus.

Mirror Elements

Some visualize the array as being made of shields. In the Hellenistic period, shields were made primarily of wood, although they might have a boss, rim, or even a very thin face sheet of bronze. Only the last would be useful in this context. Moreover, at least some shields were curved, to give some protection to the soldier's flanks.[8] They would present a convex surface to the sun, which is not what you want if you are trying to concentrate sunlight. So, Syracuse would need to have a large stock of flat sheets of bronze in order to assemble the mirror array. This itself seems improbable.

Ignition Efficiency

How quickly the mirror can ignite its prey depends on many factors: the heat flux (the rate of flow of heat energy) into the target, the target's initial and ignition temperatures, the target's specific heat capacity (the amount of heat energy needed to raise the temperature of a unit mass of the target by a single degree), the target's mass, and the target's ability to lose heat by convection, conduction, and radiation.

If combustion were not instantaneous, it would be necessary to adjust the focal length of the mirror as the enemy ship approached (and the sun moved through the sky). That is simply not possible with a conventional concave metal mirror.

Thus, Buffon's proposal—that Archimedes could have used an array of small mirrors[9]—is important. The orientations of these mirrors could have been adjusted, as needed, to keep a target in focus.

There is a loss of efficiency as a result of using an array of small mirrors, but it is surprising small for a long-focal-length mirror. Suppose one were simulating a single spherical mirror with a diameter of 10 feet and a radius of curvature of 400 feet (i.e., a focal length of 200 feet)—the specifications Hunt attributes to Buffon's experimental mirror array (see later discussion).

According to Hunt, such a mirror will produce an image of the sun, at the focal point, about two feet in diameter (this result is dependent on just the focal length of the mirror, and follows from the angular size of the sun being about 0.5 degree). "If the spherical mirror is simulated by a combination of plane elements of diameter d, then the image is increased in size by about d. Thus, if the elements are six inches in diameter then the two foot image is smeared out to two and one half feet."[10]

The Sighting Problem

The mirror has to be aimed at a point halfway between the sun and the target. If you have the only mirror in action, and your aim is reasonably good, you will see a spot of reflected light on the target, and you can adjust your aim as need be.

Confusion arises if there are many mirror-warriors, each trying to point sunlight at the target. If one mirror-wielder is pointing too high, another too low, another too far to the left, and yet another too far to the right, how will they sort out whose mirror is making which error? All the spots of light will look alike! This is why, when an observation post is sighting for artillery, one artillery piece fires at a time. An anonymous contributor to the Solar Concentrator Archives wrote:

> I tried this experiment with 20 visiting students on a field trip. We used one foot mirrors left over from the construction of a 30 foot dish. Individually we could reflect sunlight onto a cardboard box in the field. But collectively we could not. Nobody could tell which image was theirs so the reflected sunlight spots hovered about the box like a swarm of bees.[11]

If the target is fixed—a Roman ship at anchor, perhaps—then one solution is to aim the mirrors one at a time (all others being covered up) and finally, when all mirrors are pointing in the right direction, expose them to the sun. Beaty described a solar furnace constructed using an array of small mirrors. He advocated covering up all the mirrors but one initially, aiming the array so that one mirror is reflecting light to the correct spot, and then exposing the remaining mirrors, one by one, and adjusting them to reflect light to the same point. Of course, his furnace has a fixed focal point.[12] And even with an immobile target, one would still need to correct thereafter for the movement of the sun.

Claus suggested that Archimedes could have had his followers use a design similar to that of a modern signal mirror (see later description). One signal mirror design is a two-faced mirror (i.e., front and back are both reflective), with a hole near the center. The Syracusan tries to aim the mirror, as previously explained, halfway between the sun and the target. As a result, sunlight passes through the hole and a spot of light appears on the Syracusan's body or clothing. The reflected image of this spot is visible in the back-facing mirror. The Syracusan looks at the target through the hole, without blocking it, and tilts the mirror so that the reflected image "disappears" into the hole. At this point, the front-facing mirror must, in accordance with the law of reflection, be reflecting sunlight onto the target.[13]

But bear in mind that modern signal mirrors are small, typically 2 by 3 or 3 by 5 inches. I think that proponents typically visualize the use of mirrors that are the size of a heavy infantry soldier's shield, that is, several feet tall. With such mirrors, finding and maintaining the right angle of tilt would not have been easy, and not something the soldiers were trained to do.

Experimental Reconstructions

Buffon purchased over 500 flat mirrors, and of these, selected the 168 of highest optical quality. These were described as "*glaces communes*" (common glass) or as "*glaces étamées*" (tinned glass). Each of the mirrors was 6 by 8 French imperial inches in size.[14] Buffon assembled these mirrors into various arrays, and concentrated sunlight on a variety of targets at different distances.

Buffon's 1747 Burning Mirror Experiments

Number of mirrors	Distance	Target	Date; comments	Effect
40	66	cresoted beech plank	March 23, 1747, noon	ignited
98	126	cresoted and sulfurized beech plank	same day, 1 p.m.	Ignited
112	138	board covered with chopped wool	April 3, 4 p.m.; sun feeble	light inflammation
154	150	cresoted plank	April 4, 11 a.m.; sun very pale and covered by clouds	smoke produced in less than 2 minutes
154	150	mixture of sulfurized pine chips and coal	April 5, 3 or 4 p.m.; under partly cloudy skies	ignited in less than 1½ minutes

Number of mirrors	Distance	Target	Date; comments	Effect
128	150	cresoted pine plank	April 10, afternoon, clear	instantaneously ignited; spot about 16 inches in diameter
12	20	small combustible materials	April 11	ignited
21	20	beech plank		burnt in part
45	20	flask of six *livres** of tin		melted
117	20	silver chips	same day	melted

*One livre poids de marc *was 489.5 grams, and the English pound is 453.6 grams (Wikipedia).*

Edward Gibbon, in *Decline and Fall of the Roman Empire*, commented:

Without any previous knowledge of Tzetzes or Anthemius, the immortal Buffon imagined and executed a set of burning-glasses, with which he could inflame planks at the distance of 200 feet (Supplement a l'Hist. Naturelle, tom. I. 399–483, quarto edition). What miracles would not his genius have performed for the public service, with royal expense, and in the strong sun of Constantinople or Syracuse?[15]

In view of Buffon's success, Gibbon was reluctantly persuaded to give some credence to the "burning mirror" tale: "since it is possible, I am more disposed to attribute the art to the greatest mathematicians of antiquity, than to give the merit of the fiction to the idle fancy of a monk or a sophist."

Nonetheless, in 1807, François Peyrard pointed out that Buffon's array took too long to adjust to a given range. According to John Scott, Peyrard argued that even with the large number of assistants employed by Buffon, it took "about half an hour" to adjust the 168 mirrors, and "after a short interval, a readjustment would be rendered necessary" by "the varying position of the sun." (Or, I would add, by the changing the range to the target.) Peyrard suggested that each mirror could have been furnished with a telescope to make it easier to aim it properly, but the telescope was not invented until the seventeenth century.[16]

On November 6, 1973, Archimedean scholar Iaonnis Sakkas lined up seventy Greek sailors at the Palaska Training Center on the island of Salamina (near the site of the famous naval battle of Salamis), each holding a flat, rectangular mirror. The mirrors were each about three feet by five feet (1.4 square meters) and were tipped to catch the sun's rays and direct them at a wooden rowboat, 160 feet away. The reports disagree as to whether the mirrors were all bronze, or copper-backed glass. "One report says that the ship caught fire at once," and another that "in less than three minutes the boat was ablaze." The reports also vary slightly as to the number of sailors (60 versus 70), the size of the mirrors (one says with diameters of 0.7 to 1.7 meters), and the distance to the target (160 feet versus 55 meters).[17] From a photograph, it is apparent that what was set on fire was not the rowboat itself, but rather a "tar-coated plywood silhouette of a Roman galley, attached to one side."[18] (For the reasons set forth in the appendix, I do not think the combustion was instantaneous.)

On September 28, 2004, *MythBusters* conducted a test of the practicality of Archimedes's reputed burning mirror.[19] The target was constructed from tarred and waxed spruce but got quite waterlogged prior to testing. Jamie constructed a segmented concave dish mirror, with 300 mirrors in a "Fresnel" array screwed to a 400-square-foot frame. (Several different types of mirrors were tested: bronze [too expensive], bronze mylar, aluminum on Plexiglas, and a "household" glass mirror. The last was the most effective and I believe that's what they used; when the frame was dropped after the test, the mirrors broke.) The mirrors could be tilted to the desired angle by adjusting the depth of the screw from behind the frame. The frame itself was raised with a forklift.

The range was not stated, but the video shows the beam dispersed over several square feet, and only smoldering occurred. The *MythBusters* crew concluded that the design was impractical; a mirror large enough to cause quick enough ignition would be too difficult to construct and aim.

In October 2005, students at the Massachusetts Institute of Technology set up an array of 127 one-foot-square mirror tiles in two curved banks, one on the ground and the other elevated, thus forming a curved wall rather than a curved trough. The target was ten feet long and made of one-inch-thick red oak; it was positioned 100 feet away. At 100 feet, the light spot from each flat mirror was three square feet in area. (Since the solar rays striking the mirror are virtually parallel, the divergence with distance is very slow.) A targeting mirror projected an "X" on the target, and each mirror was aimed individually, then covered. With "five team members … this took 10 minutes," during which the movement of the sun caused the targeting "X" to move five feet horizontally along the target. The mirrors were then all uncovered. Once the sun came out from behind clouds, it took less than ten minutes to set the target on fire.[20]

Professor Wallace notes that the ignition time would depend on the moisture content of the wood, as the water would have to be boiled off before the temperature could further increase; he estimated ignition times of 3.5 minutes for dry oak, 4.5 for the oak used (10 percent moisture), and 11 minutes for "wood soaked in water for several weeks." If the Roman ship sides were cedar rather than oak, the ignition times would have been shorter by about 45 percent.[21]

The Discovery Channel invited Wallace's group to do a retest with two important changes. First, they would use bronze mirrors. Since bronze is less reflective than the modern mirror tiles, the group was allowed to use more mirrors to compensate (300 instead of 127). These had to be arranged in four tiers rather than two; the array was "about 110 feet long." Second, the target would be on water, not land; it was a 30-foot wooden fishing boat.[22]

The experiment aired on *MythBusters* (January 25, 2006, episode). The first trial was at 150 feet; the 300 mirrors took "10 minutes to aim," with "14 people working in pairs…. The bright spot was estimated to be 4–5 feet in diameter." The boat was charred but did not burn. The second trial achieved some "glowing, smoldering embers," and after a time "a modest open flame broke out," possibly thanks to a sea breeze fanning them. The "death ray" left a 10-inch diameter hole in the hull, and it was still burning intermittently two hours after the test ended.

Wallace blames the less impressive results of this retest on the high moisture content of the wood (the boat had been underwater for an unstated period of time), and also admits, "we did underestimate the effect of water content."

There was also an attempt made to burn a sail rather than the hull. This was completely ineffectual; Wallace suggests that both the "lighter color of the sail" and "convective cooling on both sides of the cloth" as a result of the wind were responsible.

In 2010, *MythBusters* assembled 500 high school students and equipped them with an equal number of double-sided mirrors, with one side covered with bronze Mylar film (cf. Claus's aiming strategy). Nonetheless, they were unable to set a wooden ship on fire. "Jamie noted that the reflections from the mirrors were extremely distracting and had blinded him temporarily, and that this may have been Archimedes' true intent in recommending their use in warfare."[23]

In 2011, Adam Hart-Davis, in the BBC episode "Local Heroes: Italy," proposed that soldiers could have each been given a flat mirror three feet square, assembled in ranks on a tiered stand, and then directed their reflections at the same target. Hart-Davis set up 96 horizontally pivotable, palm-size mirrors in four vertical banks (each vertically pivotable), and directed them at a piece of wood about six feet away. (We don't know how long the adjusting process took, but from the video it appears that the "spread" he accepted was about a foot.) After a few minutes, he achieved a temperature of 230°C at the brightest point, and the wood was scorched, but not set on fire.[24]

The experiments just described cast doubt on the feasibility of a "burning mirror" attack on even a stationary target. Also presumably, the Roman ships were advancing, and consequently the mirrors would have to be continually readjusted to account for the diminishing range, as well as the movement of the sun.

Conceivably, the Syracusans waited until the Roman ship had come to anchor before their walls before employing the mirrors. Or perhaps there were navigational hazards that would force the Roman ships to come to a halt, or to move extremely slowly, long enough for the mirrors to work. But even under these ideal circumstances (target at close range and stationary), it would seem that the labor that would go into making and wielding the mirrors could more profitably go into shooting fire arrows and catapulting other inflammables at the Roman ships.

Archimedes was a practical man and would not have used mirrors if the Romans could easily frustrate his designs. If they saw an array of flaring mirrors, and a deck plank started smoldering, would they not splash water on it to cool it off? The Romans could cool the deck down a lot faster than the mirrors could heat it up.

Also, the mirrors would be useful only if a large number of Roman ships could be burnt before they could retreat out of the effective range. Otherwise, the Romans would just attack next on a cloudy day, or early in the morning. (It would not have been practical for them to launch a sea attack on Syracuse from the north, as the city was bounded by water only on the west and the south.[25])

Nor would Archimedes have used mirrors if conventional weapons would be more effective. The Greek heavy artillery (catapults) had an effective range of 400 yards, far greater than that of the hypothetical mirrors, and were certainly capable of casting incendiaries at the incoming ships.[26] It is possible that the mirrors were used against ships at close range; situated on high walls, the catapults and ballistae might have been unable to depress their fire to strike targets closer than, say, 60 yards.[27] Of course, at those distances, bowfire would have been practical; its effective range was up to at least 200 yards.[28]

Burning Mirror Alternatives

All in all, I very much doubt that Archimedes actually burned any Roman ships by means of mirrors. It is conceivable that the mirrors were used as *MythBusters* suggested, not to burn the enemy ships, but to blind the sailors and soldiers aboard, making them more likely to run afoul of natural or artificial navigational hazards during their approach, and less able to accurately return missile fire. This could have been misinterpreted, even contemporaneously, as an attempt to burn the ships.

Leonardo da Vinci claimed that Archimedes built a steam-powered cannon,[29] and Cesare Rossi has proposed that Archimedes used mirrors to heat not the enemy ships, but rather water in boilers, and that the steam was then used to propel projectiles. The projectiles might have been incendiary in nature, which would have made it more likely that a commentator might jump to the conclusion that the burning was caused directly by the mirrors.[30]

MythBusters tested the practicality of a "steam cannon" and concluded that with a "flash boiler," the pressure would have been inadequate. (A flash boiler has no reserve; the steam just flows out of the boiler as fast as it is produced.) On the other hand, if the steam is allowed to build up in and thus pressurize the boiler, before being released by a valve, a cannonball can indeed be projected a respectable distance.[31] (This author observes that iron cannonballs had not yet been invented in Archimedes's time, and that it takes considerable time and skill to carve a stone into a spherical shape.)

It's also possible that a "solar-steam" gun was used to fire smaller projectiles, with people rather than ships as the main target. In 1824, Jacob Perkins obtained a British patent on a "steam gun" for rapid fire of musket balls. This used steam as the direct propellant (in a conventional cannon, the projectile is propelled by the expanding combustion gases from burning gunpowder). Unfortunately, the coal-fired boiler weighed five tons, and it took two hours to generate 900 pounds per square inch of steam.[32]

A steam-powered "centrifugal machine gun" (a rapid-fire catapult) was patented by William Joslin in 1858 and Charles Dickinson in 1859.[33] Here the steam was used to power the catapult arm.

Neither type of steam cannon saw action on the battlefield. However, around 2011, a Massachusetts Institute of Technology student group was asked by *Myth-Busters* to assess the feasibility of an Archimedes steam cannon. They were able to achieve a firing pressure of 3,500–4,000 pounds per square inch and a muzzle velocity of 280 meters per second, with the expanding steam (from just a half cup of water!) directly propelling a 0.5-kilogram spherical projectile to 1,200 meters.[34]

However, the war machines described by Polybius and Plutarch would have been capable of hurling or dropping incendiary projectiles, and that would be the simplest explanation for why Lucian said that Archimedes "burned the ships of his enemies by means of his science" and Galen said that Archimedes "set on fire the enemy's triremes."[35] And even a steam cannon could be operated without a mirror being involved.

Post-Archimedes Burning Mirrors

Zonaras also asserted that a segmented mirror "was employed by Proclus to destroy the Gothic vessels in the harbor of Constantinople, and to protect his benefactor Anastasius [Byzantine emperor, 491–518 CE] against the bold enterprise of Vitalian [who revolted 514–516]."[36] However, another chronicler explained that Proclus used some kind of incendiary or explosive weapon, with sulfur as an ingredient, against the Gothic fleet.[37]

Despite the suspect nature of the accounts, the story of the burning mirrors of Archimedes excited the imagination of many later thinkers. For example, Ibn al-Haytham, a famous eleventh-century Moslem philosopher-scientist, wrote about burning mirrors.

Roger Bacon was convinced that the Moslem interest in burning mirrors was not just scholarly; indeed, al-Haytham's *Opticae Thesaurus* had an illustration showing the use of burning mirrors against ships. Bacon warned Pope Clement IV that the Moslem would turn them to military use: "This mirror will burn fiercely everything on which it could be focused. We are to believe that the Anti-Christ will use these mirrors to burn up cities, camps and weapons."[38] I am reminded of the letter that Albert Einstein wrote to President Roosevelt, warning of the possibility of constructing nuclear weapons.

Bacon proposed to build a Christian burning mirror of great size, and remarked cryptically, "the most skillful of Latins is busily engaged in the construction of this mirror."[39]

There are also stories concerning burning mirrors in *De la Pirotechnica* (1550), a metallurgical treatise written by Vannoccio Biringuccio (1480–1539). A German acquaintance claimed:

that he had made one [a concave mirror] almost half a braccio [a braccio was literally an arm's length, and is assumed by his translator to be about 24 inches] across, which extended the clear rays of its brightness more than a quarter of a German league [each German league was equal to 4.6 modern miles] when he caught the sun with it. One day, when for amusement he was standing in a window to watch a review of armed men in the city of Ulm, he bore with the sphere of his mirror for a quarter of an hour on the back of the shoulder armor of one of those soldiers. This not only caused so much heat that it became almost unbearable to the soldier, but it inflamed so that it kindled his jacket underneath and burned it for him, cooling his flesh to his very great torment. Since he did not understand who caused this, he said that God had miraculously sent that fire on him for his great sins.

My friend also told me that with the same mirror he had often melted a gold ducat with the rays of the sun. This was held with a small pair of tongs and it melted in less than a quarter of an hour, as if it had been of lead or wax.[40]

The soldier story is very improbable. A concave mirror concentrates light at its focal point. In order to heat the soldier of the first tale, the mirror would have to have a focal length of more than a mile. A mirror with a focal length of a mile would be very gently curved; the technology of the time would not have permitted the control of the shape. Even if it could be achieved, at that distance the two-foot-diameter mirror would be ineffectual. Its angular size as measured from the focal point would be only a small fraction of the angular size of the sun (about half a degree). For energy to be concentrated, the former must be larger than the latter. Even if the German had used a larger mirror, his target was wearing armor, which would have tended to reflect the light.

It is also inconsistent with the second story; both feats could not be accomplished with the same mirror. To melt the ducat, it would have to have a focal length such that the mirror could be held in one hand, and the tongs in the other—that is, a focal length of just a few feet. To harass the soldier, it would need a focal length of more than a mile. Standing alone, the melted ducat tale is plausible.

A modern use of mirrors to burn away flammable objects appears on a website discussing "heliostats" (sun-tracking mirrors, discussed later) as "death rays":

DOUG WOOD: I thought you might like to see these 100 mirrors (40 by 80 cm) tilted up with sticks to burn the tops of too tall trees to the south. Slight sag from gravity made things hot. [Author's note: The sag would have made the mirror system concave. Wood later comments, "The mirrors flexed slightly compressing the image 2:1 from gravity sag. 100 mirrors had the temperature equivalent of 200 stiff mirrors. Because they tilt between the sun and the target a rectangle images a square."]

ERIC ROSSEN: You were right: I DID find those pictures interesting! May I ask why you did that experiment with the 100 mirrors? Did you really mean to burn that tree?

DOUG WOOD: Yes I did mean to burn those trees. There were about 12 trees that were causing problems at our concentrator site and they needed to be topped. It was not an experiment, just an application. I did this alone, it was easy. After the mirrors were tilted up the image moved from tree to tree day after day burning mostly foliage off green wet winter evergreens. They did catch on fire. The trees were green and wet so the solar fires were self extinguishing.[41]

Solar Furnaces, Engines, and Other Devices

A more mundane use of concave mirrors is for heating, or for setting of fires for civilian purposes such as rituals. The Chinese were pioneers in this area. In the fifth century BCE, they recognized that such mirrors have a focal point, which they called *zhongsui* ("central fire").[42] According to the *Chou Li* (20 CE), a Chinese ceremonial text, "the Directors of Sun Fire have the duty of receiving, with a concave mirror, brilliant fire from the sun … in order to prepare brilliant torches for sacrifice."[43]

Plutarch, in recounting the life of Numa Pompilius, comments that among the Greeks, when a perpetual holy fire was extinguished through some great misfortune, "then, afterwards, in kindling this fire again, it was esteemed an impiety to light it from common sparks or flame, or from any thing but the pure and unpolluted rays of the sun, which they usually effect by concave mirrors, of a figure formed by the revolution of an isosceles rectangular triangle," that is, a conical mirror.[44]

The Incas of Peru are alleged to have used concave mirrors to set fires. "Garcilaso de la Vega … wrote that the way the Incas made sacred fire was by using a 'highly burnished concave bowl, in the shape of a half orange'.… Concave bowls similar to the ones reported by Garcilaso can be found at a collection in Lima, Peru, made on gold."[45]

The Olmecs of Mesoamerica made many concave mirrors. Archaeologist Jose Lunazzi used a spherical mirror of 50-millimeter focal length, 15-millimeter diameter, aluminized to 80 percent reflectivity, to simulate the possible use of ancient concave mirrors to make fires. As a result of a fortuitous mistranslation of Garcilaso's description of the Incan technique, he put fire-blackened cotton at the focal point of the test mirror. Two experiments were run:

> one in Campinas city, Brazil, which is at the latitude of the tropic of Capricorn, and the other in Mexico City, both in September (not summer) days about 24° C temperature with some wind. Our results could be compared to the case of Olmec mirrors using a 20 percent reflectivity spherical mirror of 30mm diameter, or a little larger if we consider the expansion of the image of the sun by spherical aberration. It was clear that only the blackened cotton can start fire, not possible with not prepared cotton, and that by using dry leaves or wood we could only obtain smoke.[46]

Concave mirrors were also used to melt metals. Giovanni Magini (1555–1617), an Italian astrologer, astronomer, and mathematician, "presented large concave spherical mirrors to Prince Jacopo Boncompagni, Card. Farnese, and Rudolf II. The Duke of Mantua gave him 500 scudi and diamond rings worth hundreds of scudi for one of the mirrors. The gift of the mirror to Rudolf led to a memorably long struggle by Magini to get the payment that he thought had been promised."[47] Magini wrote that he had used such mirrors, which were probably less than two feet in diameter, to melt "lead, silver or gold in small quantities such as a coin held firmly with tongs."[48]

Villette, of Lyons, France, in the late 1600s, built a spherical mirror which was four feet in diameter and had a radius of curvature of 76 inches.[49] A Villette mirror was said to be "able to make tin melt in three seconds, cast iron melt in 16."

Leonardo da Vinci asserted that "the sculptor Andrea del Verrochio employed a burning mirror to solder the sections of a copper ball lantern holder for the Santa Maria del Fiore Cathedral in Florence";[50] this may well have been the first practical use of a burning mirror to melt a metal.

Burning mirrors could also be used to heat liquids. In 1515, Leonardo da Vinci conceived of a parabolic mirror concentrator for an industrial application (cloth dyeing).[51] Had his conception been brought to fruition, the dyeing vats would be brought to a boil by solar energy.[52] By 1561, alchemists knew that to make perfume, one could place a clear vase, filled with water and flowers, "at the focal point of a spherical mirror," and then heat the vase water with concentrated sunlight, causing the fragrances to be extracted from the flowers.[53]

In the early modern period, burning mirrors could not only achieve temperatures higher than those generated in furnaces; they did so without introducing contaminants. This made them useful in chemical experimentation.[54]

The Baron Ehrenfried Walther von Tschirnhaus (1651–1708) created the largest mirrors of his day by hammering copper into a thin sheet ("scarce twice as thick as the back of an ordinary knife"); his reflector was "five and a half feet in diameter." In 1687, he reported that "a piece of tin or lead three inches thick, as soon as it is put into the focus, melts away in drops…. A plate of iron or steel placed in the focus immediately is seen to be red hot on the back side, and soon after a hole is burnt through."[55]

Tschirnhaus was a native of Saxony, and its ruler, the Elector Augustus the Strong, had an insatiable appetite for porcelain (nicknamed "white gold"). In the first year of his reign he spent 100,000 thalers (about $2.4 million in today's currency) on the mysterious Oriental ceramic.[56] Tschirnhaus's goal was to duplicate Chinese porcelain and produce it for export.

The Baron became convinced that porcelain was the product of clay and glass, melted together. His burning mirrors, as well as burning lenses, were employed in the study of the melting point of porcelain ingredients such as kaolin.[57] In 1706, he received royal permission to recruit Johann Friedrich Böttger, an alchemist who had unwisely boasted of his ability to transmute base metals into precious ones and was a prisoner of the Saxon monarch. Böttger had failed to deliver, and it was fortunate for Böttger, as well as for Saxony, that he became involved in the porcelain project. Through the efforts of Tschirnhaus and Böttger, Saxony ultimately became the first European source of porcelain.

Augustin Mouchot (1825–1911) was a prolific inventor of solar-powered devices. In his solar cooker, food went inside a blackened copper cylinder. This was placed inside a glass cylinder, with a one-inch airspace separating the two concentric cylinders. A trough-like mirror, made of silver-plated wood, reflected sunlight onto the glass cylinder. This trough, possibly parabolic in cross section, was positioned to face the sun.[58] Used by the French Foreign Legion, it cooked a pound of beef in 20 minutes.[59] His solar still was of similar design. Wine was heated to a vapor in the copper vessel, and the vapor was collected in another receptacle, yielding brandy.

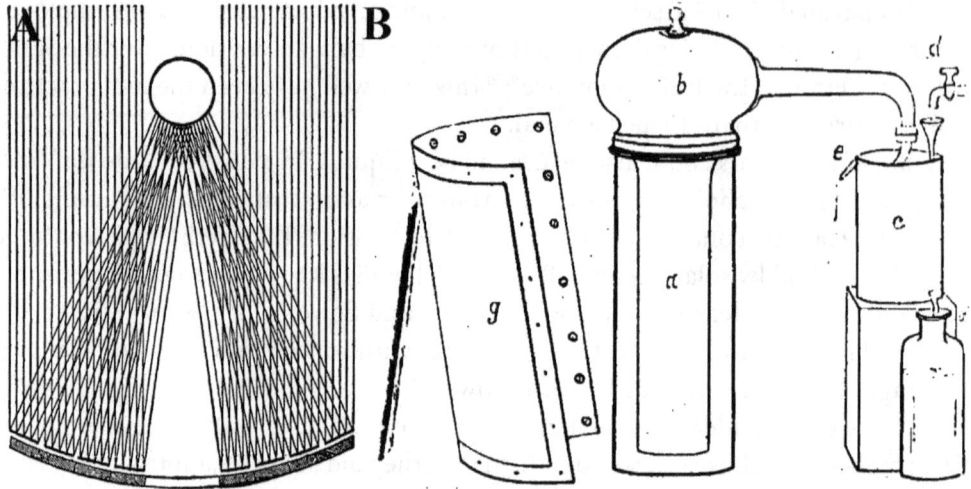

A. Schematic of parabolic reflector focusing parallel rays on cylindrical receiver. Note that the receiver shades part of the reflector. (From Pope, Charles Henry, *Solar Heat: Its Practical Applications* (self-published, 1903), 102.) B. Mouchot's solar still. From Mouchot, Augustin, *La Chaleur Solaire et Ses Applications Industrielles* (Gauthier-Villars: 1869), Fig. 20. The original white-on-black drawing has been inverted for clarity. C. Pifre's solar printing press. The steam engine is in the left foreground and the printing press in the right foreground. The tip of the central structure is called the receiver and lies at the focal point. Pope, 69.

In his solar pump, the air inside the copper cauldron expanded when heated by the concentrated sunlight, pressing down on water in a tank below. Ultimately the pressure was sufficient to cause a jet of water to shoot out of the other end of the tank.

Finally, in his solar steam engine, the parabolic trough reflector concentrated sunlight onto a copper tube one inch in diameter, boiling the water within.[60]

Another solar energy pioneer was Abel Pifre, Mouchot's assistant. In 1880, he used a dish-shaped solar collector to drive a steam engine, which in turn powered a printing press.[61]

The world's largest solar furnace is the one-megawatt facility at Odeillo, in southern France. The sunlight is captured by 63 heliostats on a hillside and then redirected to the parabolic reflector—the whole north side (40 meters high, 54 meters across) of a nine-story building! The peak temperature achieved is 7,000°F. The largest one in the United States is at Sandia Laboratories, Albuquerque.[62]

Recently, the solar furnace has drawn attention as an inexpensive replacement for wood-burning fires in equatorial countries. Steven Jones, a physics professor at Brigham Young University, designed a funnel-shaped solar cooker that weighs less than two pounds. This is a hybrid between a parabolic cooker, which achieves high heat quickly, and the box cooker, which is difficult to build. Its reflective surface was a cardboard "half-funnel" covered with aluminum foil. It concentrates the heat along a line, while a parabola directs it to a point. Inside the funnel is a black pot or jar, and it is surrounded by a clear plastic bag to help hold in the heat.

Jones says, "An estimated 3 billion people cook their food over fires. It's also estimated that two-thirds of the time you can cook with the sun instead of wood near the equator. Many women and children get sick from smoke inhalation because of the wood and other fuel sources used while they cook. Solar energy is clean, efficient and renewable." In Bolivia, during the Bolivian winter, "water pasteurization temperature was reached in 50 minutes, boiled eggs cooked in 70 minutes, and rice cooked in 7 minutes."[63]

The solar cooker also doubles as a cooler. "By pointing the funnel at the dark sky, especially at night, the pot inside cools off—infrared radiation is reflected away from the pot to the dark sky so the pot cools off. We have achieved 20 F cooling on clear nights…. We have frozen water when the overnight temperature reached about 53 F."[64]

Indirect Solar Illumination

Light enters a building through windows. The amount of light captured may be increased by an exterior light shelf, a horizontal reflective surface, typically installed outside south-facing windows. These are placed above head level so they don't reflect the light into the occupants' eyes. While an interior light shelf does not increase the amount of light entering the room, both types increases the depth of penetration, as they reflect oblique light up to the ceiling, which reflects it once again. With diffusely reflecting surfaces, "a single south-facing window can illuminate up to 20 to 100 times its unit area."[65]

Sunlight striking the roof may be directed to lower floors using a light pipe. This comprises a collector, a tube, and a diffuser. The tube is hollow, and its inner wall is made of a "highly reflective" material. Light passes down the tube by multiple reflections.[66]

Mirror Sundial

In the first century, Hero of Alexandria asked, "And will anyone not consider it remarkable to be able to tell the hour, night or day, with the aid of figures appearing in a mirror?"[67]

The traditional sundial is an outdoor one. However, mirrors have been used to move the sunlight indoors. The time of day is indicated by the position of the reflected spot of light, rather than by the shadow of a gnomon.

When Isaac Newton (1643–1727) was seventeen, he installed a "ceiling sundial" in "his grandmother's house. He placed a small horizontal mirror in a hole on the south wall of the house reflecting a point of sunlight onto the living room ceiling, on which he had painted the hour lines: an indoor sundial!"[68] Christopher Wren made a similar contraption when he was a year younger.[69]

The indoor sundial was possibly invented by Nicolaus Copernicus (1473–1543); he is said to have built the first type of this kind of sundial in Olsztyn (Allenstein), the remains of which are still preserved there.[70] By marvelous coincidence, there are two working indoor sundials on Kopernikusstraße in Bremen.

In 1637, Emmanuel Maignan (1601–1676) "constructed a catoptric sundial" for a hallway at the "Minim convent Trinità dei Monti in Rome." In 1632 Cardinal Bernardino Spada (1594–1661) purchased a "palazzo in Rome," now "known as Palazzo Spada." Spada became the Cardinal-Protector of the Order of Minims in 1642, and he commissioned Maignan to design a mirror sundial for a twenty-two-meter-long gallery in his new home. In 1644, a sundial fresco was painted by Giovanni Battista Magni, according to Maignan's design, on the vaulted ceiling, and the gallery became known as the Galleria della Meridiana.[71]

On the wall of the gallery there is an epigram written by the Cardinal that points out that the light actuating the mirror sundial is from a small, east-facing window that "hardly allows the daylight to enter." This would have made the light spot brighter and smaller than would otherwise have been the case. The incoming daylight struck a round mirror that reflected it onto the painted ceiling "so that the point of light glows in the palazzo as in the celestial axis. Numerous lines divide the vaulted ceiling into different hours; the lines that are lit in the ceiling display the day."[72] The mirror itself is set, horizontally, in a horizontal ledge immediately in front of the mirror.[73]

Maignan described his methods in *Perspectiva Horaria* (1648), and indeed this shows an engraving of his installation in the Galleria della Meridiana.[74]

A mirror was also used to reflect moonlight onto a "moondial" on "one wall of the gallery." The moondial had "movable wooden disks" and, if set for the number of days "after the last new moon," the light spot would indicate the "nocturnal time."[75]

The Galleria della Meridiana. (From Maignan, Emanuele, *Perspectiva Horaria*.... [Philippi Rubei, Rome, 1648], Plate preceding page 391.) For the Galleria today, see https://www.giusti zia-amministrativa.it/en/palazzo-spada1.

A "camera obscura sundial" operates only at noon and indicates the day of the year rather than the time of day. The camera obscura, a pinhole with or without a lens, projects sunlight onto a meridian (north–south) scale marked on the floor of a building.

The vicissitudes of life have caused the gnomonic holes on some historic buildings to have been filled in, rendering the "meridian line" inoperative. Rather than restoring the normal operation by reboring these holes, which may damage decorations or even impair structural integrity, it has been proposed to place a new "gnomonic" hole in a more convenient location and use a system of tilted mirrors to redirect noon light to the meridian line and thus restore the functionality. The mirror system does need to be adjusted, manually or automatically, for changes in the solar declination in the course of the year, so one does lose some of the elegance of the original design.[76]

Heliostats

A heliostat (*helio*, sun; *stat*, still) is a mirror guided so that it always reflects light to a fixed target. The early heliostats could be adjusted by a servant, or by clockwork. Modern heliostats are motor driven and computer guided.

Notes have been found that indicate that Giovanni Borelli (1608–1679) conceived of "a method, using a simple wheel clock, of directing the image of the sun in a plane mirror into a predetermined direction, so that it will remain fixed for the whole day."[77]

Borelli did not publish his design, and he probably didn't build it, either. So credit must be given to Willem 's Gravesande (1688–1742), who not only coined the term but built a heliostat "to facilitate the experimental demonstrations for his lectures in optics." It was described in his 1742 textbook.[78]

One use of heliostats is in conjunction with large telescopes being used to study the sun. Rather than trying to move these massive pieces of equipment along to keep up with the sun's movement across the sky, a heliostat was used to reflect the sunlight into the telescope.[79] The McMath Solar Telescope at Kitt Peak National Observatory has a heliostat with an 80-inch planar mirror.[80]

A heliostat may be used to reflect light into a window, skylight, atrium, or light pipe collector to provide better indoor illumination. One Central Park in Sydney, Australia, uses "40 motorised heliostats" (on the West Tower roof) and 320 reflective mirror panels (on the East Tower cantilever); this system "captures and redirects sunlight into retail spaces and landscaped terraces."[81] The light is reflected from the heliostat mirrors to the fixed mirrors on the underside of the cantilever.

Heliostats have also been used to provide outdoor illumination. Viganella, Italy, is "at the bottom of a very steep-sided Alpine valley," and it does not receive direct sunlight "from November 11 to February 2." In 2006, a 26-foot by 16-foot steel mirror, heliostat controlled, was mounted on the south slope to give the residents some sunbeams.[82] A similar problem was experienced by Rjukan, Norway, which installed three mirrors, each 17 square meters.[83]

Heliotrope

The term *heliotrope* refers to a genus of flowering plants, as well as to the way plants turn to face the sun. It is also given to a surveying device proposed by Karl Friedrich Gauss (1777–1855), usually considered one of the three greatest mathematicians of all time. In 1820 he was ordered by George IV of England to survey the entire Kingdom of Hanover. The problem was that in large triangulation surveys, with stations twenty miles or more apart, conventional sighting targets were difficult to spot. This prompted his invention of the heliotrope:

> One afternoon (1821) while Gauss and his son Eugene were walking along, the father, noticing the light of the setting sun reflected from the window-pane of a distant house, thought of the heliotrope; in the simple form, this instrument consists of a plane mirror 4", 6" or 8" in diameter, which may be rotated about a horizontal or a vertical axis. This mirror is at the station to be observed, the sun's rays reflected by it impinging on the distant observing telescope.[84]

The reflective target had "the brightness of a first-magnitude star at a distance of fifteen miles (24.14 km)."[85]

In the United States, Ferdinand Rudolph Hassler, the founder of the U.S. Coast Survey, had tried to solve the problem by means of target cones covered with tin foil.[86] He eventually conceded the superiority of Gauss's heliotropes. On November 17, 1837, Hassler wrote to the U.S. Secretary of the Treasury:

> Heliotropes, of which I had begun the use of last fall, have this year been used for most of the station points, and for the base points exclusively. I caused one to be constructed in our shop last winter, after the two received from Gottingen, by the kind assistance of Professor Gauss, the inventor of this instrument; and during my work this summer, I received four more; all seven are now in activity. The aim of the instrument is: to reflect the sun's image from the station-point, at which they are placed, to the observer at his station. The new instruments require a man of some intelligence to attend them, and to replace [*recte*, adjust] them about every four minutes, according to the motion of the sun…. They will show a precise luminous point, even through haze, so frequent on the eastern seashore, when the outline of the hill itself, upon which they stand, cannot be traced.[87]

In the 1880s, the heliotrope principle was adapted to nighttime surveying. The *selenotrope*, as its name implies, was used to reflect moonshine. "Selenotropes required much larger mirrors than heliotropes' 2–4 inch diameter reflectors: i.e., 6 by 6 inch mirrors for 22 mile lines, 6 by 8 inch for 48 miles, 8 by 10 inch for 70 miles."[88]

Heliograph (Mirror Telegraph)

In 1822, Gauss urged, "With 100 separate mirrors, each of 16 square feet … one would be able to send good heliotrope-light to the moon…. This would be a discovery even greater than that of America, if we could get in touch with our neighbors on the moon."[89] Gauss's plan was never put into practice. There is, however, a long history of use of reflected sunlight for communication here on Earth.

According to Herodotus, the Athenians accused the Alcmaeonidae of treason-ous dealings with the Persians shortly after the Battle of Marathon (490 BCE); they were said to "have arranged to *hold up a shield* as a signal once the Persian were in their ships." But it is unclear whether the raising of the shield was visible in its own right, or only because it reflected the sunlight.[90]

Xenophon likewise refers to a shield used as an improvised signaling device, referring to a Peloponnesian War battle of 405 BCE:

> On the fifth day as the Athenian ships sailed up, Lysander gave special instructions to the ships that were to follow them. As soon as they saw that the Athenians had disembarked and had scattered in various directions over the Chersonese—as they were now doing more freely every day, since they had to go a long way to get their food and were now actually contemptu-ous of Lysander for not coming out to fight—they were to sail back and to *signal with a shield* when they were half-way across the straits. These orders were carried out and as soon as he got the signal, Lysander ordered the fleet to sail at full speed. [emphasis added][91]

There is a similar account in Plutarch.[92] The "Chersonese" was the Thracian Chersonese, the Gallipoli Peninsula, and thus the Strait was the Dardanelles, 0.75 to 3.73 miles wide (Wikipedia). It is doubtful that the shield raising would have been visible to the naked eye from even 0.375 miles away if it hadn't reflected light.

The Romans and Byzantines both had signaling networks, but it is not known whether they used mirrors in addition to beacon fires or torches. While, in the Roman system, the number of torches displayed was a part of the signal,[93] in more modern signaling conventions, a single light source was used, and characters were conveyed by a series of short or long pulses. That meant that one had to transmit flashes of light.

Charles Babbage, a pioneer of the computing machine, devised an "occulting telegraph" in 1851: "by means of a small piece of clock-work and an Argand lamp, [I] made a numerical system of occultation, by which any number might be transmit-ted to all those within sight of the source of light."[94] While the Babbage system used a lamp, a mirror reflecting sunlight could also have been used. "The American Con-gress later appropriated $5,000 for experiments with Babbage's telegraph."[95]

During the American Civil War, the Union experimented with an mirror-based optical telegraph:

> In 1861, officers of the United States Coast Survey, at work in the Lake Superior regions, demonstrated the usefulness of the mirror, equatorially mounted, for telegraphic purposes, and succeeded in conveying their signals with ease and rapidity a distance of ninety miles. During the same year, Moses G. Farmer, an American electrician, a man of infinite invention, succeeded in thus telegraphing along the Massachusetts coast from Hull to Nantasket.[96]

In 1865, Henry Christopher Mance (1840–1926), of the British Army Sig-nal Corps, constructed a heliograph with a tripod-mounted mirror linked to a key mechanism. The key tilted the mirror in and out of position, causing it to flash on and off. This "oscillating mirror" arrangement facilitated the transmission of Morse code.[97] Mance, who was later knighted, was then a telegraph employee in Baluch-istan (western Pakistan). William Plum comments:

The Mance Heliograph is easily operated by one man, and as it weighs but about seven pounds, the operator can readily carry it and the tripod on which it rests.... During the Jowaki Afridi expedition sent out by the British-Indian government (1877–8), the heliograph was first fairly tested in war.[98]

Some versions of the Mance telegraph had two mirrors. Just one mirror was used if the sun were in front of the sender. If the sun were behind the sender, the second mirror could be positioned to reflect the sunlight onto the one facing the recipient.[99]

Rudyard Kipling (1865–1936) made several references to the heliograph in his writing. In *A Code of Morals* (1886), the heliograph operator Jones leaves for a station at the Afghan border, but not before teaching his art to his bride, "so Cupid and Apollo linked, *per* heliograph, the pair." One day, General Bangs and his staff saw the flash of their communications and hastened to transcribe. Expecting the message to be of military character, and directed to the general, they were very surprised that he would be addressed as "Dear," let alone "my darling popsy-wop." But the coup de grâce came when they intercepted the husband's warning, "Don't dance or ride with General Bangs—a most immoral man."[100]

The heliograph also made a brief appearance in H.G. Wells's *War of the Worlds* (1897): "And through the charred and desolated area—perhaps twenty square miles altogether—that encircled the Martian encampment on Horsell Common, through charred and ruined villages among the green trees, through the blackened and smoking arcades that had been but a day ago pine spinneys, crawled the devoted scouts with the heliographs that were presently to warn the gunners of the Martian approach."[101]

In 1877, Chief Signal Officer Albert J. Meyer of the U.S. Army received sample Mance heliographs and passed them on to U.S. Army Brigadier General Nelson Miles, commander of the Yellowstone Department in Montana. In 1886, Miles took command of the troops facing the Apache Indians and used the Mance heliograph in both Arizona and New Mexico. The Indian tribes had countered the electrical telegraph by cutting telegraph wires and poles; the mirror telegraph was much less vulnerable to enemy action. The signalers were typically 25 to 30 miles apart and could send messages over distances of 800 miles in less than four hours.[102] Mountain peaks were favored sites, for obvious reasons. It is no coincidence that Graham County, Arizona, has a "Heliograph Peak," 10,022 feet above sea level.[103] Bowie Peak, where demonstrations of the heliograph have been offered, was another relay point.

Miles's "network comprised 27 signaling stations.... Records from that time indicate that between May 1, 1886, and September 30, 1886, a total of 2,276 messages containing 80,012 words were transmitted over the network. The average speed of the system was reportedly 16 words per minute, or roughly 10 bits per second."[104]

The first of these stations to be established was the one at Fort Huachuca, Arizona (which became station no. 7 of the network); its operator was C.F. von Hermann. It normally communicated with Mt. Baldy, 40 miles away, but von Hermann reports that the flash could be seen even at Fort Bowie, 90 miles off. While it might have been to the advantage of the Apache to raid these stations, from von Hermann's account, the principal hazard was the sun, which was extremely fierce at the altitude of operation. Von Hermann ended up wearing a weird umbrella-like contraption, covered with white canvas, to protect himself.[105]

According to his memoirs, General Miles used a heliograph demonstration to persuade Geronimo, who had already surrendered, to persuade the recalcitrant Natchez (the son of Cochise) to bring in the last of the Apache. Miles was sure that he had "struck the savage with awe" by showing him "a power which he could not understand."

Others think that Miles overstates his case. The Indians used mica mirrors for signaling, so why should they be awed by the demonstration? One writer denigrated the heliograph system as an "expensive toy." While it appears that the heliograph only once helped direct troops against a war party, it did have a strategic importance. This arose because the Apache understood the significance of the heliographs all too well; the Apache avoided the territory crisscrossed by the heliograph network.[106]

Initially, the U.S. Army used heliographs with Mance's oscillating mirror. Later, it adopted heliographs which achieved light flashes by coupling the key to a shutter placed in front of the mirror, rather than to a mechanism that would tilt the mirror. (The antecedent of the shutter system was the "occulting telegraph" devised by Charles Babbage in 1851.[107])

On a good day, a mirror flash could be seen, even without a telescope, from thirty miles away. Binoculars or telescopes could spy the flash from much further away. Major J.D. Harris comments:

> In heliography, a mirror is used to flash light in Morse code. Using this method, eight to sixteen words could be transmitted per minute and the equipment was relatively cheap and easy to manufacture—in the absence of anything else, a simple mirror would suffice—and a "crew" of only one trained person was required. Marching distances (ranges) varied roughly, depending on the size of the reflector. For example, a 5 inch (12,7cm) reflector had a marching distance of some 50 miles (80,5 km), whilst the 9 or 12 inch (22,9 or 30,5 cm respectively) models had a range of up to 80 miles (128,7 km). Indeed, in India, marching distances of over 100 miles (161 km) had been recorded and it is interesting to note that, at least until 1975, the Pakistan Army continued to use the heliograph.
>
> The main disadvantages of the heliograph were that a light source was required, be it the sun, the moon, or limelight, and that a clear atmosphere was essential. Tactically, interception was possible anywhere along the axis of the projected light beam, but this was sometimes much reduced in diameter by the use of a narrow tube. Even a rifle barrel could be used to project the beam. The disadvantage in this was that the marching distances of very narrow beams were also less, as less light was radiated.[108]

In 1894, the U.S. Army set a new heliograph distance record. According to the 1895 annual report of the Secretary of War:

> The former world's record for long range heliographing was surpassed [by] 58 miles ... through the zealous and intelligent exertions of Capt. W.A. Glassford, Signal Corps, and a detachment of signal sergeants by the interoperation of stations on Mount Ellen, Utah, and

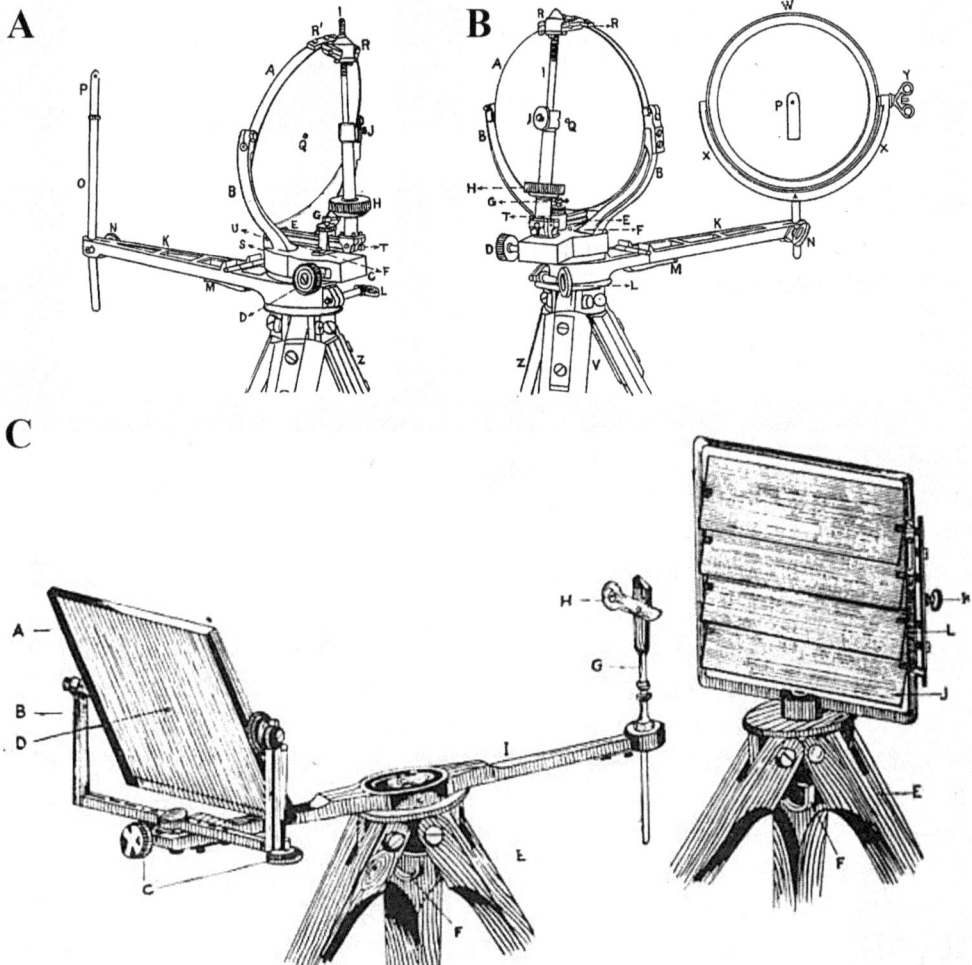

A. British Army Type One Mirror. (Fig. 98 in *Methods of Communication Adapted to Forest Protection* [T. Mulvey, 1920].) The sending key E pulls down vertical rod I and thereby tilts mirror A via claw attached to R. The unmirrored spot is Q.B. British Army Type Two Mirror. (Fig. 99.) C. American Army type. (Fig. 101.) The mirror A and the "screen" (shutter) J are on separate tripods; the screen key is K.

Mount Uncompahgre, Colorado, 183 miles apart. This unprecedented feat of long distance intercommunication by visual signals was made on Sept 17, 1894, with Signal Corps heliographs carrying mirrors only 8 inches square. It was accomplished only after much discomfort and some suffering, due to severe storms on the mountains and to the rarefied air to which the parties were subjected for ten days. The persistence, skill and ingenuity of Captain Glassford and of the signal sergeants engaged in this result are highly commendable.

Both of these mountains are more than 10,000 feet tall. If both stations were at a height of 10,000 feet, the maximum possible distance (thanks to the curvature of the earth) would be 234 nautical miles.[109]

The heliograph played an important role in military communications during the Boer War:

In the siege of Ladysmith, telegraph lines were cut off on November 2, 1899, and from then until the relieving army arrived on February 28, 1900, the heliograph was the only connecting link with the outside world. Cloudy days were tedious for the inhabitants of Ladysmith because no news could be received. One person recorded such a day in his diary, writing, "Heavy weather had settled upon us and had blinded the little winking reflector on Monte Cristo Hill."

As the relieving army, commanded by Sir Redvers Buller, approached the city, his signal officer, Capt John Cayzer, attempted to establish communication by helio. There were problems with Boer operators who intercepted the British flashes. When Cayzer finally reached a station claiming to be British, he devised a test. "Find Captain Brooks of the Gordons," he signaled. "Ask him the name of Captain Cayzer's country place in Scotland." Captain Brooks, when found, did not immediately grasp the purpose of the question and remarked, "Well, I always thought Cayzer was an ass, but I didn't think he'd forget the name of his own home!"[110]

One advantage of the heliograph over the electrical telegraph had been that it could be used even in hostile territory, where a telegraph wire had not yet been laid, or was likely to be cut. The radio, being wireless, made the issue moot. However, the fact that it was a line-of-sight device was both advantage (it was more difficult for the enemy to intercept communications) and disadvantage (friendly forces could move out of sight) relative to radio.

The heliograph was still used during World War I in some engagements in Palestine and Mesopotamia, but it was clearly on the wane. According to Lewis Coe, "Canada was the last major army to keep the heliograph as an issue item. By the time the mirror instruments were retired in 1941, they were not much used for signalling. Still, the army hated to see them go. One officer said, 'They made damn fine shaving mirrors!'"[111]

Actually, Coe's announcement of the demise of the heliograph was premature; on July 10, 1944, the crew of the USS *McCall II* (DD-400) observed a clandestine American heliograph on Japanese-held Guam and sent a whaleboat to rescue the operator.[112]

While the heliograph was abandoned by the army, its country cousin, the simple signaling mirror, was adopted by the air force. Even in modern times, downed pilots may use a signaling mirror to call for help. (Pilots are required by law to report ground flashes as distress signals.)

Signaling mirrors can also be used for fun; in "Operation on Target," teams communicate by mirror flashes with each other. In 1980–2017 it was a Varsity Scout program, but now non-Scouts may also participate. A common goal was to establish documented chains of signals from Alaska to Mexico, and from the Pacific to Denver.[113]

A modern signal mirror has a sighting aperture, and a retroreflective grid surrounding the aperture. "Small microscopic glass beads are glued to a mesh so that the light can pass through, but the returned reflection is shown as a bright spot on the mesh from the back side…. A retroreflective grid creates a 'fireball' on the grid that shows you the direction of the signal mirror reflection as you look through the back. This makes it extremely easy to line up the reflection flash and the target."[114]

Solar Propulsion of Spacecraft: Sunjammers

Mirrors may also be used to capture solar radiation, or light from a powerful moon-, asteroid-, or space station-based laser, in such a way as to propel a space-craft. Johannes Kepler was the first to propose that spaceships could use sails. He had noticed that comet tails point away from the sun. Of course, he thought this was the effect of a real wind, not of light. The existence of light pressure was proven by James Clerk Maxwell, and in 1924, Fridrickh Tsander wrote, "for flight in interplan-etary space I am working on the idea of flying, using tremendous mirrors of very thin sheets."[115]

Solar sailing has received serious scientific attention, since there is no need for fuel. The "sunjammer" has to be brought into space by other means, such as a con-ventional space shuttle. After being untethered, it deploys its sails. Some kind of sta-bilizing means is necessary so the sails don't just wrap around the cockpit. These can be rigid elements, or the sail can be spun so centrifugal force holds it open.

"In classical physics, light is characterized as an electromagnetic wave, com-posed of fluctuating electric and magnetic fields." However, it is more convenient to calculate light pressure if we, like Albert Einstein, visualize light as a stream of particle-like units (photons). The photons carry momentum, and one of the funda-mental laws of physics is that momentum is conserved. When a photon strikes an object, it is absorbed and delivers its momentum to the latter. The change in the tar-get's momentum is felt as a force. If the object reflects light, it emits a photon in the opposite direction, and, to conserve momentum, the object recoils, as if it had felt a force. Thus, light delivers one "kick" per photon to a perfectly absorbing surface, and two "kicks" to a perfectly reflecting one.

> If the sailcraft is at earth's distance from the Sun (one Astronomical Unit, A.U.), and the sail directly faces the sun, the light pressure is 4.7 newtons per square kilometer on a perfectly absorbing surface, and twice that (9.4 newtons, or two pounds) on a perfectly reflecting one. (The pressure exerted by a 12 mph wind is about 50,000,000 times the latter.)
>
> Light spreads as it moves away from the Sun, and therefore weakens in intensity, according to the inverse square law first postulated by Johannes Kepler. So solar sailing is more efficient in the inner solar system, where the light pressure is stronger.
>
> There are two solar forces acting on a solar clipper, the force of gravity (pulling it sunward) and the photonic (electromagnetic) force (pushing it outward). If the velocity of the vehicle about the Sun is constant (and nonzero), and the sail is furled (so there is no photonic force upon it), the craft will assume a fixed orbit.[116]

The photonic force is the vector sum of the forces from the incoming sunlight and the reflected sunlight. Since the angle of incidence equals the angle of reflec-tion, if all incident light were reflected, the resultant force would be perpendicular and "into" the mirror; its direction of course would be opposite that in which the sail was facing.

An orbiting sunjammer can "tack" with its solar sail to change its orbit. The photonic force would have two orthogonal components, one directly away from the sun and the other along the spacecraft's orbit. The radial component directly opposes the gravitational force. If the orbital component were forward, that is, in the orbital direction, the orbital speed would increase and the orbit would widen, moving the spacecraft away from the sun. If the orbital component were backward, opposite the orbital direction, the orbital speed would decrease and the orbit would narrow, moving the spacecraft sunward. The angle of the sail relative to the sun determines the direction of the orbital component.

"What is the best way to sail a sunjammer away from the sun?" Surprisingly, it is not to aim the solar sail directly at the sun. Then the photonic force has to overcome the force of gravity. It has been calculated that "for the power output and mass of our sun, the sail would have to mass no more than one kilogram for every 600 square metres of sail area, including the mass of payload and electronics." It is better to direct more of the force into the orbital direction. The orbital thrust is greatest when "the angle between the sun and the perpendicular to the sail is about 35.3 degrees."[117] However, less sunlight is captured, so the orbital thrust at that sail angle is only 38 percent of the total (outward only) thrust obtained with the sail facing the sun.

The sun is not the only possible source of photons. A powerful laser, mounted on a space station, or perhaps on the moon, could be used to push solar sailing ships along. The efficiency of laser light propulsion can be increased by "recycling the photons," that is, reflecting them back and forth between the laser station and the spacecraft.[118]

Whether the propulsion is solar or laser-based, the melting point, density, and reflectivity of the sail are extremely important. The higher the reflectivity, the less solar energy is converted to heat. The higher the melting point, the greater is the degree of heating the sail can tolerate. And the lower the density, the less the sail contributes to the overall mass (acceleration = force/mass).

Melting point imposes something of an absolute limit on how close to the sun a sunjammer can venture. Landis assumed the operating temperature to be "⅔ of Tmelt."

The temperature of the sail would increase until the rate at which the sail absorbed solar energy equaled the rate at which it reradiated energy as heat (this is proportional to the fourth power of its temperature and to the sail's "thermal emissivity"[119]). It can be shown that this equilibrium temperature is inversely proportional to the square root of the distance of the spacecraft from the sun.[120]

Assuming the melting point is high enough for prolonged operations at a particular distance, we may consider density. But of the metals less dense than aluminum, only beryllium has a high enough operating temperature for prolonged use at Earth's closest approach to the sun.[121]

Assuming that the absorptivity/reflectivity ratio was the same for all metals, Landis proposed in 1995 that "the figure of merit for a given material is the fourth power of the [operating] temperature divided by the density."[122]

For aluminum, and hence also for other metals, Landis assumed a reflectivity of 82 percent (absorptivity 18%) and an emissivity of 0.06.[123] On a scale where the relative figure of merit of aluminum is 1, that of beryllium was calculated as 11.1, and that of niobium, 22.7.[124] However, Landis later reported that beryllium's "optical reflectance" was just 54 percent, and its emissivity 0.02 (presumably at the "operating temperature").[125]

Niobium has been reported to have an emissivity of 0.78 at 1,350°C and 0.39 at 2,000°C; its melting point is 2,468°C.[126] Its reflectivity in the visible range is 49 to 53 percent.[127]

While an all-metal sail is possible, one in which the metal is a very thin coating on a plastic substrate is likely to be more economical and also lower in areal density. The acceleration of a sunjammer is the light pressure divided by the areal density.

If the metal is too thin, then sunlight will penetrate it, and it will also be vulnerable to degradation by solar bombardment with electrons and photons.[128]

Metals typically have a low emissivity. However, it has been proposed that the "backside" of the sail might be a material that combines a high emissivity (over 0.8) with a low areal density. That would make near–Sun round trip missions more practical.

For dielectric sails, see Chapter 7.

The first application of solar sailing was impromptu, "to save the Mariner 10 mission which had lost a large portion of its propellant margin when the star tracker locked on to floating debris instead of Canopus. The mission went on to flyby Venus and three encounters with Mercury."[129]

In May 2010, the Japan Aerospace Exploration Agency launched IKAROS. This experimental spacecraft had a four-petaled solar sail made of "two kinds of polyimide," each with a thickness of 7.5 micrometer, and a total area of about 174 square meters. The reflective surface was an 80-nanometer-thick layer of aluminum. About 5 percent of the sail was covered by "thin film solar cells" but they were "not actually relied on." The sail was successfully deployed in June 2010, and IKAROS passed by Venus that December.[130] While the IKAROS has onboard thrusters for attitude control, it was also equipped with "liquid crystal" elements at the tips of the membrane. These switch between specular and diffuse reflection, depending on whether the element is receiving power or not, and thus can create a torque if opposite tips are in the opposite states.[131]

Giant Mirrors in Orbit

The first "space mirror" may have been conceived by Hermann Oberth, in his *By Rocket into Planetary Space* (1923).[132] It was popularized by Hugo Gernsback (1884–1967), who published the first magazines devoted to "scientifiction." Oberth returned to the subject of the space mirror in *Man into Space* (1957). The mirror structure was to be 100 to 200 kilometers (60 to 120 miles) in diameter. It was not to be a single mirror, but rather an array.

The first building phase was to be the construction of the support mesh. A special rocket was to set spinning by tangential rockets. This would cause six long cables to shoot out radially. Astronauts would attach these cables to one or more external rings. Over the cables and the rings they would lay a hexagonal meshwork.

Within each of the hexagons would be a circular rail. Fitted into the rail would be two traveling "crabs," diagonally opposite each other, and connected by a shaft. This shaft would be the axis around which the inserted circular mirror "facet" could pivot. Thus, it could spin about the shaft, and the orientation of the shaft could itself be adjusted.[133] Such a mirror element could face in any direction, although it would be obstructed if it "looked" into the plane of the array. The resemblance of this structure to the mirrors of Anthemius and Buffon is evident.

In his 1957 work, Oberth just describes the mirror material as being a "thin metal foil." While it is thin to keep the mass low, it must be thick enough to reflect most of the light. Oberth assumed that the reflectivity would be perhaps 90 percent. In "The Problems of Space Flying," published in the September 1929 issue of *Science Wonder Stories*, then a Gernsback publication, Hermann Noordung explained that Oberth envisioned use of thousands of thin squares of sodium metal. Sodium is a very-low-density metal.[134]

At what altitude would the mirror operate? This is addressed only obliquely. At one point, he assumes that the mirror "floats at a height of 6000 km (3725 miles)." Later, he refers to a distance of "three earth radii" (one radius is about 6,400 kilometers at the equator). According to Gernsback, it was to orbit the Earth at an altitude of between 200 and 700 miles, making one revolution every two hours.

If the mirror were at 200 miles (320 kilometers), it would make one revolution about Earth every 91 minutes. An altitude of 700 miles (1,120 kilometers) implies an orbital period of 108 minutes. To attain an orbital period of two hours, the mirror would have to be about 1,700 kilometers (2,700 miles) above Earth's surface.[135]

In any event, the orbit of the mirror can be adjusted by using the mirror as a giant solar sail.

Oberth did not specifically state the size or number of the mirror "facets." In his Fig. 30, there are 17 facets across the diameter; if the diameter were 100 kilometers, that would imply a single facet was about 6 kilometers across. Oberth commented, "there is ... no purpose in making the facets small. Their size is limited only by the structure's static rigidity."

Oberth assumed that if the mirror structure had a surface of 12,500 square kilometers (4,825 square miles), the mirror surface proper would be 12,000 square kilometers (4,530 square miles). He expected that the mirror would be inclined at 45 degrees to the sun and would reflect 90 percent of the light down to an earthly target directly beneath it. Thus, the mirror array would be delivering the same amount solar radiation as would fall on Earth over an area of 7,700 square kilometers (2,970 square miles), if there were no atmosphere and the sun were directly overhead. If the facets were parallel, the array would be, in effect, a giant plane mirror, and its effect would be similar to that of direct sunlight over the affected circular light spot, for which the area, says Oberth, would be 5,500 square kilometers (2,125 square miles). (This last figure implies a height of about 10,000 kilometers, since the sun has an angular size of about 0.5 degree.) The facets are separately aimable and hence can be made to all focus on the same area. For that matter, the facets can be aimed deliberately at different points.

Both peaceful and military uses were contemplated. "Large towns could be brilliantly lit at night by individual facets of the mirror.... Icebergs dangerous to shipping could be melted." Cold snaps in the spring and autumn could be counteracted. Storms might be deflected.

On the other hand, Oberth declared that "munitions factories and dumps could be blown up, towns and agricultural land damaged, marching and motor-borne troops burned, etc." Gernsback claimed that the Oberth mirror "could vaporize the world's biggest cities by its titanic inherent power."

This all seems rather overblown; thermodynamically speaking, a 100-kilometer mirror which is 10,000 kilometers up cannot significantly concentrate solar radiation. (The rim half angle is too small.) However, Oberth did envision that the space mirror could alter its altitude. If it dropped down to 200 miles, the rim half angle (for a 60-mile-diameter mirror) would be about 8.5 degrees, leading to a maximum energy concentration of about 300-fold. But at the lower altitude, its movement through the sky would be more rapid, so it could not attack any one site for very long.

Gernsback suggested that a set of eight or ten larger "spatial mirrors"—each 100 miles in diameter!—could be used to create an "eternal spring." Each mirror was to be made of an array of chrome sheets; each of these "facets" was to be equipped with a solar-powered motor. The mirrors were to face the dark side of the Earth as much as possible. Gernsback explained:

> 1. It is not the purpose of the mirrors to illuminate fully the dark side of the earth. At best, with all mirrors working at maximum illumination, the night side will be in constant twilight.
> 2. The chief purpose of the mirrors is to heat the two subpolar regions sufficiently to keep the so-called temperate zones free from frost, deep snow, and ice. No attempt would be made to melt the polar caps. (That would raise the level of the oceans more than 100 feet and put all the world's coastal cities under water.)
> 3. The solar mirrors would seldom concentrate their heat on the earth itself. The mirrors would chiefly heat the atmosphere at the stratosphere level and above. It is here where the weather is created and air currents are generated, the so-called jet currents that vastly influence our weather.[136]

He added that the effect on world climate would be a gradual one. Others are less sanguine about what space mirrors would do for the environment; environmentalist Bill McKibben wrote an editorial titled "Light Up the Sky? Are We Crazy?" in 1993.[137]

One of the peaceful uses which Gernsback envisioned for the giant space mirror was "night television." Gernsback commented that 25 to 30 miles below the mirror could be a 100-foot projector. This would project an image onto the main mirror, which in turn would display it to the entire world. Gernsback asked, "What would advertisers not pay to get on a monster sky display sign, measuring 60 miles across, with letters from up to 25 to 30 miles high? And to top it all, an ad that would be seen nightly (clear skies permitting) by nearly the entire country, plus the rest of the entire hemisphere?"[138] Some of the viewers might prefer to just be vaporized.

Another use of space mirrors would be to extend the growing season. If farmers in Kansas wanted to plow their fields at night by the "light of the full moon" every night, whether the moon was full or not, could this be accomplished by placing a mirror in space? Yes, to provide the equivalent of full moonlight (only 1/640,000th the brightness of the sun) to a circular area with a diameter of 330 kilometers (the east–west length of Kansas), the mirror would have to have a diameter of just 410 meters (1/800th the target diameter). The mirror would be slightly diverging, so as to spread the light out over the whole farm area.[139]

On February 4, 1993, a Russian spacecraft unfurled an aluminized mylar mirror which was about 65 feet in diameter. When "Znamya-2" deployed, "it directed a beam of light about two or three times as bright as the moon and two-and-a-half miles wide down" to Earth. Because of the mirror's orbital motion and the rotation of Earth, observers at a given point on the ground would have only seen a "pulse" as the beam traveled over their position.[140]

At the time, the Russians touted "the idea of a space reflector as a method to light up emergency zones where all power is out and whole northern cities where dark reigns throughout the winter. They say that several reflectors together would produce a spotlight strong enough to illuminate points on Earth."[141] However, their earlier concept was more ambitious: to "increase the length of a day with the goal of boosting productivity in farms and cities in the then Soviet Union."[142]

A second attempt, in 1999, was unsuccessful; the 25-meter *Znamya 2.5* mirror was fouled by the Mir station's cargo ship's antenna.[143] Astronomers, fearing light pollution from space mirrors, rejoiced. The mirror was expected to create a light spot 3 to 4 miles in diameter, with a brightness of 5 to 10 full moons, and to be able to "direct its beam." Moreover, it was a steppingstone to a proposed "200 m reflector as bright as 10–100 full moons over a footprint of 15–45 km."[144]

〰〰

CHAPTER 5

More Mirrors at Work

Flashlights and Spotlights

The coupling of a light source with a "concave glass" to form a flashlight was proposed by della Porta in 1589: "By this very glass, we may in a tempestuous night, in the middle of the streets, cast the light a great way, even into other men's chambers. Take the glass in your hand, and set a candle to the point of inversion, for the parallel beams will be reflected to the place desired, and the place will be enlightened above sixty paces, and whatsoever falls between the parallels, will be clearly seen. The reason is, because the beams from the center to the circumference, are reflected parallel, when the parallels come to a point. And in the place thus illuminated, letters may be read, and things done conveniently, that require great light."[1]

Note that della Porta states that the reflected beams are "parallel"; this is possible only if the mirror is not merely concave, but paraboloid.

Della Porta furthermore proposed spotlighting, by setting "two or more" of these illuminators within a chamber: "In temples, watches, and nightly feasts, any man may thus with a few lights make a great light." However, for this purpose he recommends the use of lamps rather than candles.[2]

The Magic Lantern

A *laterna magica*, with a light source, parabolic reflector, lenses, and slide stage, was described in 1659 correspondence between Christiaan Huygens and Pierre Petit, and its optics were "essentially identical to that of the modern slide projector." (Some authorities assume that the magic lantern was invented by Athanasius Kircher, but he did not describe one until in the second edition of *Ars Magna Lucis et Umbrae* [1671], and his device would not have worked because the lens was incorrectly positioned.[3])

Lighthouse Lights

The Seventh Wonder of the Ancient World, the Pharos of Alexandria, was a great lighthouse, built in the third century BCE, which survived until 1326. Modern scholars believe it was a three-tiered edifice with a total height of about 100 meters, based on the travelogue of the twelfth-century Arab traveler el-Andaloussi, who traipsed over its ruins and made detailed measurements. That made it, excepting the pyramids, the tallest manmade structure in the world in its day. The lighthouse bore the inscription "Sostratus of Cnidos, son of Dexiphanes, on behalf of mariners, to the Divine Saviors" (a reference to Castor and Pollux).[4]

It was not the first lighthouse, but it may have been the first to be equipped with a large mirror, probably made out of bronze. The mirror, if it existed, would certainly have been used to reflect sunlight during the day, and may also have been used to reflect the light of a wood or oil fire at night (in antiquity, ships didn't sail at night).[5] According to Josephus, the light could be seen at least 300 stadia (34 miles) offshore.[6] Chugg suggests that the mirror was a gilded bronze sphere, reflecting but not concentrating sunlight.[7]

There were fantastic tales about the mirror, such as that it could be used to burn enemy ships, or to spy on faraway places. These rumors may well have been encouraged by the Ptolemaic rulers. Or they may have been embroidery on the part of later Arab storytellers.

If we ignore the suspect Pharos Lighthouse, the first definite use of mirrors to project light out to sea was by Liverpool privateer Captain William Hutchinson (1715–1801) in 1763.[8] "He took a large metal bowl and embedded it with small pieces of mirrored glass. The light source was placed at the focal point of the bowl. The light was then reflected from the bowl and projected outwards as a parallel beam of light."[9] (The Hutchinson mirror looks like half a disco mirror ball, turned inside out.[10])

His experiments were conducted at the Bidston Hill signal station, and one of his original reflectors is preserved at Trinity House, London. The signal station was replaced by a lighthouse in 1771; there, "the light is from oil, with one reflector of silvered glass 13 feet and 6 inches in diameter with a four foot focus. The immense reflector is lit by one large cotton wick which consumes one gallon of oil every four hours."[11] The Bidston Hill lighthouse guided shipping into the proper channel of the River Mersey; its light could be seen from 21 miles away.[12]

In 1777, Trinity House replaced the coal fire at the Lowestoft lighthouse with a glass lantern seven feet high and six feet in diameter; inside this lantern was a large cylinder covered with 4,000 small mirrors. This "spangle light" was used until 1796.[13]

In 1781, Swiss physicist and chemist Aimé Argand (1755–1803) invented the Argand lamp, which provided a steady, smoke-free light, vastly superior to a naked flame. Reynaud, a director of the French Lighthouse Service, claimed that Argand also was the first to propose combining his lamp with a parabolic reflector.[14]

An inferior version of the Argand reflector was promoted by Winslow Lewis (1770–1850) in the United States, beginning in 1810.[15] The Lewis reflectors were made of a thin copper which warped easily, losing the parabolic shape, and their silver coating was so thin that it was worn off when cleaned. Nonetheless, Lewis was the lowest bidder on the lighthouse contracts, and his lamps were operated in all U.S. lighthouses, save three, up until 1852.[16]

In 1823, a more efficient light projector, the Fresnel lens, was developed. The Fresnel lens was rapidly adopted in Europe, but it did not completely supersede the catoptric system in the United States until 1859.[17]

Sextants and Octants

Various instruments have been used by sailors to determine latitude (their position north or south of the equator) by measuring the "height" (really an angle) of the sun or the Pole Star above the horizon.

At first, the sun's altitude was determined by sighting on it directly and measuring the angle from this sight line to the horizon. One of the devices used for this purpose was the "quadrant"; Columbus had one. The first recorded mention of this instrument was in 1450.[18] The measurement process was called "shooting the sun," possibly because the quadrant looked somewhat like a crossbow.

In 1594, John Davis (1550–1605) modified the quadrant design so you did not have to look at the sun. Instead, the navigator could stand with his back to the sun and line up the edge of a shadow with the horizon.[19]

The octant described by English mathematician John Hadley (1682–1744) in 1730 was similar to the Davis quadrant, but used two mirrors, so that one looked at a reflection of the sun, instead of at a shadow that it cast. When an object is seen through a double reflection, its angle from the eye is twice the angle between the two mirrors. This allowed a scale of only 45 degrees in width (one-eighth of a circle) to represent 90 degrees of altitude. According to Hadley's presentation to the Royal Society, his reflecting octant allowed "both bodies (for lunar distances) or the body and the horizon (for altitudes) to be seen by the observer simultaneously, making observations in a moving ship practicable."[20] The Hadley octant proved its utility in sea trials in 1731.

"It was later discovered in Halley's papers that Newton had a similar idea [a double reflection instrument] in 1700 but Halley had told him it was not practical."[21]

"Quite independently of Hadley, Thomas Godfrey (1704–49), a Philadelphia glazier, had devised an altitude measuring device based on the same principle. The Royal Society recognized the equal contributions of both men and awarded them a prize of £200 each. Godfrey also received a prize from the Board of Longitude (of chronometer fame) for his work. However it was Hadley who generally received credit for the invention."[22] Indeed, in 1752 he had a constellation named after him and his octant, *Octans Hadleianus*.[23] Appropriately for a constellation honoring navigational invention, it lies near the South Celestial Pole. (Sadly, modern astronomers just call it *Octans*.)

The radius of the octant had to be made quite large (eighteen to twenty-four inches), since the scales were drawn by hand and hence were more accurate if the separations of markings were larger. This, in turn, made it prohibitively expensive to make the instrument out of metal, so it was made of wood. But wood warped, which could make the readings inaccurate. When Jesse Ramsden (1735–1800) in 1768 invented the dividing engine, which produced accurate scales mechanically, the radius of the octant could be reduced. Brass became a practical material for the frame and warpage a problem of the past.

To find one's exact position at sea, knowledge of latitude was not good enough; one needed to know the longitude (east/west position) as well. The sextant was developed to facilitate the calculation of longitude by the lunar distance method. "The method uses the fact that the Moon moves fairly quickly against the background stars, moving through its diameter in about one hour. If tables are given for the distance of the Moon from certain stars then it is possible in principle to determine an absolute time for the place of observation and thus to determine the longitude by comparing the absolute time with the local time."[24]

"The first determination of longitude by lunar distance is variously attributed to Regiomontanus in 1472, Amerigo Vespucci in 1497, and John Werner in 1514; however, for centuries it was very little used, because of lack of accurate ephemeral data on the Moon, poor instruments, and the complexity of the necessary computations."[25]

The octant could not always be used to measure these lunar distances because they could exceed the octant's 90-degree limit. But in 1757, John Bird and British Captain John Campbell (1720–1780), who had conducted the first sea trial of the Hadley octant, broadened the arc of the Hadley octant from one-eighth of a circle to a sixth (sextant) so that angles up to 120 degrees could be measured.[26]

To determine longitude, one needed this sextant and tables of lunar distances. Tobias Mayer sent lunar tables to the Board of Longitude in 1756. Astronomer Nevil Maskelyne (1732–1811) helped popularize the lunar distance method, publishing *The British Mariner's Guide* (1763).[27] Soon thereafter, he was appointed Astronomer Royal. Under his auspices, the first official British Nautical Almanac was published in 1766 (for 1767). This (and its successors) provided predicted angles ("lunar distances"), "for every three hours of the day" (Greenwich Local Apparent Time), between the Moon and the Sun, as well as the "lunar stars" Hamal (Aries), Aldebaran (Taurus), Pollux (Gemini), Regulus (Leo), Spica (Virgo), Antares (Scorpius), Altair (Aquila), Fomalhaut (Pisces Austrinius), and Markab (Pegasus). (These are the bright stars that are near the ecliptic, the Moon's apparent path across the sky.) Beginning in 1834, it switched to tabulation of lunar distances according to Greenwich Mean Time, and it provided lunar distances for the planets Venus, Mars, Jupiter, and Saturn, too.[28] Lunar distance tables were published up until 1905 (Britain) or 1912 (United States), but individuals interested in nautical history have developed software for generating them.[29]

To use the tables, the observed lunar distance had to be corrected ("cleared") to obtain the true distance, as tabulated, and this required that one simultaneously

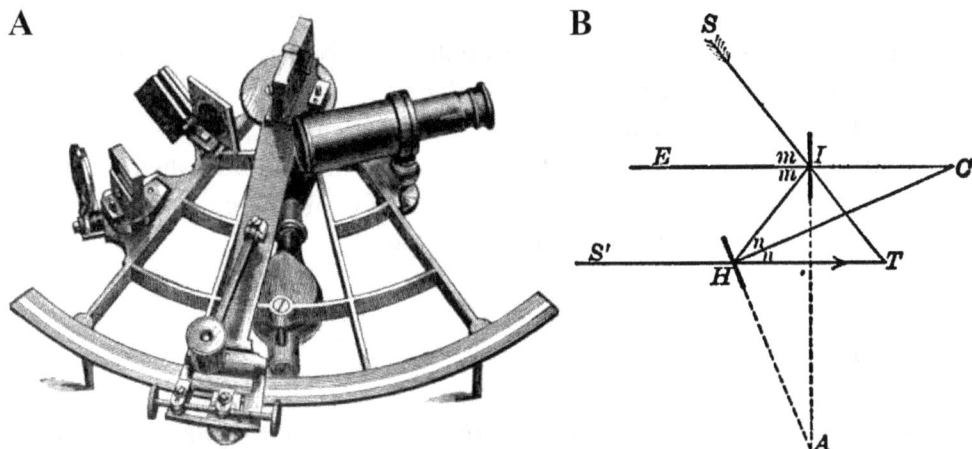

A. A late-nineteenth-century sextant. The small telescope points at the fixed "horizon glass," of which the upper half is transparent and the lower half silvered. An arm, moving along a graduated arc, turns the index glass, another mirror. There are also various filters which may be interposed if the light of one of the objects is too bright. (From Fig. 9, Greene, Dascom, *An Introduction to Spherical and Practical Astronomy* [Ginn, 1891].) B. The operating principle of the sextant. (From Fig. 10.) The telescope T points at horizon glass H and through the upper half sees object S'. The arm A is moved, turning index glass I, until by double reflection off I and the lower half of H the telescope also see object S. The angle (AIT) through which the arm moved is equal to the angle ECH between the two normals and thus to half the angle STS' between the objects.

measure the altitudes of the Moon and the reference body and then do various calculations. The navigator then took the corrected lunar distance and found "two predicted lunar distances … 3 hours apart, which bracketed his observed lunar distance." Interpolation then provided the Greenwich time corresponding to the observed distance, and that in turn could be compared to the local time to determine the longitude.[30]

The accuracy of the lunar distance method is limited by both the accuracy of the prediction of lunar motion and the accuracy of the sextant measurement; the latter depends on sea conditions. Under ideal conditions, one might expect "a time accuracy of 1 minute, which is sufficient to find Longitude within about 15 minutes of arc [about seventeen miles at the Equator or nine at 60°N]."[31]

The sextant was used to determine lunar distances, and hence longitude, on several famous expeditions, including Captain James Cook's first voyage to the Pacific, 1768–1771 (using Hadley sextants),[32] and the Lewis and Clark expedition.[33] However, once a clock that could keep accurate time at sea was invented, the lunar distance method of deriving longitude became obsolete.

Nonetheless, the sextant remained in use for obtaining the ship's latitude. In a modern nautical sextant, the observer looks through a small telescope at a fixed glass (the "horizon glass") with a silvered top half and a clear bottom half. The horizon is visible through the lower half. A second, fully silvered mirror is mounted on a moveable arm pivoted at the top. The arm is moved until the reflection of the sun off the pivoted mirror is seen on the upper half of the fixed mirror (the sextant may come with filters to reduce glare). The angular separation of sun and horizon can then be determined by comparing the position of the arm to a scale.[34]

Some sextants have a provision for reading the sun's altitude when there is no visible horizon. One such scheme involves a pool of liquid mercury (effectively, yet another mirror), which acts as an artificial horizon.

Rangefinders

As its name implies, a "rangefinder" is a device for determining the distance to some faraway object without going to the trouble of pacing it out.

The first use of the rangefinder was in the military. "No serious attempts to obtain ranges by instrumental methods were made before 1770. That year saw the short base method put to limited use, e.g. for siege purposes, but it was a time-consuming process. Development of efficient optical rangefinders did not commence until 1860, at the start of the rifled era, and culminated in the introduction of the well-known Barr & Stroud types 20 years later."[35]

A simple rangefinder uses two mirrors, mounted at opposite ends of a straight base piece of wood or metal. One mirror is fixed at a 45-degree angle to the baseline. This stationary mirror is silvered on the bottom half but clear on the top half. The other mirror is a normal mirror and is mounted so it can rotate. Look through the clear half of the fixed mirror at the target, then turn the rotating mirror until the object is visible in the mirrored half of the fixed mirror, too.

The optical rangefinder works because if you know the length of one side of a triangle, and the values of two of the angles, you can determine the lengths of the other sides. In the rangefinder described above, one angle is 90 degrees (you're looking through the clear half of the fixed mirror on a sightline perpendicular to the baseline), the length of the baseline is known, and the angle to which the rotating mirror is set is read off a scale.

Our eyes and brain actually perceive distance in much the same way. That is, the brain compares the difference (stereopsis) in the angles of view from the left and right eyes and deduces whether the object is close or far away. "For an individual having a stereoscopic acuity of 20 sec arc and an inter-ocular distance of 6.5 cm, the stereoscopic range is 670 m." That's the limiting distance for detecting "differences in depth," not the maximum range at which range can be accurately estimated by the unaided eye.[36] Visual range estimation considers cues other than just "stereopsis."

The optical rangefinder, by use of mirrors, effectively increases the separation of our eyes so that we can perceive much smaller differences in range at longer distances and also provides a means for quantifying the observed range.

Before you can use this rangefinder, you must be able to convert the angle of the

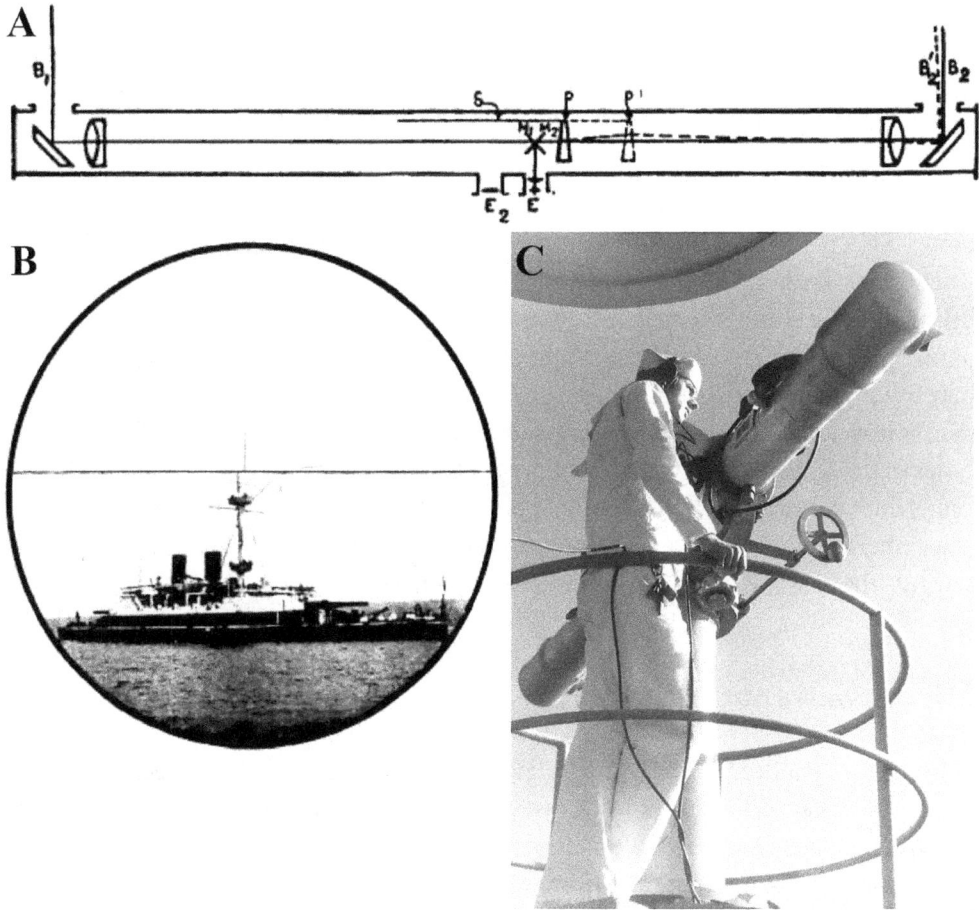

A. Schematic (simplified) of a short-base, single-observer, coincidence rangefinder. Light beams B1 and B2 (the views of the target) are reflected by reflectors R1 and R2, pass through lenses L1 and L2, and are reflected to eyepiece E by mirrors M1 and M2 so one beam is in the upper half and the other in the lower half of the eyepiece field of view. Reflector R2 is pivotable so that the target can be seen in both M1 and M2. The beams travel slightly different distances, depending on target range, and thus, without further intervention, the images are displaced from each other. A refracting prism P is interposed in the second light path, and its position is movable along scale S, e.g., to P'; it deflects the first reflected beam from R2, and when the deflection is sufficient to bring the images into coincidence, the operator reads off the range from the scale, which is visible in eyepiece E2. (Fig. 4 from Editors, "Barr and Stroud Range Finder," *J. United States Artillery*, 1908; 30: 243–49.) B. Eyepiece view from a coincidence rangefinder. Note the lack of coincidence between the upper and lower views of the mast. (Appendix Fig. 4 from Gleichen, Alexander, *The Theory of Modern Optical Instruments* [HM Stationery Office, 1918].) C. A rangefinder (type unstated) in operation on an escort carrier during World War II. (National Archives. Naval History and Heritage Command, Catalog No. 80-G-K-16163.)

rotating mirror into a distance to the target. There are two ways to do this. The more precise one is to use trigonometry to calculate the distance. Since gunners aren't keen on looking up sines and cosines in the middle of a firefight, rangefinder manufacturers could create a curved scale for the rotating mirror that was marked off in distances instead of in degrees. The other approach, for the math-averse, is to actually measure the distances to various objects, and record the corresponding angles. That works pretty well for a gun emplacement in a fortress but plainly is not of much use for a gun turret on a battleship.

The accuracy of the distance measurement is dependent on the distance (baseline) between the mirrors and thus the length of the base piece; the longer it is, the better. So, on a battleship, you would want your sight lines to come from opposite sides, or better yet, the opposite ends of the ship, and use a system of mirrors to bring the images to your eyes. You may be combining two reflected images, instead of one reflected image and one direct view, as in the simple rangefinder. And you may use prisms instead of mirrors.

While rangefinders were first used by the military, there are civilian uses of rangefinders. With cameras, it is important to bring the image into focus. That is sometimes easier said than done. It was logical to couple the focusing mechanism of a camera to an optical rangefinder so you could tell whether you were at the right setting. The first use of an optical rangefinder on a camera was in 1916.[37]

Mirror Galvanometer

Mirrors—more precisely, the mirror galvanometer—played a decisive role in one of the great technological achievements of the Victorian era, the laying of the world's first transatlantic cable. This accomplishment was as astounding in its day as the landing of a man on the Moon was in the twentieth century. Moreover, it can be argued that it had the greater practical importance, as the ability to communicate essentially instantaneously across the ocean had great and immediate commercial, military, and political implications.

For example, "three underwater telegraph cables spanned the Atlantic Ocean on 4 September 1870, when Elihu B. Washburne, the U.S. minister to France, witnessed the next French revolution. On 5 September, Washburne reported by telegraph from Paris that a republic had been proclaimed at the city hall. He asked his superiors for instructions. The next day, Acting Secretary of State J.C.B. Davis cabled that Washburne should 'not hesitate to recognize' the provisional government as soon as it appeared to be the de facto source of authority. Washburne officially recognized the Third Republic one day later, on 7 September 1870."[38]

A "galvanometer" is a device for measuring electric current. The electrical current is converted into a mechanical movement, which can be read on a scale. In 1826 Johann Christian Poggendorff (1796–1877) invented a magnetic declinometer (a device for measuring Earth's magnetic field) in which the motion of the compass needle pivoted a mirror, causing a spot of light to be deflected on a scale.[39] (This is called a "moving-iron" galvanometer because the needle is made from soft iron. The needle is not a permanent magnet.)

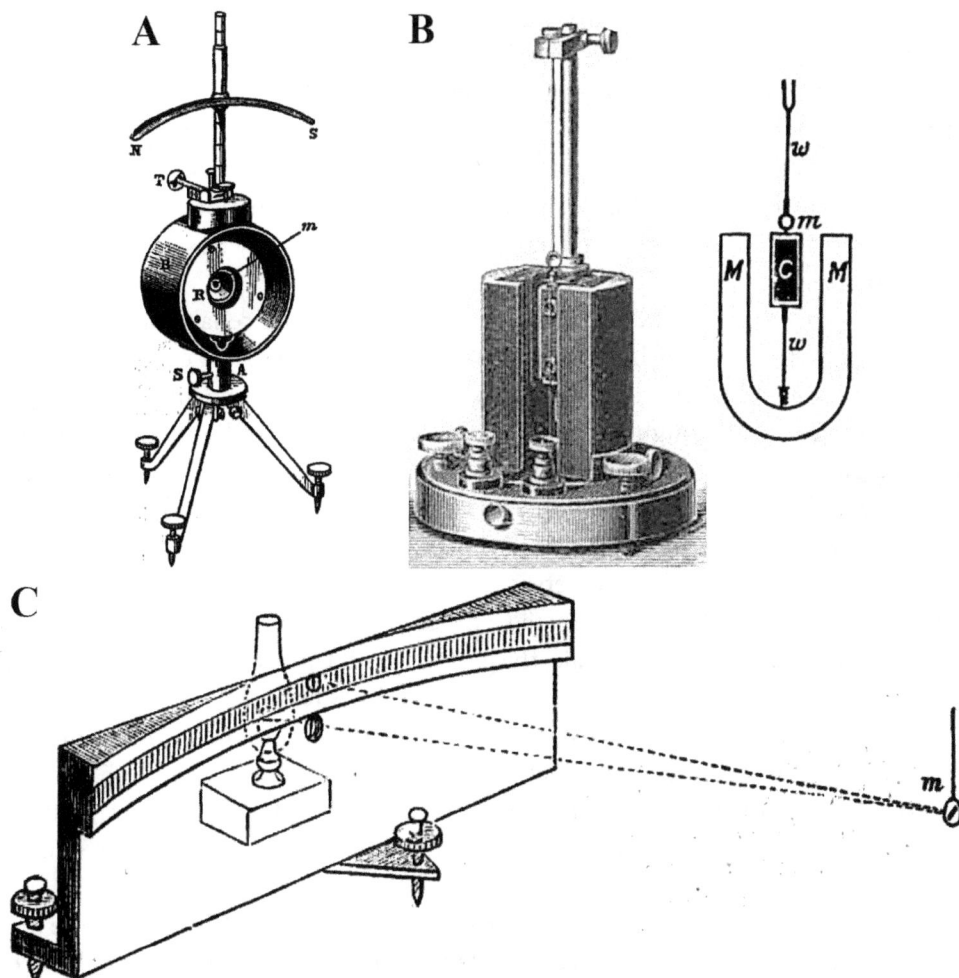

A. A "moving iron" mirror galvanometer, without the scale. R is the reel around which the fixed coil is wrapped. A small concave mirror m, with a "magnetized strip of steel" cemented to its back, is "suspended by a single silk fibre at the center of the coil." The wires of the coil go out the back and are not visible in this illustration. A large permanent magnet (N-S) is mounted above the coil; it has various uses, including adjusting the sensitivity of the galvanometer, and negating the influence of the geomagnetic field so only the current running through the coil is relevant. (Stewart, Robert Wallace, *Second Stage Magnetism & Electricity* [Clive, 1906], Fig. 183.)
B. A "moving coil" mirror galvanometer, perspective (left) and schematic (right) views. The rectangular coil C, which encloses a soft iron cylinder, is suspended by a fine wire w between the poles of a strong horseshoe magnet MM. The tiny mirror m is above the coil. (Stewart, Fig. 185.)
C. The readout system for a mirror galvanometer. A lamp shines through a hole that is bisected by a vertical wire. (A lens, not shown, may be used to focus light on the hole.) The mirror M reflects onto the arcuate scale "the image of the aperture," including the wire. After the mirror galvanometer is adjusted so, with no current, the image of the wire falls on the zero of the scale, and any current in the coil will deflect the mirror and thus the light spot. (Stewart, Fig. 51.)

In the 1830s, Karl Friedrich Gauss and William Weber used a mirror in the "readout" system for an experimental electrical telegraph. When the current changed direction, the "moving magnet" pivoted, causing the mirror to also pivot.[40]

Despite these earlier designs, it was the mirror galvanometer of William Thomson (Lord Kelvin) (1824–1907) which made the transatlantic cable a success. His 1851 "moving iron" design was "far superior to the earlier ones in that the oscillation period of the moving components is shorter, and the damping more rapid. His galvanometer has two small magnetic needles placed inside two circular coils. The needles are suspended by a torsion thread and move in unison with a small flat mirror. The small size and reduced mass of the needles greatly lessens its period of oscillation."[41]

By 1858, Thomson had developed a highly sensitive and reliable mirror "moving magnet" galvanometer in order to detect the small electrical currents that, in his opinion, were all that could practically be carried by submarine cables. Light from a lamp passed through a slit, producing a thin pencil of light. This light was reflected off a mirror, striking a scale. The mirror was suspended, on a silk thread, in the heart of a copper coil, and four small magnets were fastened to the back of a mirror. When a current passed through the copper coil, inducing a magnetic field, a force was exerted upon the magnets, causing the attached mirror to pivot. This in turn deflected the light so it illuminated a different spot on the scale. The longer the distance from the mirror to the scale, the greater was the movement of the light spot. (This was an opto-mechanical arrangement sometimes termed an "optical lever." If the angular movement of the mirror were the angle theta, the change in the angle of the reflected beam would be twice theta.)[42]

When used to send Morse code, the rightward shift of the light beam signified a dot, and the opposite movement, a dash.[43] The initial speed of transmission of messages was about eight words per minute.[44]

Unfortunately, Thomson's subtle approach was scorned by Whitehouse, the first leader of the submarine cable project. Whitehouse proposed to solve the problem of cable communications by a brute-force approach, using high voltages to achieve a more detectable current level. These harsh voltages (almost 2,000 volts) hastened the demise of the first submarine cable. Even with high voltages, Whitehouse found it difficult to communicate over this cable using the conventional galvanometer and soon was secretly using Thomson's mirror galvanometer. He had a clerk read the transmissions using Thomson's device and then key them into Whitehouse's own apparatus to create the appearance of success. However, the cable eventually failed utterly.

In 1861, the Submarine Telegraph Committee blamed Whitehouse for the failure and transferred control to Thomson. On July 28, 1866, the first successful transatlantic cable was completed, under his supervision. The current needed to activate the mirror galvanometer was supplied by "a toy battery made out of a lady's silver thimble, a grain of zinc, and a drop of acidulated water."[45] On November 10, 1866, Thomson was knighted for his achievement.[46]

The mirror galvanometer did have its limitations. "First it took two men to operate it, one to watch the beam of light and call out the letters while a second operator wrote the letters down. Second, there was no permanent record of the message received."[47] The mirror galvanometer was replaced by the siphon recorder, invented by Thomson in 1869, in which the deflection of the magnet resulted in the redirection of the outlet of the "siphon."

In the 1858 Thomson mirror galvanometer, the coil was fixed, and the magnet moved. This was reversed in the very sensitive 1882 mirror galvanometer of Marcel Deprez and Jacques-Arsène d'Arsonval.[48]

Paired mirror galvanometers ("scanners") are still used in laser light shows. One scanner deflects the laser beam horizontally, and the other vertically. The galvanometers are usually of the moving iron or moving magnet types.[49]

To "write" separated characters on the screen, "blanking" is needed; one method of achieving this was to "send the beam through a third scanner which [first] deflected it to ... a corner reflector. The corner reflector sent the beam back to the scanner and from there it was deflected into the X-Y scan pair. This arrangement formed an optical switch as any movement of the [third] scanner caused misalignment of the beam through the corner reflector, turning the beam off."[50]

Measurements of the Speed of Electric Currents and of Light

In 1834, Charles Wheatstone (1802–1875) used a double-sided rotating mirror in an experiment to measure the velocity of the electric current in a copper wire. The test wire was half a mile long, with three spark gaps, but the gaps were brought near each other so they could be observed simultaneously. The mirror "revolves 800 times in a second; and during this time the image of a stationary point would describe 1600 circles." He believed that an angular movement in the image of the spark as small as half a degree could be seen, and that would correspond to a duration of $\frac{1}{1,152,000}$th of a second. The spark at the middle gap occurred at a later angular position than those at the ends, leading him to conclude that the speed was 288,000 miles per second. The speed of light in a vacuum is actually about 186,000 miles per second, and the speed of an electric current is actually somewhat less. Wheatstone later acknowledged that while he could perceive the difference in timing, he could not measure the angular change with this apparatus, and thus the speed was high but uncertain.[51]

The speed of light in air was first successfully measured by Hippolyte Fizeau (1819–1896) in 1849, using a setup employing two fixed mirrors, several miles apart, and a rotating cogwheel. Léon Foucault (1819–1868) later used a light source, a slit, a rotating flat mirror, and a stationary concave mirror. The light was triply reflected (off the rotating mirror, off the stationary mirror, and finally off the rotating mirror once more), but since the rotating mirror turned while the light traveled between it and the stationary mirror, the triply reflected light spot was deflected. By also placing a diagonal beam splitter glass between the light source and the rotating mirror, the observer could be enabled to see both the single reflection off the beam splitter and the aforementioned triple reflection, allowing measurement of the deflection. In 1862 Foucault reported a speed of 298,000 kilometers per second, which is nearly identical to the modern value.[52]

Early Motion-Picture Devices

The essence of the motion picture is the creation of the illusion of movement through a succession of still images. Our eyes merge these images as a result of what is called "persistence of vision."

The praxinoscope, an early movie-making device, was invented by Charles-Émile Reynaud (1844–1918) in 1877. It is, in effect, a double cylinder. There is a series of mirrors in the center of the device, on the outside of the first cylinder, facing outward. There are small spaces between the mirrors. On the outer cylinder, facing inward, are the still scenes which are to be merged. When the praxinoscope is spun, your eye receives the images reflected from each mirror in quick succession.[53] A light source can be placed on top of the central axis to provide additional illumination. The Praxinoscope Theatre added a background, and the Projecting Praxinoscope combined the praxinoscope with a magic lantern.

Reynaud later developed the first nonlooping motion picture system, the "Theatre Optique." Hand-drawn images on

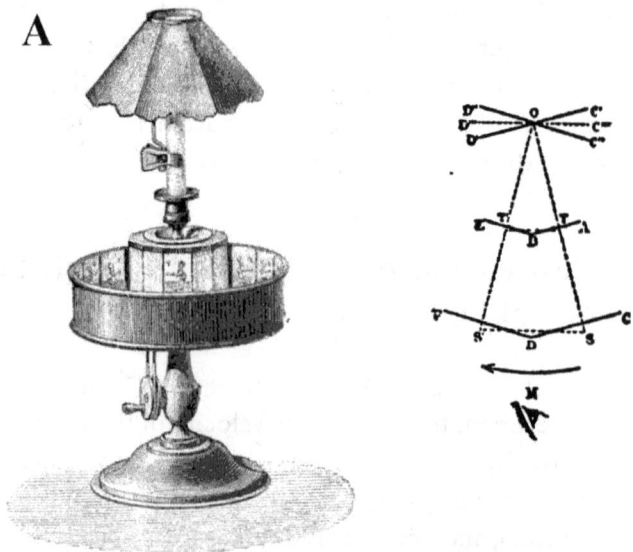

A. "Praxinoscope" (Figs. 128–129, p. 126). B. "Praxinoscope theatre" (Fig. 130, p. 127). (From Tissandier, Gaston, *Popular Scientific Recreations in Natural Philosophy....* [W.H. Steele, 1882].) Compare with the illos in Wikipedia, https://en.wikipedia.org/wiki/Praxinoscope.

celluloid frames were joined into long ribbons and run from one spool to another. A projection system, using both lenses and mirrors, projected the images on a screen. These images were superimposed on a still background image provided by a second projector. The images all used black backgrounds to hide the superposition. The films were 10 to 15 minutes long and were accompanied by music and sound effects. He applied for a patent in 1888; it issued in 1889.

"From 1892 until 1900, with some breaks, the Théâtre optique gave 12,800 shows at the Musee Grevin which were attended by over half a million spectators." Unfortunately, after his show was canceled, in a fit of outraged artistic temperament, he threw all of his apparatus, and all but two of his films, into the River Seine. He died in a poorhouse.[54]

Mirrors as Artists' Aids

Mirrors may aid in developing artistic perceptions. In his *Discourse on Painting*, Leonardo da Vinci says, "When you wish to see whether the general effect of your picture corresponds with that of the object presented by nature, take a mirror and set it so that it reflects the actual thing, and then compare the reflection with your picture and consider carefully whether the subject of the two images is in conformity with both, studying especially the mirror."[55]

"Alberti recommends looking at a painting in a mirror to expose its weaknesses. In his *Della pittura* he writes, 'A good judge for you to know is the mirror. I do not know why painted things have so much grace in the mirror. It is marvelous how every weakness in a painting is so manifestly deformed in the mirror.'"[56]

The earliest use of mirrors by artists was to facilitate a self-portrait. According to Pliny the Elder (23–79 CE), at Naples there was, in his time, a self-portrait by the Roman painter Iaia of Cyzicus, "taken by the aid of a mirror. There was no painter superior to her for expedition; while at the same time her artistic skill was such, that her works sold at much higher prices than those of the most celebrated portrait painters of her day."[57]

Of course, if a conventional mirror were used, the resulting picture would be left-right reversed. "An interesting example is preserved in Rembrandt's self-portrait of 1668.... Evidently realizing that he appeared left-handed, Rembrandt painted out and changed over his hands and arms. This is clearly visible in an X-ray of the painting."[58]

Heather Champ's "Mirror Project" website accepted self-portraits made by photographing oneself on a reflective surface; "you, or some part of you, and the camera should be visible within the reflection." The site (presently inactive) received over 34,000 images; favored surfaces (per "Themes") include bathroom mirrors, rear and sideview mirrors, Mylar balloons, water, windows, Christmas ornaments, chrome auto bumpers, convex safety mirrors, spoons, knives, toasters, garden globes, guitars, hand mirrors, sunglasses, tools, and TV screens.[59]

Filippo Brunelleschi (1377–1446), the architect and engineer who gave Florence's cathedral its famous dome, experimented with the use of accurate perspective in painting. According to his biographer Antonio Manetti, in 1415 Brunelleschi painted a now-lost picture of the Baptistery in Firenze, Italy. Manetti's account is sketchy. There is a dispute as to how the initial, true-perspective picture was drawn. The simplest explanation is that he just painted the picture on a mirror facing, and therefore reflecting, the Baptistery. Alternatively, he may have used some combination of architectural plans, surveying devices, geometric calculations, and graphical constructions.[60]

In any event, in this picture he made a small conical peephole, most likely at the "vanishing point" of the painting. He set up the picture so that it faced the actual Baptistery. In front of the picture he placed a second mirror so that, looking from the back of the picture through the peephole, you could see the reflected image of the painted scene. If the second mirror was taken away, you could see Baptistery directly, and therefore how closely the picture painted on the first mirror mimicked the actual perspective.[61]

Having confirmed that his original picture (painted mirror or not) was accurate, Brunelleschi could mathematically relate the perspective view to the plan and elevation views of the Baptistery. This allowed him to devise a system of horizon lines and vanishing points whereby an artist could show perspective without resort to mirrors.[62]

Gerrit Dou ["Gerard Douw"] (1613–1675) has been credited by Jean-Baptiste Descamps with the use of an optical device that is referred to interchangeably as "*un miroir concave*" and "*verre concave.*"[63] While the latter alone could refer to a concave lens, the context makes it clear that this is a concave mirror of the lead-on-glass or tin-on-glass type. If the object is far enough away, its image will be reduced (and inverted). Steadman translates the text as follows:

> I do not know whether it is to him that we owe a rather ingenious invention—although one with various drawbacks—for reducing large objects into a small space. He made use of a type of screen on a stand, in which he had fashioned and framed a concave mirror [*miroir concave*], on the level of his eye when he was seated. This screen was a sort of enclosure [*cloison*] between him and the object to be represented. The object formed a reduced image of itself [*se traçoit en petit*] in the concave mirror, and the painter needed to do no more than imitate the outline [*trait*] and the colour.
>
> Once his composition was laid out, he brought to his canvas—divided into an equally spaced square grid—the objects that he needed. This division was repeated with threads on a little framework whose size was that of the circumference of the concave mirror, in such a way that when he fixed the framework on the mirror, it represented a square inscribed in a circle.[64]

Descamps went on to criticize this method as one "contrary to harmony and elegance."

The Camera Obscura

The term "camera obscura" was coined by the astronomer Johannes Kepler in 1604; it is Latin for "dark room." If there is a pinhole in one wall of such a room,

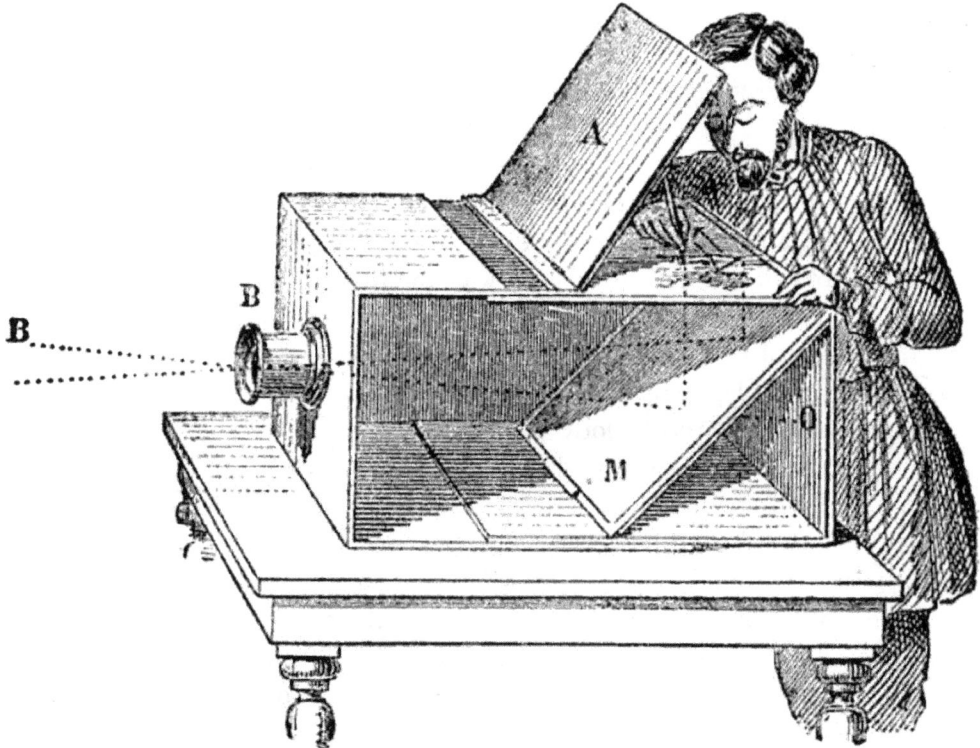

The camera obscura. Light enters through lens B and is inverted and reversed. It strikes mirror M, which corrects the inversion, and the image is displayed on the unlabeled ground glass screen. The artist has laid a sheet of tracing paper over the screen and is drawing the image. The image is focused by sliding the lens back and forth. Lid A covers the screen when the camera is not in use. (From "The Camera Obscura," Fig. 4, in Lardner, Dionysus, *The Museum of Science and Art*, Volume 8 [Walton & Maberly, 1859], 203.)

an inverted (and reversed) image of the outside would be projected onto the opposite wall. We do not know when the first camera obscura was constructed, but the astronomer Reinerus Gemma Frisius used one to observe a solar eclipse in 1544.[65]

A "reflex" camera obscura is one that includes a mirror. In 1585, Giovanni Battista Benedetti (1530–1590) proposed the use of a mirror angled at 45 degrees to the direction of the light coming from the lens in order to right the image.[66] However, that would have caused the image to appear on the ceiling of the room. The use of mirrors inside a camera obscura was also described by della Porta. He suggested putting a convex mirror near the hole, and a concave mirror at a distance, arranged to obtain an erect image.[67]

The first attempt to make the "camera obscura" portable was by Friedrich Risner (1572), who proposed use of a "lightweight wooden hut," with lenses in place of the pinholes.[68] However, one still had to observe the image from inside the hut, and the need to fit an observer inside placed a limit on the degree of miniaturization.

The key innovation was to use a box with a translucent screen opposite the lens (or pinhole). This appears to have been proposed by several writers in the mid-seventeenth century, including Gaspar Schott (1657) and Robert Boyle (1669).[69] Not only did the translucent screen allow the observer to stand outside the camera obscura; it also corrected the left-right reversal. Johann Sturm has been credited with adding in 1676 a 45-degree mirror, so the image would be projected upright, onto a sheet of oiled paper on the top rather than in the back of the box. (Later, the oiled paper was replaced with ground glass.)[70]

Given the faintness of the projected image, if the device were used outdoors, one would probably want to cover one's head and the top of the device with a cloth hood.[71]

The use of the camera obscura as a drawing or painting aid was proposed by Giambattista della Porta in 1558.[72] In Venice, both Antonio Canal ("Canaletto") (1697–1768) and Bernardo Bellotto (1722–1780), and probably also Paolo Caliari (1528–1588), "made accurate reproduction of buildings using a camera obscura on the site"—so accurate that modern scientists have used them to determine the contemporary height of the "green-brown algal belt" on a building and compare it to its height on the same building today, thereby quantifying the rise in mean sea level.[73]

Superficially, the image on the obscura's screen appeared to be "a painting in miniature." But since the camera obscura used a lens, its image lacked some verisimilitude. Objects in the periphery of the scene were distorted and of diminished brightness.[74]

The Camera Lucida

Another artist's aid, the "camera lucida," was proposed by Johannes Kepler in *Dioptrice* (1611),[75] but there is no evidence that he built it. The device was reinvented by William Hyde Wollaston (1766–1828), who received a British patent on it in 1806.[76] Wollaston was a physician who left medicine in 1799. He became wealthy as a result of developing a process for refining platinum ore and then cornering the British market on platinum. As a chemist, he is remembered for the discovery of the elements palladium and rhodium, and as a physicist, for his study of the solar spectrum and the invention of the meniscus camera lens, the reflecting goniometer, and the camera lucida.

Unlike the camera obscura, the camera lucida does not project an image into a darkened room or the inside of a box. The purpose of the device is to project a scene on a paper in such a way that it is easy to trace the scene on the paper.

The simplest way of viewing the paper and the outside scene is to look down at the paper through a sheet of plain glass, inclined 45 degrees. The image of the scene is then reflected upward. However, there are several problems. The reflection is relatively weak; the image is inverted; and it is difficult to focus on both the image and the drawing.[77]

One may instead use two transparent glass reflectors and a planoconvex lens. The light from the scene strikes the lower reflector first; it is inclined at an angle of 22.5 degrees to the horizontal, so the angle of incidence is 67.5 degrees. The reflected light then strikes the upper reflector, which is at an angle of 135 degrees to the lower one, or 22.5 degrees to the vertical. The twice-reflected light thus travels vertically upward, to the eye, which thus sees the outside image. Because the light is reflected an even number of times, the image is upright.

Since the upper reflector is transparent glass, the artist also sees through it, down to the paper below. A planoconvex lens is placed in that optical path, between the upper reflector and the paper. This is to adjust for the disparity between the distance of the eye to the subject, and that of the eye to the paper.[78]

A disadvantage of this system is that relatively little light is twice reflected. The reflections are "external" (air-to-glass) and with an angle of incidence of 67.5 degrees and a glass refractive index of 1.52, each reflection is of 14.6 percent of the incident light,[79] and thus 2.1 percent is doubly reflected.

In his 1806 patent, Wollaston suggested that the upper reflector could be "partly silvered." However, this didn't mean "lightly silvered," as in the case of a conventional beam splitter. Rather, this

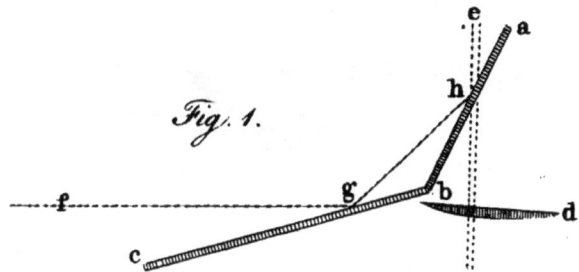

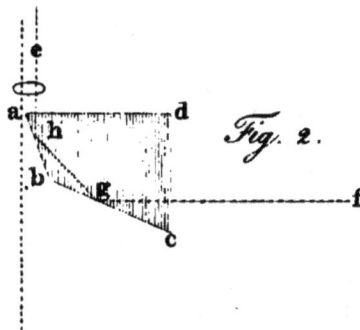

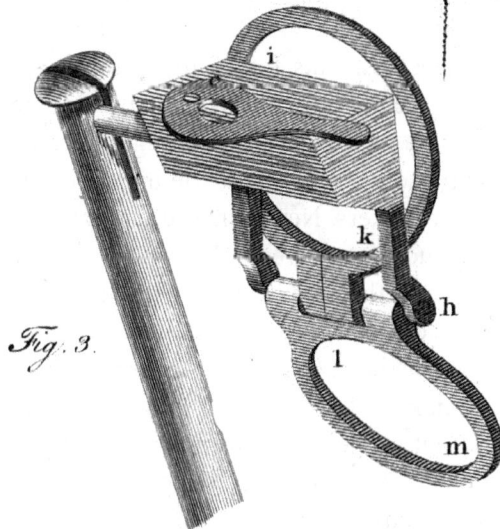

The camera lucida of Wollaston (1807). A. Schematic (Fig. 1) of the simple two-mirror form. The light from scene f is reflected at g off glass cb and at h off glass ba, striking eye e. The eye also sees the paper through planoconvex lens d and glass ba. B. Schematic (Fig. 2) of the prismatic form. The light from scene f enters the prism abcd, and is reflected at g off surface bc and h off surface ba, striking eye e. The eye also looks past corner a of the prism at the paper. The corrective lens is not shown. C. Perspective view (Fig. 3) of the prismatic form. A moveable eyehole c is used to adjust the relative intensity of the light seen through the prism i or directly from the paper (not shown).

A

B

The prismatic camera lucida in use. A. At a worktable, drawing a small figurine. Note how the support rod for the prism is clamped to the table. (From *Scientific American Supplement* [1879, Jan. 11], https://upload.wikimedia.org/wikipedia/commons/9/90/Camera_Lucida_in_use_drawing_small_figurine.jpg. See also Stanley, William Ford, *A Descriptive Treatise on Mathematical Drawing Instruments....* [Spon, 1878], 131.) B. Drawing while resting the camera lucida on a drawing pad. (From Dolland, George, *Description of the Camera Lucida* [1830], Beinecke Rare Book & Manuscript Library, Yale University, https://collections.library.yale.edu/catalog/2012328.)

reflector allowed "the paper to be seen through an opening in the silvering."[80] (Or one could just look "past the edges of the same.")

In yet another disclosed embodiment, a quadrilateral prism replaced the two reflectors. Two of the sides correspond exactly in their orientation to the reflectors of the first embodiment. For plate glass (refractive index 1.52) to air, the critical angle is about 41.1 degrees. Thus, total internal reflection is experienced at both of these sides. However, since the reflection is total, the drawing paper cannot be observed through the prism; one must look "past the edge," such that the edge bisects the pupil.[81]

"Wollaston arranged for the manufacture of camera lucidas by two firms of instrument-makers, Newman and Dollond, and put them on sale from 1807. They were a big success and were taken up by artists and amateurs for making landscapes and portraits. The camera lucida was much lighter and more easily carried than the camera obscura, and remained in wide use until the 1840s, when it was largely superseded by the photographic camera."[82]

Opticians in Europe also thought that the camera lucida would be a hot sales item and put versions of it into their catalogs. The physicist Giovanni Amici sold both books and instruments. "The Camera Lucida rather suddenly … eclipsed the Camera Obscura…: Amici's workshop delivered 269 Camera Lucidas between

1815 and 1832, whereas only 2 Camera Obscurae were sold in this period." Fiorentini suggests that the camera lucida was also more useful to surveyors, as well as to artists.[83]

In 1830, Captain Basil Hall warned that the camera lucida has "no means of supplying taste," and will not "enable people who are totally ignorant of the use of the pencil … to make good drawings, without very considerable practice." Hall also anticipated the concern that "there is less merit attaching to a sketch made with the Camera, than to one made in the ordinary way." But he urged that "the wish to gain credit for making sketches" should be "subordinate to the wish to represent natural objects correctly."[84]

Steadman reports that the camera lucida is "not an easy tool to master. There is a narrow boundary at which the eye must be placed so that one sees the subject superimposed on the drawing surface. The field of view is small, and it is difficult to see enough of the subject at once to trace long lines smoothly."

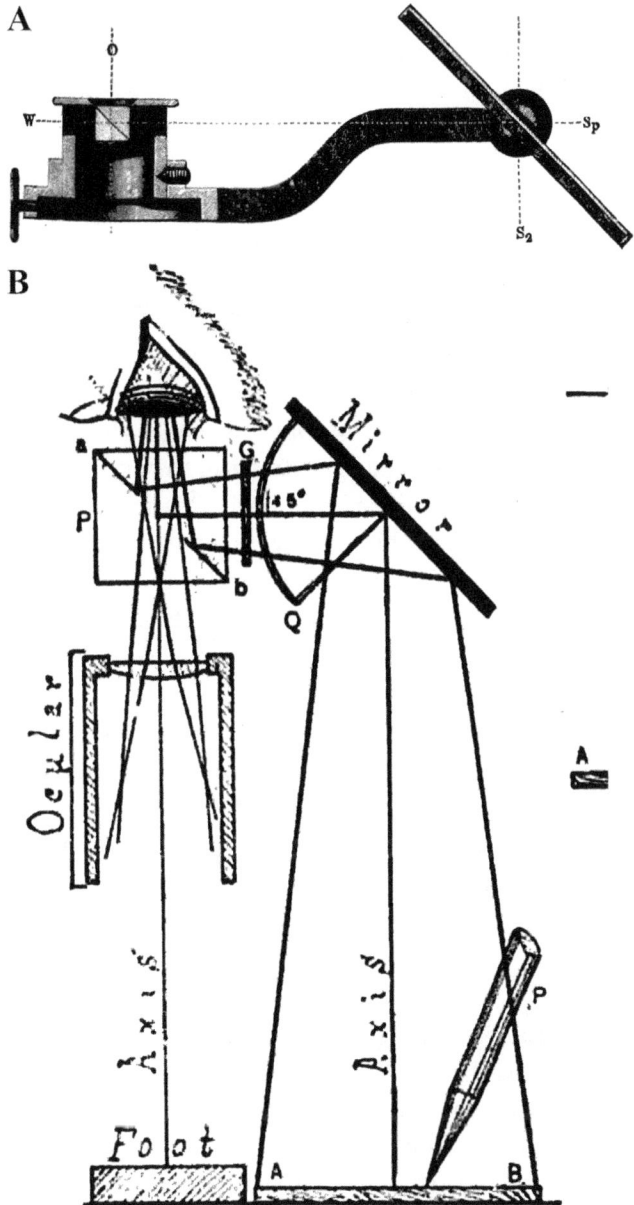

Abbe camera lucida. A. Fig. 220, Carpenter (1901). B. Fig. 100, Gage, Simon Henry, *The Microscope: An Introduction to Microscopic Methods and to Histology* (Comstock, 1920).

Wollaston's "prismatic" camera lucida competed with others using one or two ordinary glass reflectors. In 1877, Terquem proposed that "for the usual camera lucida with either one or two reflections [off ordinary glass] a glass semi-silvered by Foucault and Martin's process can be substituted with great advantage."[85]

The camera lucida is used by microscopists as well as artists. Amici patented "a catoptrical microscope with a camera lucida ocular."[86] In 1857, the microscopist Lionel Beale favored a simple system that was a single piece of tinted glass inclined at 45 degrees and held in front of the eyepiece, thus reflecting the magnified image upward.[87] The purpose of the tinting was to obscure the "ghost reflection" from the second surface of the glass.[88]

The microscope tube had to be horizontal. That in turn meant that the glass slide was vertical and thus one couldn't use a "Beale reflector" to draw a specimen in a well slide. (The liquid would escape.) This was also a problem with the use of the Wollaston camera lucida for microscopy.

This problem was solved by the late nineteenth-century Abbe camera lucida. In this device, the microscopic subject is viewed directly, while the drawing surface is doubly reflected. The drawing surface is seen through a special prism and a large, 45-degree mirror offset horizontally from the microscope tube. The prism is made by slicing a cubical glass diagonally into two triangular wedges and silvering the diagonal surface of the upper wedge except for a small circular aperture in the center. The wedges are then cemented together to reform the cube. Light transmitted through or reflected off the subject passes up the microscope optics and through the clear aperture.[89]

With the Abbe camera lucida, the microscope tube may be vertical. The drawing surface must be parallel to the microscope stage, so if the tube is tilted, the drawing surface must also be tilted. In later refinements, the prism could be replaced with another prism having a narrower or wider aperture, and smoked glasses could be inserted into the optical paths to adjust the relative intensities of the two images.[90]

The Hockney–Falco Thesis

The artist David Hockney and the physicist Charles Falco argue that many Renaissance artists used optical aids to make their paintings more realistic. Hockney contends that the paintings show detail that could not readily be achieved without the aid of an optical device like a camera lucida; that there are out-of-focus areas that are optical artifacts; and that there are changes of perspective from one part of the picture to another that are suggestive of an optical device being re-aimed or refocused.[91] He also claims that at the end of the sixteenth century, the artists seemed to have painted more people as left-handed than would have been expected. The last he sees as evidence that a mirror was used.[92]

Hockney also constructed an artists' aid of his own. The subject, strongly lit, was seated beside a hole (a large one, not a pinhole as in the camera lucida), opening into a darkened room. On the opposite wall Hockney hung a concave mirror. Beside the hole, facing the mirror, Hockney hung drawing paper. The subject's reflection was projected onto the paper, and Hockney drew over it. (To make this work, the mirror would have to be tilted slightly to one side.) Hockney reports that the image was upside down but not reversed left-and-right.[93] He says that the usable image is never much more than a foot across.[94]

Christopher Lüthy differentiates the strong (optical aids were used directly by the artist) and weak (optical aids merely inspired the artist) versions of the thesis. Which version applies depends on which artist (and painting) is under discussion. In particular, Lüthy concludes that Gaspare Vanvitelli (1652–1736) used a camera obscura in creating his cityscapes, but "several stages of artistic transformation separate the camera obscura projection from the finished painting." But he concedes that Vermeer is "Hockney's strong thesis turned flesh."[95]

The Hockney–Falco thesis has been challenged from several different perspectives. The first is for allegedly giving insufficient acknowledgment to prior hypotheses that particular Renaissance artists used optical aids. The second is for seeming to belittle the knowledge and skills of the Renaissance artists, that is, by allegedly asserting that they needed those optical aids to create their works and deliberately concealed this "cheating" (in the critics' view) from their patrons. The third is for relying primarily on perceived optical artifacts in small portions of a limited number of paintings as evidence of use of optical aids.[96]

With regard to the specific issue of use of a concave mirror, David Stork has argued that the paintings in question would have required use of a mirror with a much longer focal length than would have been feasible to make by blowing and cutting a glass globe.[97]

Vermeer's Hypothetical Use of Optical Aids

The paintings of Johannes Vermeer (1632–1675) have been viewed by some as photorealistic in character. "In 1891, the American lithographer and etcher Joseph Pennell" declared that it was "extremely likely that ver Meer used the camera lucida, if it was invented in his time, for it gives exactly the same photographic scale to objects."[98] However, given the history of the camera lucida (already described), Philip Steadman concluded that it was "improbable that any such instrument would have been available to Vermeer."[99] Instead, he proposed that Vermeer used a camera obscura.

The film *Tim's Vermeer* documents the techniques by which Tim Jenison, a non-painter, duplicated Vermeer's *The Music Lesson*. He first "built an exact full-scale model" of the original drawing room, and then painted it, using an optical device to guide his brushwork. This device consisted of "a 4-inch diameter lens with a focal length of 28 inches, a concave mirror with a focal length of –12 inches, and a small first-surface mirror mounted on a stick." "It took another seven months to actually paint the picture…. My experiment doesn't prove that Vermeer worked this way, but it proves that he COULD have worked this way."[100]

Tim's optical device was, essentially, a form of camera lucida. Jenison rejected Steadman's hypothesis that Vermeer used a camera obscura: "The light projected by the lens obscures the color of the paint you are applying to the canvas. It makes the paint look too dark and too colorful. You must constantly turn on the light to see what color you have actually painted. There is simply no way to accurately compare the paint color to the projection. They interfere with each other."[101]

However, Steadman proposed in a 2003 interview that Vermeer painted "in monochrome directly over the projected image in the camera [obscura]" and added the color later. "Vermeer's first paint layers are in monochrome, either dark brown or blue over a light ground."[102]

The Claude Lorraine Mirror

Some nineteenth-century romantics would, upon encountering a beautiful vista, turn their backs to it and view it indirectly by its reflection in a mirror, preferably a tinted (black, blue, or brown was favored), slightly convex one. In this way they sought to see the landscape as if it were painted by some master, like the seventeenth-century French painter Claude Lorraine (1600–1682), in subdued colors.

This pastime was sufficiently popular that mirrors were specially designed for the purpose. An 1870 catalogue advertised, "Obsidian, Claude Lorraine, or Landscape Mirror … for condensing landscapes in true perspective view, and imparting a soft shady tint to the natural colors; a most valuable acquisition for the artist." Martin Kemp comments, "It was a compact-looking fold up mirror, that was slightly convex. Either hand tinted or with a black backing. The overall effect for the artist's purpose was to allow the subtlety of the middle greys to emerge, while suppressing the overwhelming highlights. The darks [sic] were still preserved with detail. The convexity of the glass compacted the overall scene and aided in the area of perspective."[103]

Essayist Gail Hamilton, describing a trip down the St. Lawrence in her *Gala Days* (1863), wrote, "In the sunny day all things are sunny, save when a Claude Lorraine glass lends a dark, rich mystery to every hill and cloud."[104] William Dean Howells said of Mark Twain's home in Hartford, "The windows of the library … showed the leaves of the trees that almost brushed them as in a Claude Lorraine glass."[105]

A black "Claude mirror" was used by the meteorological pioneer Luke Howard in his study of clouds, as it increased the contrast between the clouds and the sky.[106]

In 2016, there was an "art installation at Wimpole Hall" (Cambridgeshire, England) that took "the form of a series of large-scale easels…. Each easel is composed of a Claude mirror and they come in different shapes, are tinted in different colours and have different landscape emphasis."[107]

Opaque Projector

An opaque projector, once ubiquitous in schools, is essentially a portable camera obscura that projects on an external screen (or wall). A strong light source, usually concentrated by concave mirrors, shines on the document. The latter reflects some of the light and that light is projected through a lens, creating a magnified image on the screen. If the light is reflected upward from the document, a diagonal mirror is placed between the document and the lens so the image appears on the wall rather than on the ceiling.[108]

The early camera obscura was usually used to observe an outdoor scene. However, della Porta advised that one could use it to help one copy a picture set outside,[109] or to project text written on the surface of a plain mirror receiving the sunlight from the pinhole.[110] In 1646, Athanasius Kircher described a technique he called "catoptric steganography," which involved "methods for projecting texts using both sunlight and candles, with the aid of both flat and concave mirrors, and a convex lens." Gorman characterized it as "the early modern version of the Powerpoint presentation."[111]

A. A megascope used to project the demonstrator's hand. The chimney on top is for ventilation. (Dolbear, Amos Emerson, *The Art of Projecting* [Lee & Shepard: 1887], Fig. 24.) Dolbear comments, "minerals, crystals, shells, bright-colored beetles, bugs, butterflies, etc. may all be exhibited and appear, with the shades and shadows, like real objects." B. Plan of the Wonder Camera. (Fig. 68, Hopkins [1898].)

Euler described a type of opaque projector in a 1762 letter to Princess Friederike Charlotte of Brandenburg-Schwedt and noted that he had presented the device to her six years earlier. This used the mirror only for concentration of the light of the lamps. He pointed out that his "magic lantern," unlike the common type, could be applied to "objects of all sorts," as opposed to just "figures painted on glass."[112]

However, to avoid confusion, it came to be known as an "opaque lantern," a megascope (because the lens-to-screen distance was typically much greater than the lens-to-object distance, and consequently the projected image was greatly magnified), or an episcope (because it was often constructed so as to shine down on the object). The greater the desired magnification (twenty-fold or more was possible[113]),

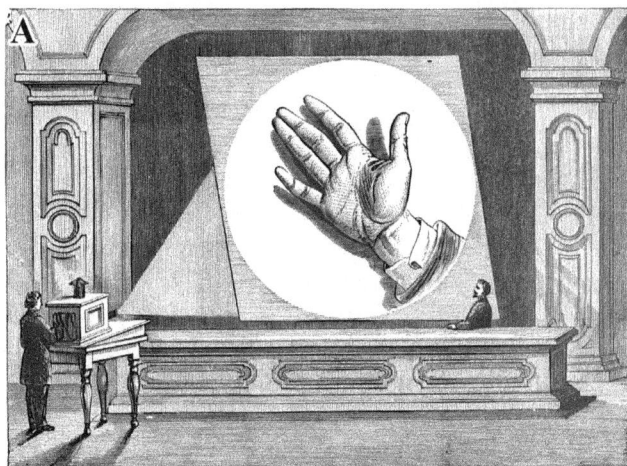

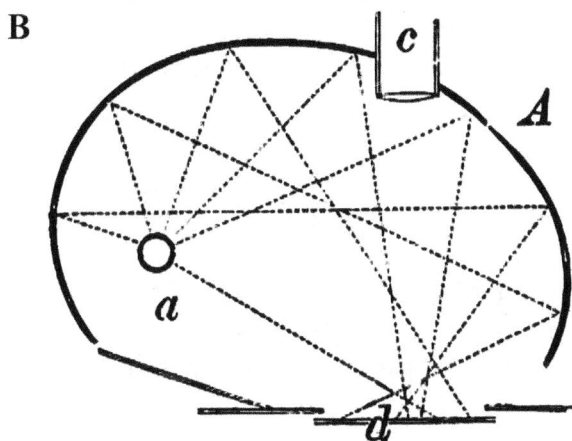

the more intense was the necessary light source, and measures had to be taken to dissipate the heat and protect the object (and operator) from it. By 1888, an electric megascope had been introduced by Trouvé.[114]

In the nineteenth century, the opaque projector was used to project images of three-dimensional objects, not merely documents. It was even "possible to enclose large objects," such as a "human skull," inside the lamphouses of some megascopes. "Due to the limited depth of field, the focused parts of the projected image appear to be situated between other parts that are either more or less in focus. This creates an illusion of depth when projected on a two-dimensional screen…. Changing shadows and reflections, caused by the manipulation of the object, are equally realistic on the screen as when seen directly. This manipulation during projection heightens the illusion of depth."[115]

The "Wonder Camera" had an ellipsoidal reflector, with an Argand lamp at one focus and a holder for the object at the other. There was a perforation in the reflector immediately opposite the holder, and through this the light reflected off the object was projected.[116]

Photographic Enlarger

The traditional photographic process produced a negative image, because the silver halide particles on the glass or film were darkened by light. Positive prints could be made by passing light through the negative, and the light then striking a photosensitive paper. Early prints were contact prints, without any enlargement of the image. For enlargement, the lens-to-paper distance must be greater than the negative-to-lens distance.

In 1768, Martin Ledermüller published an illustration showing the marriage of a solar microscope (see Chapter 3) and a camera obscura to project the microscope image onto a wall, where he could trace it. In 1807, Thomas Wedgewood and Humphry Davy reported use of a similar setup to project the image of a painted glass onto photosensitive paper. However, at the time there was no means to fix the silver print. In the 1830s, William Talbot, who had devised a fixative, duplicated their experiment, but this time produced a permanent photomicrograph.[117]

However, "transparent negatives were not available [to photographers] until about 1850."[118] A few years thereafter, Woodward patented (1857) a "solar camera," essentially a portable camera obscura which used a flat mirror to reflect sunlight through the negative and then through a projection lens and, enlarged, onto the paper. Or, he taught, it could be used with an internal artificial light and a concave mirror.[119] It did not have a separate dark chamber for the paper, so the room had to be darkened for printing.

Long exposures were often required, and Gale received a patent (1864) on a sun-tracking system for a solar camera (cf. heliostats, Chapter 4).[120]

Digital Video Projection Systems

In the late twentieth century, movies were shown by using an intense light source, a parabolic reflector, and a lens to project light through film onto a movie screen. Nowadays, movie projectors are primarily digital projectors. Digital projectors are also found inside digital television sets, which have built-in screens.

One type of digital projector, the "DLP" ("digital light processing"), uses a "digital micromirror device." The latter is an array of very small mirrors, for example, 13.65 microns wide for XGA display (768 × 1,024) or 10.8 microns FullHD (1,920 × 1,080). Each mirror has two positions, one that reflects the light through the objective lens, and the other that reflects it to an "absorbing site" (essentially a heat sink). Thus, each pixel may be turned on or off. Thanks to persistence of vision, shades of gray are possible by switching rapidly between the two states within a single "frame period," with the appropriate proportion between the "on" and "off" states.[121]

Color may be rendered in one of two ways. One may use three DLP chips, each receiving light of a different primary color. Or one may use a single chip but shine white light through a rotating color wheel carrying the appropriate filters, with the micromirror array synchronized to the color wheel to arrange the micromirrors to correspond to the appropriate component color image.[122]

LCD and LCoS digital projectors are explained in Chapter 7.

Optical Discs

Optical discs (CDs and DVDs) are removable data storage media. The optical reader directs laser light to a particular position on the disk and determines whether that light is reflected back or not. The reflective layers used are "aluminum, gold, silver, and silver alloys." For data to be read from both data layers of a single-sided, double-layer DVD ROM disc, the first reflective layer must be only semireflective; the second is fully reflective.[123]

On CD-ROM and DVD-ROM discs, the reflective layer is also the data layer, with the data being encoded by stamping pits into a polycarbonate substrate and then depositing metal upon the substrate. On recordable discs, the data layer is separate. In the write-once type, the data layer is a photosensitive organic dye and the laser beam of the writer causes a chemical change in this dye. In the rewritable type, the data layer is a reversible "phase-changing metal alloy film." All of these expedients create a mosaic of areas that absorb light and areas that transmit it.

Decorative Mirrors in Indoor Architecture

The first use of mirrors in architecture may have been by the infamous voyeur Hostius Quadra. Lucius Annaeus Seneca [the Younger] (4 BCE–65 CE) wrote, "Mirrors faced him on all sides in order that he might be a spectator of his own shame."[124] It is unclear whether they were actually fastened to the walls of his bedroom; if the mirrors were magnifying, as Seneca states, they would have to have been concave in shape, which in turn would have made it more awkward to use them as wall coverings. But it is possible that he used both free-standing concave mirrors and wall-based flat mirrors.

Ultimately, Hostius Quadra was murdered by an abused slave; Seneca sneers, "he ought to have been immolated in front of a mirror of his own." While the Roman custom was to put all the slaves in a household to death if one murdered his or her master, Augustus declined to take any action to avenge Hostius's slaying.

As advancing technology made economical (at least by the standards of the nobility) the production of large mirrors, mirrors became popular as wall coverings. This led, perhaps inevitably, to "mirrored rooms" and "halls of mirrors."

The first chamber of mirrors appears to have been at Torgan Castle, on the Elbe River, 75 miles south of Berlin. According to Schiffer, as early as the twelfth century the castle boasted a "mirror chamber, full of mirrors, formed in the most diverse manners, such that one could see above on the ceiling and on the walls around the table, in the room or in the bed, in the chamber, everything ... in the courtyard, in the street, in the country, and on the Elbe river."[125]

We do not know whether these were metal or glass mirrors, or a mixture of the two. But in the first decade of the seventeenth century, architect Robert de Cotte introduced large glass mirrors facing each other. The secretary to the Swedish ambassador to France wrote back to Stockholm on March 22, 1697, "by reflection, a room with two to four candles is brighter and gayer than another with twelve."[126] It is reported that "by 1667 Sir Samuel Morland ... had installed in his house at Vauxhall a room of which the walls were covered entirely in looking-glass."[127] Two of Charles II's rival mistresses, Nell Gwynne and Louise de Keroualle (later made Duchess of Portsmouth), had mirror-walled rooms. No doubt once one had it, the other found it a necessity. The Duchess's hall was visited by the Moorish ambassador, who commented that "he much wondered at the room of glass where he saw himself in a hundred places."[128]

The most famous Hall of Mirrors is the "Galerie des Glaces" at the Palace of Versailles. It was constructed during the period 1678–1684, at the urging of Louis XIV, the Sun King of France. The Sun King used it as a council room, but it was also a site for grand public ceremonies. These included "extraordinary audiences to receive the Doge of Genoa (1685), the Envoys of the King of Siam (1686), the Ambassador of the Shah of Persia (1715), the Ambassador of the Sultan (1742), among others. Balls for the princes' weddings were also held there, such as those of the Duc de Bourgogne (1697) and of the Dauphin, Louis XV's son (1745)."[129] There, in 1871, William I was crowned emperor of Germany, and there, in 1919, the First World War officially ended (with the signing of the Treaty of Versailles).[130] Additionally, the Hall of Mirrors was "the direct way for the sovereigns to reach the Chapel each morning. Crowds of people gathered there to admire the procession going by."

The Hall of Mirrors, originally known as the Grand Gallery, was located on the west side of the palace, so it received afternoon sunlight, and, on occasion, the light of firework displays in the gardens. The official website offers the following description:

> The hall measures 73 meters long, 10.5 meters wide, and 12.3 meters high. At one end is the Salon of War, at the other is the Salon of Peace. Seventeen windows overlooking the garden are matched by seventeen arcaded mirrors along the wall. These exceptionally large mirrors were made in a Paris workshop founded by Colbert to compete with Venice's glass factories. The arches are set on marble pilasters whose gilded bronze capitals are decorated with the symbols of France—the fleur-de-lys and the Gallic cockerel—according to the new "French order" of architecture invented by Le Brun.[131]

Each of the seventeen mirrors was actually made of twenty-one mirror panes, fitted together as seamlessly as possible. Fifteen of the panes were of identical size, "twenty-six by thirty-four inches." These were arranged in five rows, each with three panes abreast. Each mirror also had three additional rectangular panes, forming a narrow row between the two top rows of large panes. (These aligned with trim above alcoves containing statuary.) And above these eighteen panes, forming a seventh and topmost row, there were three more panes, which were cut to match the arch of the windows on the opposite wall. Thus, the hall had a total of 357 panes.[132]

From the date of construction, it is obvious that they were made "by a glass-blowing process and not by casting on an iron table." Based on the glass composition, it is thought that they were made at a plate glass factory in Tourville, Normandy, which had the right to produce small mirrors like the Versailles panes.[133] Nonetheless, in later years, the Hall of Mirrors may have served as advertisement for the wares of the royal manufactory in Saint-Gobain.

The architect, Jules Hardouin-Mansart, is credited with the idea of creating a mirrored wall; there is speculation that he did it to minimize the space which his rival, Le Brun, would have for painting.[134] However, Barter insists that both the windows and the mirrors were intended to maximize the illumination of the ceiling paintings.[135] In any event, Le Brun found plenty of scope for his talents on the ceiling of the great hall.

Mad Ludwig II of Bavaria, a great admirer of the Sun King, just had to have his own Hall of Mirrors. It is in Herrenchiemsee Palace and is almost 100 meters long. Like the Hall at Versailles, it has seventeen windows.[136] Another noteworthy mirrored room is in the Amalienburg Hunting Lodge of Nymphenburg Castle in Munich. "The Amalienburg was begun in 1734 by architect François Cuvilliés, an architect from the Spanish Netherlands, for Prince Elector Carl Albrecht. It is a delightful little building with silver, blue and yellow Rococo decor and a central circular Hall of Mirrors (Spiegelsaal)."[137] Another famous Hall of Mirrors is the one at Schönbrunn Palace, summer palace of the Habsburgs, where a six-year-old Mozart played for the Empress Maria Theresa in 1762.[138]

In London, at the Colosseum in Regents Park, there was a Hall of Mirrors, opened to the public in 1835. Not only were there floor-to-cornice mirrors at the ends of the hall, but the windows shutters were mirrored, so at night the lit chandeliers were repeatedly reflected. An 1839 account stands witness to the "fairy enchantment" evoked by the brilliantly lit hall:

> The endless reduplications of reflection from the mirrors give it an appearance of interminable extent in every direction; and the various coloured dresses of the company, which assumes the appearance of a countless multitude in constant motion, produce an impression of grandeur, magnificence and beauty.... The whole scene is one effulgent blaze of splendour, perpetually changing as the spectator varies his position, and presenting new combination of elegance and beauty in endless succession.[139]

There are prominent halls of mirrors outside Europe. One is at the Aina Mahal in India, created by Ram Singh Malam, under the auspices of Maharao Shri Lakhpatji, around the middle of the eighteenth century. The walls are white marble, covered with mirrors.[140]

A different approach is taken in the Shish Mahal, or Palace of Mirrors, of Shah Jehan's Lahore Fort in Pakistan. "It consists of a row of high domed rooms, the roofs of which are decked out with hundreds of thousands of tiny mirrors in the fashion of the traditional Punjabi craft of 'Shishgari' (designs made from mirror fragments). A fire-brand lit inside any part of the Palace of Mirrors throw back a million reflections that dizzy the eye and seem like a galaxy of far-off stars turning in an ink-blue firmament."[141]

Photophones

"I have heard articulate speech produced by sunlight! I have heard a ray of sun laugh and cough and sing.... I have been able to hear a shadow, and I have even perceived by ear the passage of a cloud across the sun's disk."[142] With those words, Alexander Graham Bell triumphantly announced his June 3, 1880, creation of a photophone, a device that transmitted sound by light. I describe here what Bell considered to be the "best and simplest" of "about fifty forms of apparatus for varying light" in order to carry a signal; the description is based on Bell's lecture "Selenium and the Photophone," delivered at the Boston meeting of the American Association for the Advancement of Science (AAAS) in 1880.[143]

His transmitter, which converted sound into light, took advantage of the flexibility of a mirror made from a mica sheet. The mirror vibrated in response to sound waves. These vibrations altered the path of light reflected from the mirror. While Bell noted that any powerful light source could be used, he experimented mainly with sunlight. This was concentrated by means of a lens onto the diaphragm mirror. Another mirror and lens were used to collect the reflection off the diaphragm mirror and send it in the direction of the receiver.

The heart of his receiver was a selenium photocell; light increases the electrical conductivity of selenium. The selenium cell was at the focal point of a parabolic reflector, which captured the light signal and directed it onto the selenium.

The selenium cell, in turn, was connected in a local circuit to a telephone-type receiver and a battery.

Bell described one trial to AAAS as follows: "Mr. Tainter operated the transmitting instrument, which was placed on the top of the Franklin schoolhouse in Washington, and the sensitive receiver was arranged in the windows of my laboratory, 1325 L Street, at a distance of 213 metres. Upon placing the telephone to my ear I heard distinctly from the illuminated receiver the words, 'Mr. Bell, if you hear what I say come to the window and wave your hat.'"[144]

Bell characterized the photophone as "the greatest invention I have ever made; greater than the telephone."[145] Four of his thirty patents were for this invention.[146] In his AAAS lecture, Bell commented, "there seems no reason to doubt the results will be obtained at whatever distance the beam of light can be flashed from one observatory to another." Of course, in this confident remark was concealed an inherent limitation of the invention; it was strictly and literally a "line of

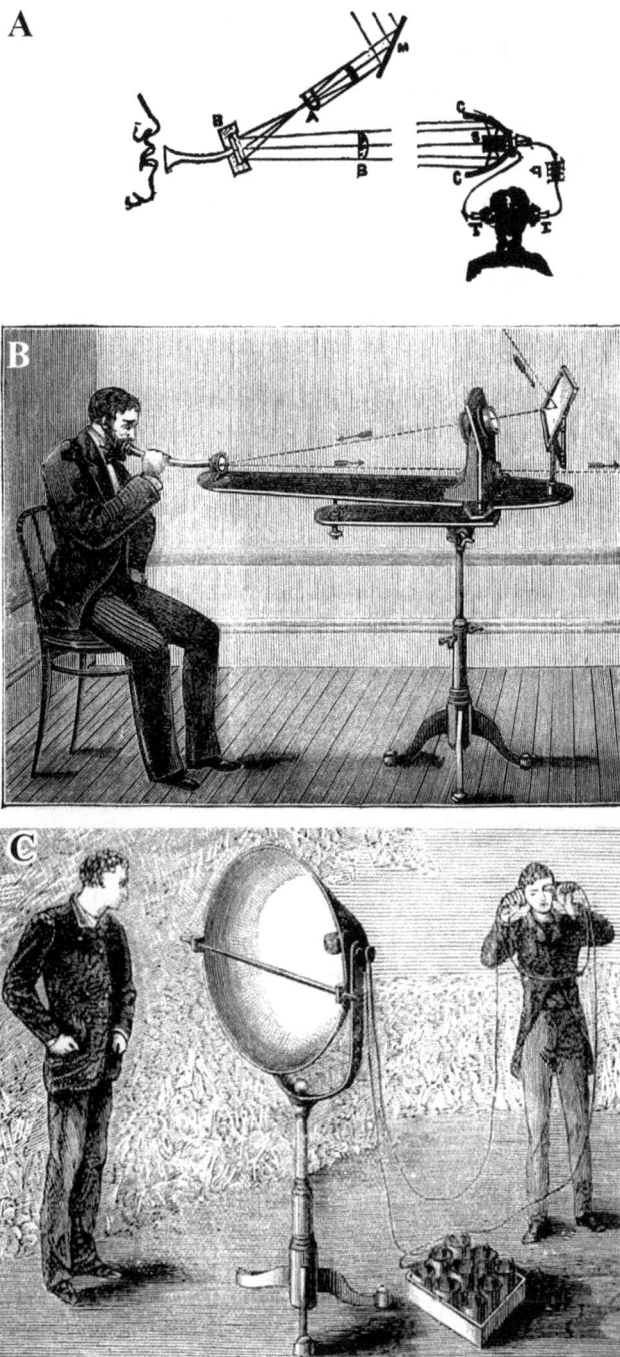

A. Theoretical diagram of the articulating photophone. B. The transmitter. C. The receiver. (All from Munro, J., "The Photophone," *Quarterly J. Science*, 18: 13–14 [1881].) Munro's diaphragm was of "thin flexible glass, silvered on the outside," rather than mica.

sight" communications device. Of course, in theory this limitation could have been overcome by use of relay stations; as had been done for heliograph communications.

Not everyone was enamored of the photophone concept: "'Does Prof. Bell intend to connect Boston and Cambridge with a line of sunbeams hung on telegraph posts, and, if so, what diameter are the sunbeams to be?' sniffed an editorial in The New York Times. 'Until (the public) sees a man going through the streets with a coil of No.12 sunbeams on his shoulder, and suspending them from pole to pole, there will be a general feeling that there is something about Prof. Bell's photophone which places a tremendous strain on human credulity.'"[147]

While the photophone did not revolutionize the communications industry, it was nonetheless prescient in one respect: Light was to become the basis of long-distance transmission of information. However, it was necessary to guide and protect that light so communications were not line-of-sight limited, and not subject to interference by atmospheric conditions. The solution came in the form of fiber optics.

Lissajous Figures

Jules Antoine Lissajous (1822–1880) was a French physicist who developed several optical methods of studying vibrations. He obtained what are now called "Lissajous figures" by "successively reflecting light from mirrors on two tuning forks vibrating at right angles. The curves are only seen because of persistence of vision in the human eye."[148]

Lissajous figures are produced by two sine waves acting at right angles to each other. For example, equal amplitude, equal frequency sine waves which are 90 degrees out of phase produce a circle. If the waves are in phase, but with a 2:1 frequency ratio, a figure-eight is traced.

Suppose you had a bag of sand with a tiny hole at the bottom. If this bag were suspended by a string, it could swing to and fro, letting out a string of sand as it did so. The deposited sand would trace out a Lissajous figure.

The pattern shown on *The Outer Limits* when you are warned "Do not attempt to adjust your picture—we are controlling the transmission" is a Lissajous figure. A simpler one is the ABC logo, with three loops.

Reflective Eyeglasses

Like one-way mirrors, mirrored sunglasses have a thin metallic coating to reflect some of the incoming light.

"Thin metallic films, whose effect is principally reflection rather than absorption, have been used for the protection of the eye since 1859." In welder's glasses, gold films were used, since they could reflect 98 percent of infrared and transmit up to 75 percent of visible light.[149]

Popular Science (1937) reported that "lenses in novel sunglasses recently introduced act as mirrors as well as eye protectors. The glasses are treated with a very thin coating of silver which reflects much of the brilliant sunlight, so that the light transmitted through them has a glareless, bluish hue." It added that "women have found the unusual glasses useful as make-up mirrors."[150]

Another early type, "Kool Krome," was designed by optician Monroe Levoy. It had a chromium coating that reflected 30 percent of solar infrared. A 1948 article said that they provided "complete ocular privacy" (an onlooker can't see the wearer's eyes).[151]

I have wondered whether it had any connection to wartime research. The USAF School of Aviation Medicine reported, "a combination reflective-absorptive filter was produced during the war. It consisted of a sage-green lens, on the upper surface of which a thin graduated film of chrome-nickel alloy was deposited."[152]

CHAPTER 6

Mirrors at Play

Mirrors are not just tools; they are also toys. Their use to entertain and mystify goes back to antiquity. In the first century, Hero of Alexandria noted that they were useful in "affording diverting spectacles." More particularly, he said that "with the aid of mirrors" we may "see ourselves inverted, standing on our heads, with three eyes, and two noses, and features distorted, as if in intense grief."[1] As mirror making technology improved the quality of the image, the illusions achievable with the aid of mirrors became more convincing and elaborate.

Magic of the Orient: "Light-Penetration" Mirrors

According to Shen Gua [Kuo], an eleventh-century scholar, "when the sun shines on a light-penetrating mirror, the characters on the back appear in the image reflected on the wall." He noted that his "family has three mirrors like this and I have seen others treasured in my friends' homes. They are antique." The first "magic" or "light-penetrating" mirrors were made in China during the Western Han period (206 BCE–24 CE).[2]

Bright areas in the projected image match up with raised areas on the back. The projected image is a consequence of tiny variations in relief on the mirror side. If light strikes a flat area, it is reflected normally, and a light area appears on the wall. If it strikes a convex area, the reflected rays diverge, and the corresponding section of the wall is darker. And if it strikes a concave area, and the mirror is just the right distance from the wall, then you get a very bright feature flanked by darkness.

What is less certain is how these variations in relief were achieved so as to match up with the design on back. Replicas have been made by various methods and compared with the surviving originals. Researchers concluded that in the historical Chinese magic mirrors, the variation in relief was primarily attributable to differences in cooling speed, and therefore in casting stress, between thin areas and thick ones. The thick ones correspond to the raised designs. The thick rim of the mirror compressed the disk, "and thin areas respond by becoming more convex, thick areas are less flexible and change less."[3] The thin areas were just 0.5 millimeters thick.[4]

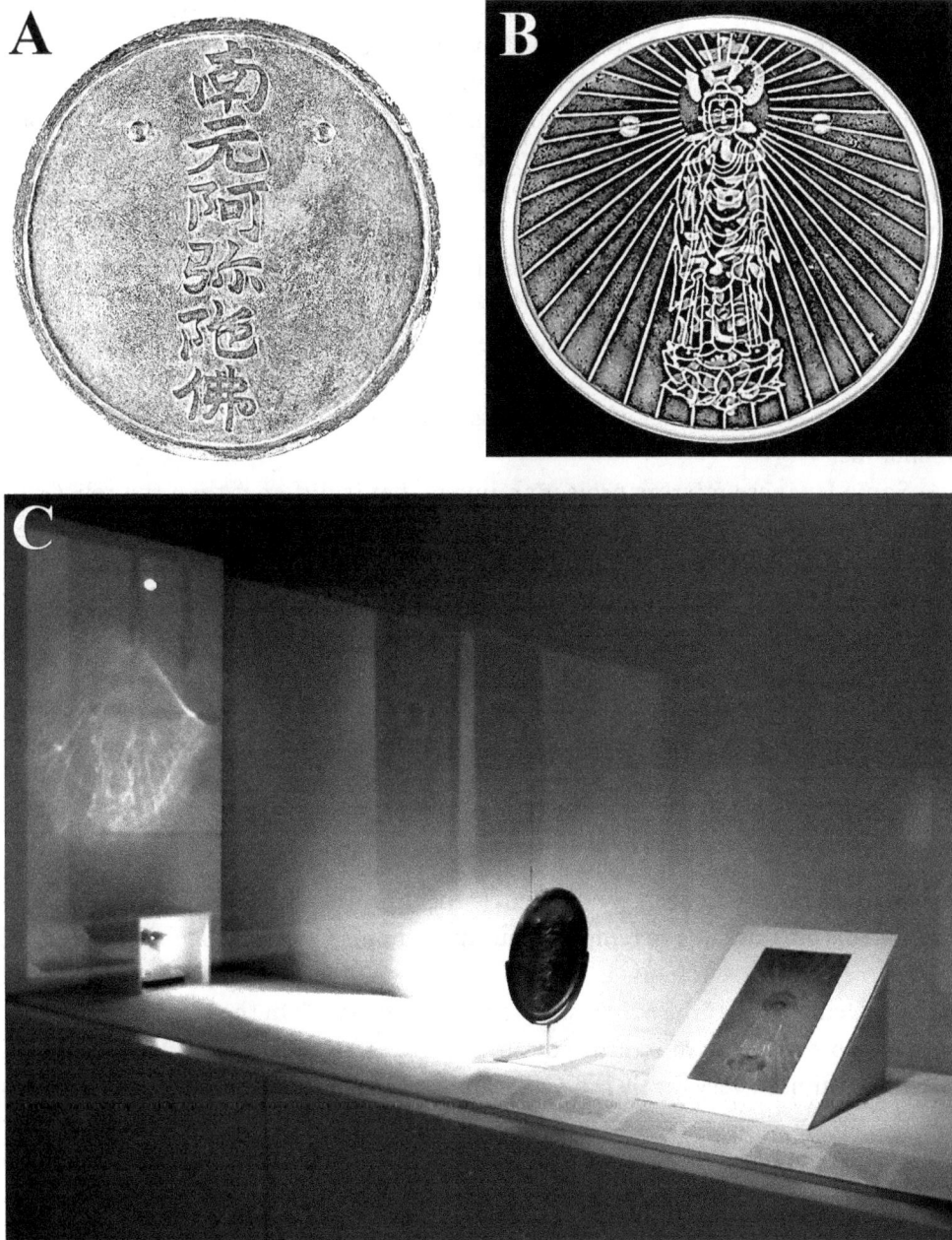

A. Nineteenth-century bronze "magic" mirror back, with six-character inscription *"Namu Amida Butsu"* ("Hail the name of Amida Buddha"). B. The projected image (Amida Buddha "emanating forty-eight rays of light") from the mirror. C. Mirror illumination setup. (The Rogers Fund, 1909, accession 09.62, The Met.)

Modern ("commercially produced") Chinese magic mirrors were found to have "a top reflective transparent layer" as well as a subsurface layer exhibiting "ups" and "downs" ("+700nm to -500nm"), detectable by "phase measuring reflectometry."[5] Mak suggests that the subsurface layer was formed by electroplating the bronze with an acid-labile alloy and then acid etching the design.[6] That would allow the reflected image to be one without any correspondence to the back relief.

Japanese magic mirrors first came to the attention of Western scholars in 1832, and it is not clear when they were first made. In contrast to the Chinese mirrors, which offer flat and convex relief, the Japanese ones exhibit flat and concave relief. This "front relief pattern is formed during fabrication by elastic and plastic mechanical transfer of the back pattern. The making of these mirrors involves first casting the mirror with the back relief, thinning by filing, then polishing the front surface while exerting a strong pressure on the mirror plate using a tool called 'distorting rod,' going repeatedly over the whole surface…. [A] back-face protrusion will induce a corresponding front depression focusing the reflected beam…." The mirror thickness is on the order of millimeters while the "undulations" are on the order of microns.[7]

Some Japanese mirrors project an image that is different than the relief pattern on their back. The explanation is that this is a "false" back; there is a hidden back that created the front relief. Four such "two plate" mirrors, dating to the Edo period (1603–1867), have a back design showing a "turtle, crane, pine and bamboo," but project an image of "Jesus Christ on Cross." Another has a back relief showing Mount Fuji but projects an image of Amida Buddha.[8]

The last remaining magic mirror (*makkyo*) maker in Japan is Yamamoto Akihisa in Kyoto, who characterized himself as a fifth-generation *makkyo* craftsman.[9]

Opaque Mirror-Based Illusions

Mirrors have long been used to baffle the uninitiated. The earliest use of mirrors for this purpose was in temples. Pausanias, in his *Description of Greece*, tells of a visit to a temple of Artemis in Arcadia. He saw a gigantic block of stone, carved to show Demeter and "the Mistress" sitting on a throne. "On the right as you go out of the temple there is a mirror fitted into the wall. If anyone looks into this mirror, he will see himself very dimly indeed or not at all, but the actual images of the gods and the throne can be seen quite clearly."[10] Since Pausanias's description is rather vague as to the layout of the room, it is difficult to say exactly how this effect was achieved.

One possibility, suggested by Bur,[11] is that it used the two-mirror arrangement described by Hero's *Catoptrica*. He introduces it as follows: "To place a mirror so that one approaching it sees neither his own image nor that of another but only the image which we select." The diagram shows a person who sees, in an inclined mirror above his head level, the doubly reflected image of a statue behind a wall below that mirror. There is a second inclined mirror, parallel to the first, hidden by the wall. Hero specifically suggests "the placing of the mirrors in a temple."[12]

Another possibility is that it was a "one-way mirror" (i.e., the mirror is "unsilvered" glass or mica) and the statue was strongly lit and the visitor in darkness.

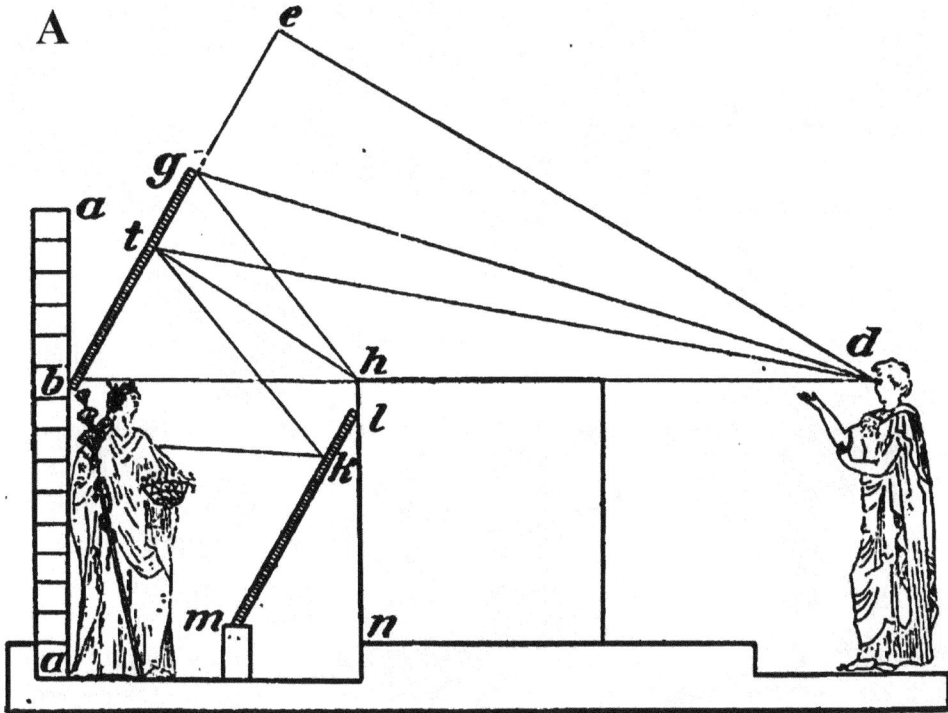

A. Hero's two-mirror apparition. The statue is reflected in mirrors lm and bg, so her shoulder appears to be along the line dt. Fig. 91a (358), in *Catoptrica* XVIII. (From Nix, L., and W. Schmidt, *Heronis Alexandrini Opera Quae Supersunt Omnia, Vol. II, Fasc. I, Mechanic et Catoptrica* [G. Teubneri, Leipzig, 1900].) B. Marion proposed that an arrangement similar to Hero's was used by Nostradamus to trick Marie de' Medici into believing that his "divining mirror" showed that she would share the throne of France with Henry of Navarre. (Fig. 56 in Marion, Fulgence, *The Wonders of Optics*.)

The same possibilities exist with respect to della Porta's cryptic instructions for "How to make a Glass that shall show nothing but what you will,"[13] although, like Hero, he says nothing about lighting.

The life story of Andrew Oehler (1781–?) offers an object lesson in the dangers of demonstrating conjuring to the superstitious. He gave a séance for the governor of Mexico and other high-ranking Mexican officials. He led them into a dark chamber where they were treated to displays of lightning and thunder, candles that went out and relit themselves, and a phantom that appeared in the smoke from a pan of hot coals. His distinguished guests left without a word, and the next day he was arrested as a sorcerer and incarcerated. He was freed months later as a result of the intervention of a more worldly Spanish marquis. After other mishaps, he abandoned his unexpectedly perilous profession in 1809. In the supplement to his 1811 autobiography, *The Life, Adventures and Unparalleled Sufferings of Andrew Oehler*, he explained that "the ghost was projected by a hidden magic lantern, then reflected by a slanting mirror on the rising smoke from the pan of coals."[14]

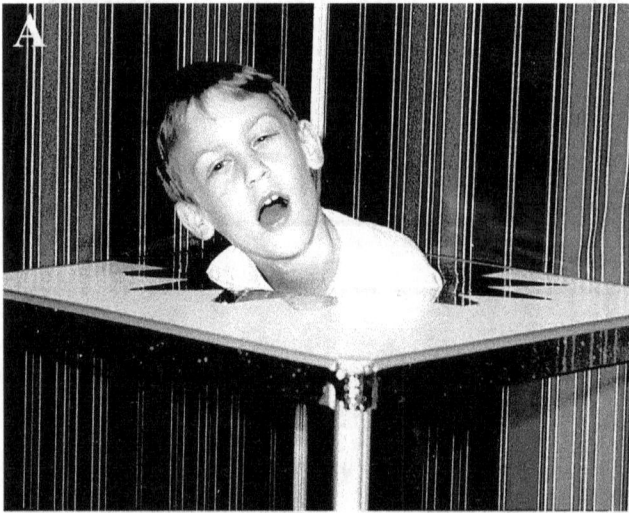

A. Author's photograph of his son (at elementary school age) in "talking head" mode. B. The divided telescope illusion. (Fig. 141 from Tissandier [1882].)

Stage conjurors find mirrors useful to misdirect the audience. In the Colonel Stodare (Alfred Inglis) "Sphinx" illusion, as presented in London on October 16, 1865, a disembodied head, resting in a box on a table, opened its eyes, smiled, and recited poetry. The audience was sure that there was no one under the table, as they could "see" beneath it to the rear curtain of the stage. The underlying trickery was created by Thomas W. Tobin; two mirrors met at a 45-degree angle under the three-legged table, reflecting the side curtains in such a way that they mimicked the rear curtain. The body of the "Sphinx" was concealed under the table, behind the two mirrors.[15]

A favorite of "Haunted House" displays during Halloween is the "Talking Head." It appears that a disembodied head is lying on a table, talking to you (at my son's school, it was on a dinner plate). The principle is the same as for the "Sphinx" illusion. One source provides a few tips for success:

> The illusion is most effective if the mirror is very clean and its edges are concealed. The table should be placed well away from other obstructions that would be eclipsed by the mirror. If a carpet is used, it must be aligned carefully so that there is no discontinuity at its edge where it goes behind the mirror. One should be careful not to stand directly in front of the table lest the reflection of one's legs be seen in the mirror. The table should not be placed too close to the audience to avoid reflections of the audience. The illusion works best if the audience is seated slightly above the level of the top of the table. The proximity of the audience to the head would seem to favor discovery of the trick, but, on the contrary, it is indispensable to its success.[16]

An extremely elaborate form of this illusion was inspired by the H. Rider Haggard novel *She*. A lady stood on a table with candles beneath it. A cylindrical screen was lowered over her. On a signal, the lady was set on fire, as shown by smoke billowing out from over and under the screen. The screen was raised, and on the table the audience saw "a few smouldering embers and a pile of bones surmounted by a skull."

As in the "talking head" illusion, there were mirrors at a 90-degree angle to each other beneath the table, with the "V" pointing at the audience. These at first reflected the sides of a folding screen surrounding the table, and then the cylindrical screen. There was a trap door on the top of the table into which the lady descended, and under the table were the bones and fireworks, which she placed on the table and set off.[17]

John Nevil Maskelyne (1839–1917) used a concealing mirror in a levitation; a woman was "raised on a pedestal hidden behind a looking glass."[18] (I presume that he actually used two mirrors, in the "V" arrangement, about the pedestal.)

Jean-Eugène Robert-Houdin's magic cabinet used swinging mirrors. The assistant entered the cabinet and, with the door closed, pulled toward herself two mirrors that had been flush against the sides of the cabinet. She was thus enclosed in the "V" formed by the mirrors, and when the door to the cabinet was opened, the audience saw that she had vanished. This is, of course, merely a dynamic version of the "Talking Head" illusion, with the "V" being formed during the trick.[19]

This was reversed by Charles Morritt, who seemingly materialized women, inside the inner of two cages, out of thin air. "Two mirrors met behind the bar at the far right corner of the inner cage. One ran back to a bar of the outer cage; the other extended diagonally to the far right corner of the larger cage. From the front the cages seemed to be empty. Each time Morritt lowered a canopy around the cage and raised it, a girl appeared. The canopy only covered the front and two sides of the

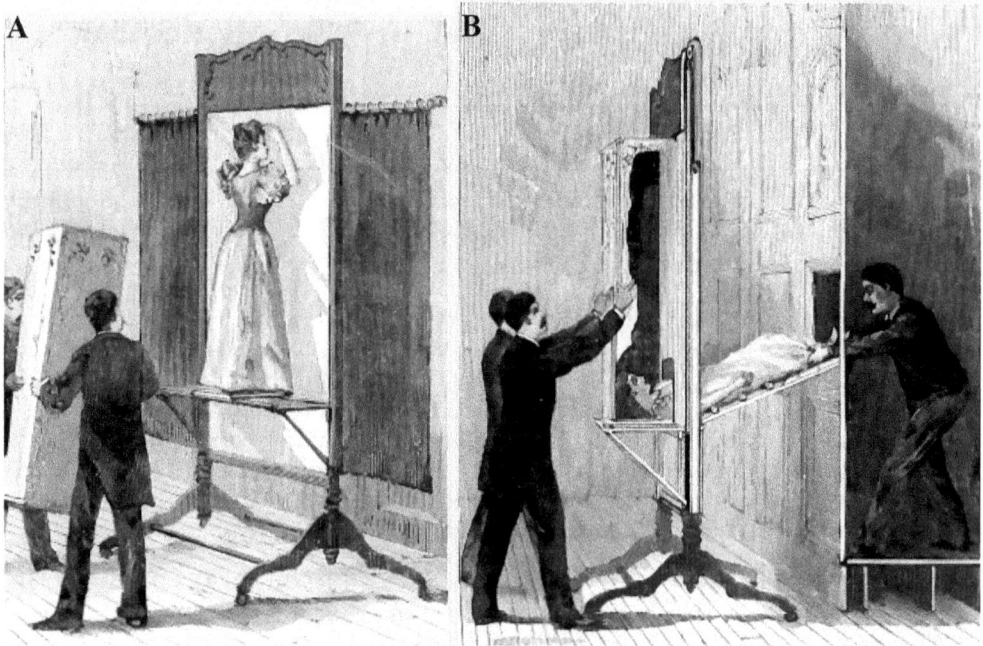

A. The "Vanity Fair" illusion, performed by Alexander Herrmann and his wife. The lady stands in front of a large pier glass mounted on a portable frame. After a bit of patter between her and the magician, a screen (the box held by the two assistants) is placed over her. The screen is removed; she has vanished! The glass is wheeled away. B. The secret. The mirror is raised; the upper part is concealed by the upper part of the frame. This exposes an opening in a second, stationary mirror behind the movable one. A stagehand opens a secret door in the back of the stage and extends a plank. She is pulled out by her feet; the plank is withdrawn; and the movable mirror drops back down. (Hopkins, Albert A., *Magic: Stage Illusions and Scientific Diversions...* [Sampson Low, 1897], 27–30. https://www.gutenberg.org/files/45235/45235-h/45235-h.htm.)

inner cage. The back of the cage and the first mirror were spring-hinged. The assistants came one by one from their V-shaped hiding place between the mirrors."[20]

While in the previous illusions the mirrors reflected the sides, in a cabinet illusion performed by Maskelyne, a man disappeared inside by the simple expedient of climbing up on a shelf and pulling down "a hinged mirror that cover his hiding place and reflected the ceiling."[21]

Double-sided mirrors were used by Kellar in his "Queen of Flowers" illusion. This had, as a background, a helter-skelter arrangement of flowers and bushes. In front was a four-poled stand, which he closed off with a curtain. Spectators on the extremities could still (they thought) see behind the frame, but in fact they were seeing the reflections of the background on the mirrors, which ran from the back of the stand to the background. When Kellar pulled back the curtain, the "Queen" (who had walked from behind the stage, between the double-sided mirrors, and onto the stand) was revealed.[22]

In the "divided telescope" illusion, the spectator can see through the instrument even when an opaque object, such as a brick, is placed between the two halves of the telescope. The trick is that the scene is reflected in four diagonal mirrors, one in each half of the telescope and two in the base.[23] Marion says that this trick could "often be seen in the streets of London,"[24] with the device being highly portable.

At a late-nineteenth-century exhibition in Paris, a statue showed "Goethe's Marguerite standing" with her back to "a mirror," but the reflection in the mirror was of Faust. The trick was that the back of the statue was carved with Faust's likeness.[25]

There were also illusions which used a single mirror, inclined forward 45 degrees. In one, a seated assistant put his or her head in a hole in the mirror; the mirror reflected the ceiling, making it appear to be the back of the stage, and the assistant's head to be disembodied.[26] My favorite is "the Spider and the Fly." A person descending a staircase finds it blocked by a giant web, with a chimeric monster (woman's head, spider's body) at the center, and hastily retreats. The top edge of the mirror has a notch in which the female assistant has positioned her head, her body lying on an inclined support. The spider's body is held by the web (ropes). Her body is hidden by the mirror, but "the mirror reflects the lower steps, so that this reflection really appears to be a continuation of the steps, and the entire flight seems unbroken."[27]

"Transparent" Mirror Illusions

Pepper's Ghost

Imagine attending a performance of Dickens's *A Christmas Carol* or of Shakespeare's *Macbeth* in which the ghosts aren't men in white sheets, but "real"; you can see them but still look through them. This marvel was made possible by the ingenuity of John Henry Pepper (1821–1900). He was a professor of chemistry at the London Royal Polytechnic Institute. This institution was the nineteenth-century equivalent of the San Francisco Exploratorium. The institute, for which the goal was to entertain using science, opened its doors in 1838. A giant induction coil was used to generate thirty-foot sparks. Organisms in a glass of drinking water were magnified and projected onto a screen. There was no "IMAX" back then, but the institute had its own "magic lantern" theater. Images could be projected on the back or front of a screen; with multiple projectors, one image could be dissolved into another.[28]

In 1862, Pepper published "Wonders of Optical Science," which included what has come to be known as the "Pepper's Ghost" illusion. In essence, this is a one-way mirror effect, made possible by the development of inexpensive but large panes of unsilvered plate glass. The glass was placed at a 45-degree angle to the audience. One scene would be in the direct line of sight from the audience to the glass, and view by transmitted light. The other would be at right angles and viewed by reflected light. The glass could be stood vertically, so it reflected a scene on the left or right, or at a 45-degree inclination, so it reflected one above or below stage level. Since glass is both reflective and transparent, by adjusting the relative lighting of the two scenes, one could make one scene or the other visible. An abrupt change in lighting would result in a sudden transformation, while a gradual one would result in a slower metamorphosis.

The effects that could be achieved with this simple arrangement were limited only by the imagination (and the availability of a large enough piece of flat, clear glass): Appearances, disappearances, and transformations were the stock in trade. A figure could appear to glide around, fly, or swim. A candle could appear to be underwater (candle on one side of glass, beaker of water on the other).[29]

There were a number of technical limitations. The audience side of the theater had to be in darkness, which was not then customary. The "average width of a provincial Victorian theatre stage was about thirty feet," and the largest pane of glass produced at the time was about nineteen by eleven feet. It is likely that the glasses actually used only covered "around a third to a half of the stage," and thus the location of the apparition would have been limited. The performers could not see it themselves, so the stage had to be marked to show that location. Their actions had to be timed according to musical cues.[30]

"Due to the thick glass separating the performers from the audience, all of the theatrical skits in which the Ghost appeared had to be performed in pantomime, since the actors couldn't be heard!"[31] However, narrators could be employed.[32]

In the original form—with the glass tipped forward toward the audience—the concealed actor had to also "be tipped ... so that his image would appear to be standing upright on the stage." He was given a support. He could not walk, but he could be moved by means of a "trolley on a track." And his alcove was called the "oven," because of the heat generated by the light.[33]

Pepper did not actually invent the illusion. "In 1862, a Liverpool civil engineer named Henry Dircks [1806–1873] constructed a miniature working model of the effect.... But it was John Henry Pepper ... who, having seen Dircks' model, built the first practical full-size version and exhibited it on stage."[34]

Dircks had been unable to interest any theaters in scaling up the illusion. The problem was that since his version used a "vertical sheet of glass," a "two-level theater" would have had to be custom built to exploit it. Pepper proposed the "angled sheet," which made it possible for existing theaters to adapt themselves to the illusion.[35] In 1863, Dircks and Pepper received a patent for "projecting images of living persons in the air."[36]

A phantasmagorical version of *A Christmas Carol* was staged on Christmas Eve, 1862, at London's Royal Polytechnic Institute. "'The apparitions," wrote Thomas Frost, "not only moved about the stage, looking as tangible as the actors who passed through them, and from whose proffered embrace or threatened attack they vanished in an instant, but spoke or sang with voices of unmistakable reality."[37]

The ghost illusion was an immediate sensation. At the Royal Polytechnic, it "drew crowds for a full fifteen months," and "earned 12,000 pounds"—implying

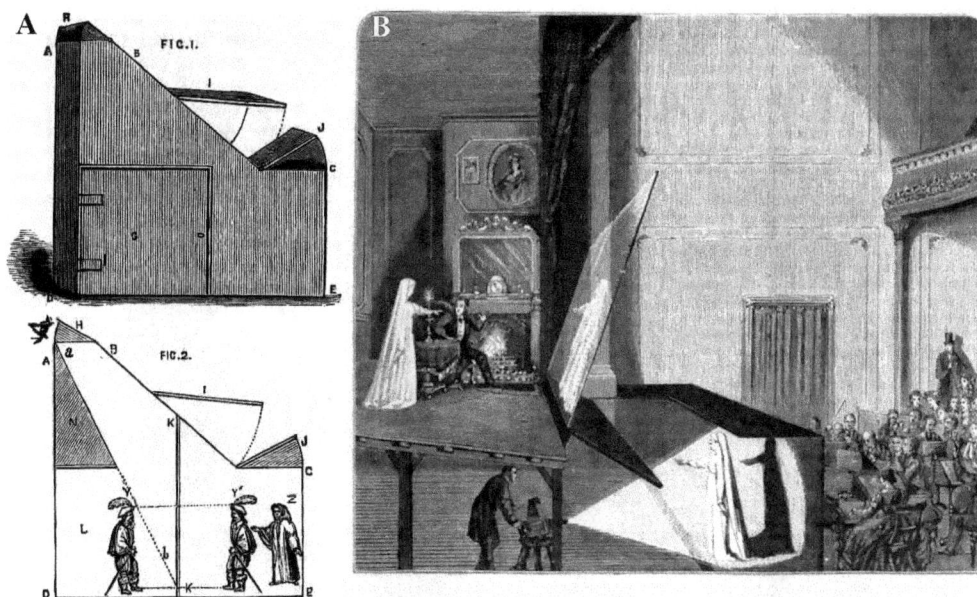

A. Dircks's phantasmagoria. Top: external perspective view; bottom, vertical section. Note that the intervening glass K is vertical, not diagonal as in Pepper's stage version. (From Dircks, 46.) B. Pepper's Ghost on stage. The actor portraying the ghost is under the stage, illuminated by a strong oxy-hydrogen light. A glass sheet, near the front of the stage but behind an opening in the floor, is inclined so the "ghost" is reflected in it, and the reflection is visible to the audience. The ghost appears to be on the stage behind the glass. (From "The Wonders of Light," *The World of Wonders*, 21: 201 [1868].)

the attendance of "nearly a quarter of a million visitors.... Pepper also licensed the effect to various theatres and music halls."[38] Dircks reports that this caused a shortage of large plate glass.[39] However, the illusion primarily appeared at "fairground ghost shows" and in performances by "spectral opera companies" that rented their venues.[40]

The Cabaret du Néant was a particularly impressive Parisian form of the illusion, as it was able to seamlessly incorporate audience members into the three-act performance. In the second act, a volunteer was asked to enter an upright coffin and stand on blocks of wood so his head touched the top of the coffin. He was then wrapped in a white sheet. "Now, as the spectators watch him, he gradually dissolves or fades away and in his place appears a skeleton in the coffin. Again, at the word of command the skeleton in its turn slowly disappears, and the draped figure of the spectator appears again. The illusion is perfect to the outer audience; the one in the coffin sees absolutely nothing out of the common."

The key to the illusion was that standing on the blocks brought the volunteer into perfect registration with the skeleton in the hidden coffin, with the wrappings hiding any minor discrepancies. The operator slowly dimmed the Argand burners illuminating the volunteer, while intensifying those directed at the skeleton, thus effecting the first transformation, and then reversed the process.[41]

A

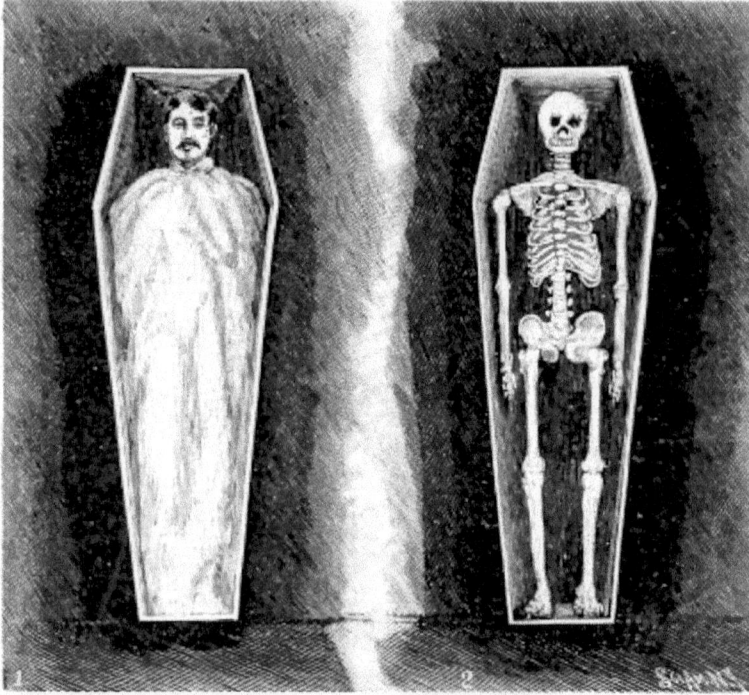

B

Cabaret du Néant, Act 2. A. The spectator and the skeleton in their respective coffins; note the positions of their heads. B. A perspective view of the stage, showing the positions of the coffins, lights, and glasses. At the moment shown, we are in mid-transformation; the illuminated skeleton's reflection is visible on the diagonal glass, but the spectator is still illuminated, too. (Hopkins, 56–59.)

The Pepper–Dircks collaboration went sour. As so often happens, the estranged pair went public with their complaints about each other.[42] Just to complicate the picture further, several other individuals have been touted as creators of this illusion. The earliest of these was della Porta, who somewhat cryptically described "how we may see in a Chamber things that are not." The chamber in question was to be fronted by a large, polished glass window. As a result, "what is without will seem to be within," thanks to the reflection in the glass. However, there was no teaching that the glass be placed diagonally, to reflect what was in a room hidden from the spectator's direct view. Nor was there any teaching of control of the lighting.

Next was the stage magician "Henri Robin," who claimed (in an 1863 book) to have conceived the illusion in 1845 and employed it in a European tour in 1847. By his own admission, in the form he devised, "it produced little effect" and was also "very costly."[43] His only proof in 1863 was a playbill referring to the title of the illusion, "Living Phantasmagoria," but that was a "generic term for a performance of projected lantern images on smoke or gauze curtains." The illusion builder and historian Jim Steinmeyer has expressed doubt that "Robin would have toured with such large pieces of glass."[44]

In 1852, Pierre Séguin patented a toy, the "polyoscope," in France. (Robin accused him of plagiarism.) According to Pepper, "it consisted of a box with a small sheet of glass placed at an angle of forty-five degrees, and it reflected a concealed table, with plastic figures, the spectres of which appeared behind the glass, and which young people who possessed the toy invited their companions to take out of the box, when they melted away, as it were, in their hands and disappeared." This was, Pepper conceded, "substantially the ghost apparatus and produced that illusion."[45] But of course it lacked the lighting control of Pepper's stage version.

The Pepper's Ghost illusion has been used in film, such as the nickelodeon scene in Coppola's *Bram Stoker's Dracula* (1992), in which a woman in Victorian dress turns into a skeleton.[46]

A number of theme parks use Pepper's Ghost effects. If the Pepper's Ghost effect is combined with motion detection, "as is done in one popular theme park, you have a ghost that can travel alongside the visitor."[47]

The largest Pepper's Ghost illusion in the world is probably the Grand Ballroom at Disney's Haunted Mansion. One fan describes the scene: "While ghostly couples waltz as they disappear and reappear, a mysterious birthday party is taking place with transparent guests. Ghosts flutter in and out as the reverie continues with an evil organist playing a haunting refrain on a massive pipe organ. With each note, wraiths fly out of the pipes and vanish like wisps of smoke."[48] There are actually two ballroom sets, one with the animatronics, and the other with the furniture. It is claimed that "The Imagineers forgot to take the 'mirror image' aspect of the Ballroom effect into account when designing the animatronics. As a result, the ladies lead the men in the dancing!" (Actually, while waltzing couples normally turn clockwise, a counterclockwise turn is a standard variation, readily leadable by the fellows. I speak from experience.)

Illusions Using Metallized Glass

While the original Pepper's Ghost illusion used a stationary and unsilvered glass, an 1879 patent awarded to Pepper and James Walker proposed mounting the glass so it can be "rapidly and noiselessly moved diagonally across the chamber," and using a glass with a "graduated" deposit of silver. The latter, as it advanced, would permit a gradual change of appearance without any change in the lighting.[49]

The graduation helped to conceal the edge of the glass as it moved against the background. It was achieved not by controlling the thickness of the silver deposit, but rather by depositing it so the glass had narrow, widely spaced silvering on the leading edge, but progressively becoming wider and more closely spaced as you moved along the glass. In addition, the leading edge of the glass was stepped rather than flat.

In one routine, a servant takes apart and reassembles a suit of armor (#1) on a stand in the back, stage left, of a chamber. Then, undetected by the audience, the mirror advances diagonally, hiding the first one and reflecting a second (#2) in the front, stage right. An actor dressed in a suit of armor (#3) hides suit #1 and takes its place. The mirror is withdrawn, and the servant comes over to dust the suit (which the audience thinks is #1 but is actually #3). The suit animates, chases the servant, beats him up, and returns to its original position. The mirror advances again, and the actor puts suit #1 back into place and leaves the stage. The mirror withdraws again, and the master of the house demonstrates that the suit of armor (#1) is empty.[50]

In the late nineteenth century, semireflective platinized glass (see Chapter 2) was used in an illusion in which a person could see his image transformed to that of a horned devil. The picture of the devil was separated from the glass by a shutter. With the shutter closed, you saw your own reflected image. With it open, the light penetrated and illuminated the devil picture. Electric lighting was used to ensure the clarity of both images.[51]

In David Devant's 1916 routine "The Magic Mirror," it appeared to the audience that a bright red spot appeared in the mirror, grew larger, and revealed itself as Satan. Satan gestured to the audience, then faded away. The mirror was a lightly silvered one; concealed lights built into the rear of the mirror frame made the mirror effectively transparent, revealing an assistant dressed as the devil.[52]

Motion Picture Trickery

The Pepper's Ghost Illusion can fool not only the eye but also the movie camera lens; in the movie industry, the illusion is called the Schüfftan process.

Just as Pepper's Ghost was preceded by magic lantern shows, the Schüfftan process evolved from rear projection methods. This early special effect allowed cinematographers to use a studio to make an indoor set appear to have been filmed outdoors. A previous filmed "background" still was projected onto a translucent screen placed behind the actors, to create the impression that they were filmed "on location." The first use of this technique was in *The Drifter* (1913), by Norman O. Dawn.[53]

Of course, rear projection of a still image did not accomplish much more than what could be done with a painted backdrop. The next improvement was more significant: It was to project a background movie onto the screen, blending it with the live action in the foreground. That required synchronizing the projector with the camera. The dynamic rear projection scheme was devised by Hans Koenekamp, and first used by George Teague, in *Just Imagine* (1930).[54]

One advantage of rear projection was that it avoided the costs of going on location in faraway places. The car-chase scene in *Doctor No*, a low-budget action movie (the entire film had a $900,000 budget), was filmed with back projection.[55] However, the real impetus toward the widespread adoption of this technique in the movies came earlier, as a result of the transition from silent films to "soundies." In the 1930s, sound recording techniques were so primitive that it was impractical to film on location.[56]

Unfortunately, the background image was relatively dim, because the light had to pass through a translucent screen. Farciot Edouart (1897–1980) solved this problem by synchronizing three projectors, one projecting directly onto the screen and the other two being aimed at the screen using mirrors.[57]

Front projection of the image would also have made the background image brighter, but the actors would have cast shadows onto the screen. This problem was solved by another application of the "one-way" mirror. A diagonally oriented mirror was placed between the camera and the actors. Behind the actors was a special screen, made of many tiny glass beads, each of which acted as a retroreflector (that is, each bead reflected light back in the direction from which it came). A projector was placed at 90 degrees to the main light path, so it cast an image onto the "one-way" mirror, which was then reflected onto the screen. While the actors would cast shadows onto this screen, their shadows would be masked out by their own bodies. While variations of the system were developed independently by several inventors in the 1940s and 1950s, its first cinematic use was in Stanley Kubrick's *2001: A Space Odyssey*.[58]

In 1923, Eugen Schüfftan invented a form of back projection which allowed full-size actors to appear in miniature sets. The camera was aimed toward the set (behind which was the rear projection screen), but in the light path there was a diagonally oriented silvered glass mirror with part of the silver scraped away. The image of the actors was reflected to the lens by the silvered portion, but the miniatures could be seen through the exposed glass. (Naturally, the background behind the actors had to be chosen to match the appropriate region of the miniatures and the rear projection screen.) If the reflected light path to the actors was longer than the transmitted light path to the miniatures, the actors would look smaller. (Naturally, the position of the actors and the miniatures could be reversed. Also, in theory, the

Schüfftan miniaturizing effect could be accentuated by use of a curved mirror.)[59] "The mirror is positioned quite close to the camera's short focal (wide-angle) lens, which is focused on infinity. As a result, the reflection of the model is in focus, whereas the mirror's surface itself is blurred."[60]

Further borrowing from Pepper, in Fritz Lang's *Die Nibelungen: Siegfried* (1924), Schüfftan used two mirrors to implement the petrefaction of the dwarfs. The fixed, fully reflective mirror miniaturized the live actors. The second mirror "was partially transparent and could slide up and down," and in the "up" position it reflected their stone replicas. The result was that the dwarfs were petrified "from the bottom up."[61]

That was the first use of the Schüfftan process, but its more famous early use was in Lang's *Metropolis* (1927). "The transition between the Moloch miniature and the full-size stairs … is remarkably difficult to spot. This is a result of Schüfftan's broad jagged zones where the reflective surface gradually peters out."[62]

Fractured Light: Kaleidoscopes and Other Mirror-Based Toys

The Prehistory of the Kaleidoscope

The effect of combining two mirrors at an angle was apparently first studied by Mo Zi (468–376 BCE), who wrote, "(If two mirrors are used) the larger (the angle formed by the mirrors within the limit of 180 degree) the fewer (the images)."[63] However, I do not agree with Tolansky[64] that this means that the Chinese invented the kaleidoscope. They did not put the mirrors into a tube and use it to create radially symmetric images.

In the second century CE, Hero of Alexandria's *Catoptrica* described how to fashion a "multiview" mirror, which "shows many faces" and "makes one finger many." It consists of two plane mirrors that can "revolve about their common side" (they are joined by a hinge).[65] His illustration shows that if two mirrors were placed at a particular angle, two images would be formed,[66] but he does not explain the relationship of the angle to the number of images.

Giambattista della Porta wrote that by joining two mirrors so that "they may be shut and opened like a book," and narrowing the angle between them, instead of seeing just one face, "you shall see many in them both." And he described arranging mirrors in a polygonal shape, to likewise elicit multiple reflections.[67]

Athanasius Kircher (1602–1680), in his *Ars Magna Lucis et Umbrae* (1646), describes, like della Porta, two mirrors hinged together like a book. However, he tabulates the number of images formed for different angles between the two mirrors.[68]

As pointed out by Brewster, neither Porta nor Kircher required the eye to be in a particular position to the object, or angles that would assure the formation of symmetrical reflections, or to provide lighting by which these would be uniformly illuminated.

Brewster's Kaleidoscope

The kaleidoscope (*kalos*, beautiful; *eidos*, form; *scopos*, watcher) was patented by Sir David Brewster (1781–1868) in 1817.[69] Brewster constructed his first telescope at the age of ten. At twelve, he began studying for the ministry. Unfortunately, "the first day he mounted the pulpit was the last"[70]; he suffered extreme stage fright when confronted by a congregation. Religion's loss was science's gain: He went on to an illustrious career as a physicist, winning all three medals offered by the Royal Society.

According to Brewster, the kaleidoscope "is an instrument for creating and exhibiting an infinite variety of beautiful forms, and is constructed in such a manner as either to please the eye by an ever-varying succession of splendid tints and symmetrical forms, or to enable the observer to render permanent such as may appear most appropriate for any of the numerous branches of the ornamental arts."[71]

His patent[72] goes on to describe a two-mirror system. The long edges of two rectangular mirrors were placed together at an angle, such as 18 degrees (1/20th of a circle). He explains that "if any object, however ugly or irregular in itself," is placed before one end, the part of it that is seen through the other end will be successively reflected by the two mirrors to produce a symmetrical form. With mirrors at 18 degrees, there is 20-fold symmetry, so one sees a ten-pointed star.

Although intended as a tool for artists (Brewster called it an "ocular harpsichord"), the kaleidoscope found its niche as a toy. It was the Rubik's Cube of its day. In 1818, Dr. Roget wrote:

> In the memory of man, no invention, and no work, whether addressed to the imagination or to the understanding, ever produced such an effect. A universal mania for the instrument seized all classes, from the lowest to the highest, from the most ignorant, to the most learned, and every person not only felt, but expressed the feeling, that a new pleasure had been added to their existence.[73]

The standard kaleidoscope is tubular. The object cell, at one end of the tube, contains translucent, colored objects, such as beads or colored oils. These objects can move about. A set of mirrors runs down the length of the tube. There can be two, three, four, or more such mirrors. The mirrors are what create the magic of the kaleidoscope. The image that you see in a kaleidoscope is a combination of a direct view of a pie-slice sector of whatever lies at the far end, and multiple, symmetrically arranged views of reflections of the direct view.

"In the two-mirror system, the mirrors are arranged in a 'V' with a third side that is blackened. The angle of the 'V' determines the number of reflections.... The most perfect symmetry and best images occur when the angle between the mirrors divides equally into 360 degrees." For example, a 60-degree angle results in sixfold symmetry and a three-pointed star.[74]

"The three-mirror system can be arranged in any form of triangle.... It produces a continuous field of honeycomb-like patterns." Typically, the angles are 60–60–60, 90–45–45, or 30–60–90. As previously mentioned, the 60–60–60 arrangement produces sixfold symmetry. With a 90–45–45, the 90-degree angle creates fourfold

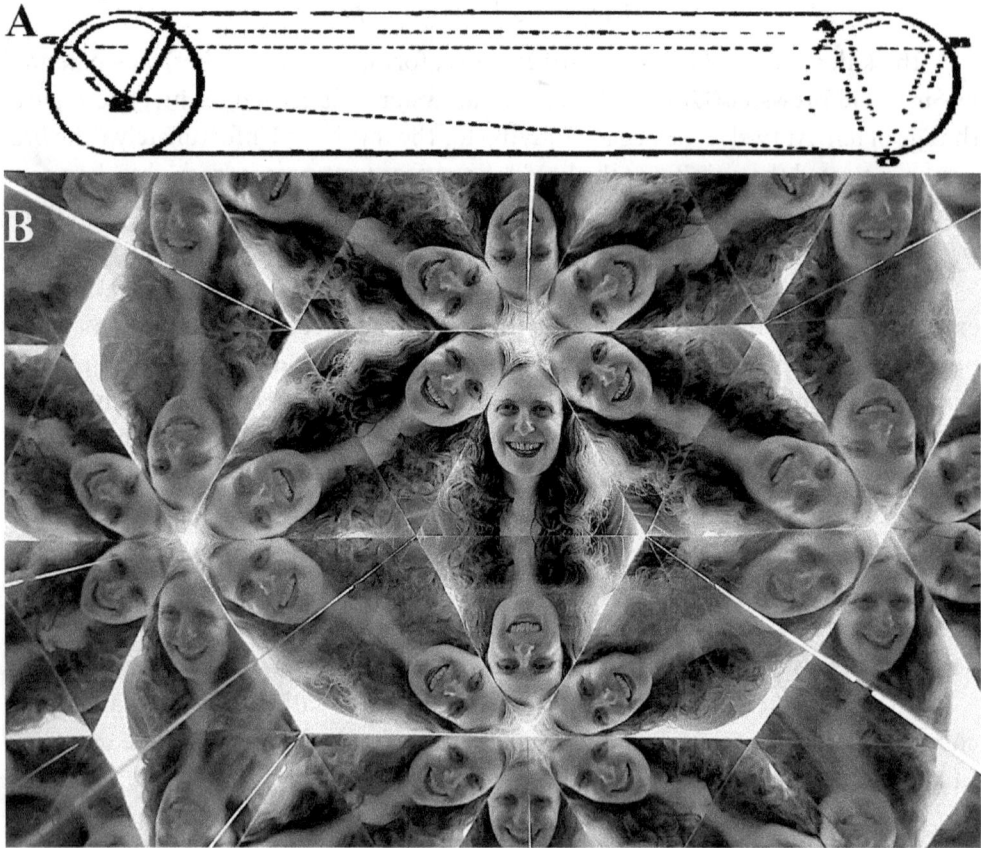

A. Two-mirror Brewster kaleidoscope, schematic perspective view. (Plate B, Fig. 13 from Brewster, David, *A Treatise on the Kaleidoscope* [Archibald Constable, 1819].) B. Teleidoscope image of author's daughter; note the sixfold symmetry. Photographed by author at Museum of Illusions, Orlando, Florida.

symmetry while the 45-degree angles produce eightfold symmetry. A 30–60–90 kaleidoscope has three kinds of symmetry: fourfold, sixfold, and 12-fold (the last from the 30-degree angle).[75]

The conventional mirror is plate glass with a reflective metal coating on the rear surface. This has the advantage of protecting the coating. However, it yields two reflections, one from the front surface of the glass, and another from the coating to the rear. If a rear surface mirror is used in a kaleidoscope, the double reflection results in images with fuzzy edges. Hence, the better kaleidoscopes feature front surface mirrors. Such mirrors also yield a brighter image.

Charles G. Bush, a Prussian immigrant to a Boston, was the first major American kaleidoscope designer. One of his innovations was the use of a liquid-filled object box. In a preferred form, this object box contained two or more immiscible (nonmixing) liquids of different colors, as well as air bubbles.[76] Another was the use of an object box with an opening and a cover, so the contents could be changed readily.[77] Still another contribution was combining a rotating object box with a means for illuminating the objects so that opaque objects could be visualized.[78]

A *teleidoscope* is a variant on the kaleidoscope: Instead of an object cell, it has a lens. Hence, any indoor or outdoor scene can be fractured and multiplied in a kaleidoscopic fashion.

Multiphotography

In the nineteenth century, the term "multiphotography" was applied to the process of obtaining, in a single photographic image, multiple views, from different angles, of a subject. Woodbury says that this was "at one time quite popular." For example, with the mirrors set 75 degrees apart, one obtains a back view, two three-quarters views, and two profile views. It was used not only for portraits, but also for product photography. "In France it is used for photographing criminals."[79]

A. Setup for multiphotography of a sitter. (Woodbury, Fig. 3.) B. The corresponding multiphotograph (Fig. 5). C. Author's multiphotograph of his daughter playing cards (at the Museum of Illusions, Orlando, Florida).

The "Mirage"

Mirrors have been used in other toys, too. An example is the flying saucer-shaped "Mirage" or "Mirascope" toy.[80] An object appears to be on top, but there is nothing there, just an image. The real object is hidden inside. The secret of the device is that the image is formed by two concave mirrors that face each other and indeed meet at the edges. In the center of the top mirror (which faces down), there is a small transparent window. The object is placed underneath this window, sitting on the bottom mirror.

What is the underlying physics? Light comes in through the window and is reflected off the object. The object is sitting at the focal point of the upper mirror.

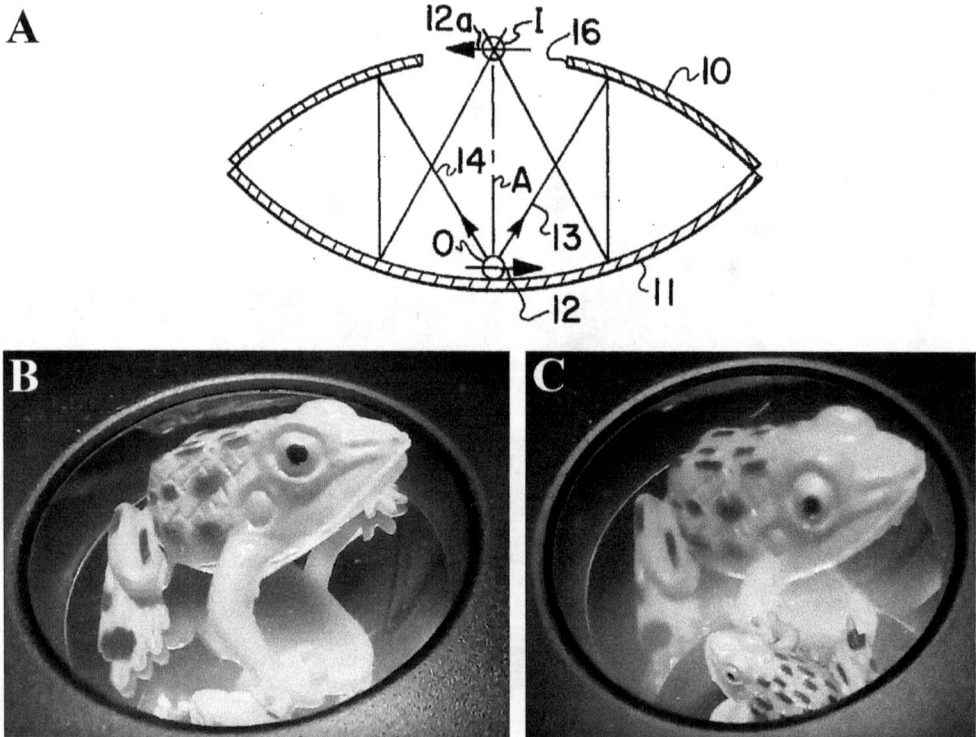

A. Fig. 1 from Elings, Virgil B., and Caliste J. Landry, Optical Display Device, USP 3,647,284 (issued March 7, 1972), showing the working principle of the "Mirascope" toy. Mirrors 10 and 11 have "their concave side facing each other"; the top mirror 10 has a "central aperture." The object O is placed on or near the surface of the bottom mirror 11. Light striking the object is reflected off the object, then of the surface of the top mirror 10, and finally off the surface of the bottom mirror 11, converging to form a real image at point I. This image is reversed, as shown by the arrows 12 and 12a. In other embodiments, the image is not reversed (Fig. 3); there's a transparent barrier to prevent easy removal of the object (Figs. 8, 9), or there is internal illumination of the object (Fig. 10). B. The image of a toy frog appears just above a "Mirascope." C. Looking more directly into the Mirascope, the real object can be seen. Note that the image is inverted relative to the object. (B and C are frame grabs taken with permission of Autistivision from its video, "3-D Mirascope-Instant Illusion Maker (Toysmith/DaMert Company)," https://www.youtube.com/watch?v=mbBQ1vK0ELo.)

The reflected rays strike the upper mirror, which, being parabolic in shape, converts them to rays parallel to the optical axes of the upper and lower mirrors. These now strike the lower mirror and are focused at the focal point of the latter, which lies just on top of the window. The image looks real and is upside-down.

The two mirrors have the same radius of curvature, and hence the same focal length. It can be shown by matrix analysis that with this two-mirror system, the focal lengths can equal either half the radius of curvature (resulting in inverted image), or 1.5 times the radius of curvature (yielding an upright image). The mirrors are fitted so that the focal lengths meet the former condition. This means that if you slowly lift the upper mirror, at the right height the right-side-up image of the object will appear.[81]

The thirteenth-century natural philosopher Roger Bacon would have been pleased by these effects. He suggested that mirrors could be used to produce "strange apparitions."[82]

Nonreversing Mirrors

The first-century engineer Hero of Alexandria claimed that with the aid of catoptrics, "mirrors are constructed which show the right side as the right side, and, similarly, the left side as the left side, whereas ordinary mirrors by their nature have the contrary property and show the opposite sides."[83]

Hero was no doubt aware that if two mirrors are set at right angles to each other, like a partially open book rested on its bottom edge, the subject is doubly reflected, and thus a secondary image is formed that does not appear to reverse right and left. (The subject's image is actually reversed twice.)

In the sixteenth century, della Porta likewise observed that with two mirrors at an angle, "the right parts will show right, and the left to be the left, which is contrary to Looking-glasses."[84]

Despite this centuries-old prior art, John Hooker was able to get an 1887 patent on a compound mirror with the two surfaces meeting and rigidly connected at one-half, one-quarter or one-eighth of a right angle.[85] Nor was this the last patent directed to a variation on this ancient invention; I know of at least four others, the latest of which emphasized that its "True Mirror" provided a "center split which is virtually invisible."[86]

Funhouse Mirrors

"Hey, Dad! Look at me!" Your child is delighted as he or she is successively squashed or stretched by the funhouse mirror. These mirrors have a wavy surface. They were traditionally found at amusement parks, carnivals, state fairs, festivals and the like. However, they can also be found in daycare centers, doctor's offices, and restaurants.

The distorting mirror is ancient. Pliny the Elder (23–79 CE) says, "Mirrors … have been invented to reflect monstrous forms; those for instance, which have been consecrated in the Temple at Smyrna."[87]

Della Porta discussed in detail the effect of convex cylindrical, concave cylindrical and "pyramidal" mirrors on one's appearance and proposed putting mirrors of different shapes in the same room: "If a crooked be set in one place, in another a Concave, and a plain one in the middle, you shall see great diversity of images." The distortions clearly amused him:

> If that part of the glass, that is set against your mouth, shall stick forth before like a wreathed band or a Boss-buckler, you mouth will appear to come forth like an ass's or sows snout. But if it swell forth against your eyes, your eyes will seem to be put forth like shrimps eyes. If the angle be stretched forth by the length of the glass, your forehead, nose, and chin, will seem to be sharp, as the mouth of a Dog.[88]

An unusual collection of funhouse mirrors appeared at the Pavilion of Fun of Coney Island:

> In a variation of the usual row of fun house mirrors that made people fat and squat, tall and thin, or even crooked, Tilyou's Funny Mirrors were set in a circle and revolved so that mirrors paraded past the viewer. A six inch wide row of nails encircled them and when a patron leaned on the rails to view the amusing and distorted images of themselves, they received a mild shock. And those who decided to take a seat were surprised when the seat collapsed four to five inches. Nearby was a stereoscopic viewer that had a blowhole at its pedestal. When a distracted woman looked at the image, she became embarrassed as air blew her dress upwards for all to see her undergarments.[89]

It is not necessary that a mirror actually distort the image in order to create an unusual visual effect. The "anti-gravity mirror" is a favorite of "hands-on" science museums. You stand at one end of a large, vertical plane mirror, so that you are bisected by it. Adjust your position so that someone at the other end sees a composite image (half direct view, half mirror image) that looks normal (because your mirrored half, reversed, looks like your hidden half). Now raise your exposed leg. The reflected image is also raised, and you seem suspended in the air.

Mirror Mazes

"Mazes go back in history at least 4,000 years. For the first 3,000 years they were entirely in the form of unicursal Labyrinths, consisting of a single convoluted path, without junctions. These labyrinths were not puzzles, but instead were for ritual walking, running and processions."[90] The most famous maze was, of course, the Labyrinth which King Minos of Crete was said to have built to house the Minotaur. "Whatever the truth of this myth, it is a fact that the 7-ring Cretan labyrinth design was used on Cretan coins in the 1st century BC."

Gustav Castan, the owner of the Panopticum in Berlin (a wax museum), received a French patent on a mirror maze in 1888, and his "Moorish Maze" opened on the third floor of the Panopticum in 1889.

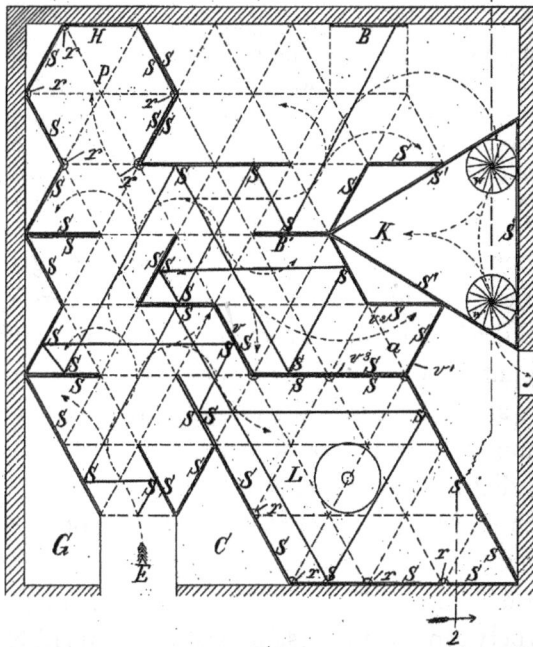

A. Mirror maze, from Castan patent Fig. 1. The entrance is at E and the exit at A; the kaleidoscope K is actually above the main floor. B. Mirror maze at Mall of America, Bloomington, Minnesota. (Photograph by Michael Ocampo, posted to Flickr under Creative Commons Attribution 2.0 Generic license. Converted to monochrome for this publication. https://www.flickr.com/photos/coolmikeol/4156970741.)

His American counterpart patent states:

> The primary object of my invention is to provide such an arrangement of mirrors in a room or inclosure as shall cause them, by their reflection of objects suitably located with relation to the mirrors, to present to the vision of a person in the apartment the illusion of a labyrinthian device composed of seemingly endless passages, which appear to him to be freely traversable until he is stopped in his course by an obstructing mirror, from which long passages seem to extend to the right and to the left.[91]

Many mazes are constructed on a rectangular grid, or no grid at all. However, mirror mazes use a triangular grid, in which adjacent mirrors are at a 60-degree angle (or an integer multiple thereof) to each other. The triangular grid is plainly disclosed by Castan's patent. Castan also teaches that "the mirrors and passages shall present a uniform appearance in width and height, as well as in the shape, size, and design or decoration of the frames or borders."

Castan's figure 1 is believed by this author to be an actual plan of his "Moorish maze." It contains a "rhombic compartment L in imitation of the Lion Court of the Alhambra." At P there is "a tropical garden, the background H of which appears as the entrance to a Moorish temple." There are "palms and exotic plants … placed on the architectural lines between the mirrors," as well as "pillars r." Near the exit of the maze, visitors had the option of entering a "kaleidoscope K, represented in the shape of a regular tetrahedron, having three sides formed of the three mirrors S….' The base of the chamber K is supported above the plane of the floor of the mirror-maze proper," and was pierced by two spiral staircases, one for entry and the other for exit.

There are two other early American mirror maze patents. Perry points out that the passages may be formed by panes of clear glass as well as mirrors, and also addresses the danger of a person inadvertently walking into either of them.[92] Palm was likewise concerned with "protecting the mirrors or transparent glasses from breakage," and also with rendering the maze "readily portable and with interchangeable parts so constructed and arranged as to be readily put together … and as readily taken apart."[93]

One mirror maze was created in 1896 for the Swiss national exhibition in Geneva and has been at the Glacier Garden, in Lucerne, Switzerland, since 1899. It has ninety mirrors, and the arches are in an "Alhambra" style.[94] Another is at Wookey Hole Caves, in Somerset, England. There are over forty mirrors, each eight feet high and set at precise 60-degree angles, to create the illusion of long passageways. The illusion lasts until you bump your nose into one of the mirrors. A musical fountain, positioned at the end of the maze, is reflected back through the mirrors.[95]

A mirror maze was one of the features of the Pan-American Exposition in Buffalo, New York, in 1901. According to the Official Catalogue and Guidebook, "At the west end of the Midway is a building occupied by Dreamland, or the Mirror Maze. Behind mirrors is a large amount of fun for those who attempt to explore its recesses. No illusion on the Midway is more confusing and amusing."[96] The exposition opened in May 1901; in September, while visiting the Exposition, President William McKinley fell to the bullet of an assassin.

At Coney Island, during the 1920s, the park management added an attraction for young children called "Noah's Ark," a funhouse that looked and rocked like a boat. "Inside was a mirror maze, moving floors that nearly knocked one off their feet, and hidden compressed air jets that blew woman's skirts high in the air. There was also a menagerie of animals that Noah and his wife saved from the Flood."[97]

During Prohibition, the Stotesburys of Philadelphia installed a mirror maze in their mansion, currently the home of the Catholic Philopatrian Literary Institute. It is unclear whether the purpose of the maze was to entertain the guests, or to make it more difficult for the authorities to find their way to where the guests were drinking. "The mirror maze continued to be effective as a burglar device until 1958 when it was disassembled to become the hall of mirrors that currently exists."[98]

Mirror mazes were equally popular on the West Coast. "In 1956 CBS and the Los Angeles Turf Club (Santa Anita) acquired the lease on the Ocean Park Pier and they proposed to build a $10,000,000 nautical theme park to compete with Disneyland.... The 28 acre park was decorated throughout in a sea-green and white art moderne look.... Visitors entered the park through Neptune's Kingdom where they descended in a submarine elevator to the oceanic corridors below. Across from the elevator was an enormous sea tank where it appeared a shark and its prey shared the same tank.... A standard glass mirror maze was part of the fun house experience."[99] The park closed in 1967; this may now be the only mirror maze open to fish.

A mirror maze figured prominently in Orson Welles's film noir *The Lady from Shanghai*. The mirror maze in the sequence used 2,912 square feet of glass—eighty mirrors, each seven feet high, and another twenty-four distorting mirrors, all rigged as one-way mirrors, so they could be filmed through.[100]

Mirrors were also used effectively in Bruce Lee's epic *Enter the Dragon* (1973). The arch villain, played by Shih Kien, fled through a rotating wall section into a mirror maze. It took hours to set up the maze for filming, as the production team had to make sure that none of the camera or lighting equipment would be visible. Eerie music played as Bruce Lee sought out his opponent among the 8000 mirrors[101] used to construct this maze. He was frustrated until he remembered the words of the head of the Shaolin Temple: "the enemy has only images and illusions, behind which he hides his true motives. Destroy the image and you will break the enemy." The moment in which a slow-motion kick appears in a dozen images at once is particularly memorable.

Anamorphic Art

An anamorphic image is one which when viewed conventionally appears distorted, but when viewed in a special way (such as at a particular angle, or in a curved mirror) appears natural.

Here, we are concerned with catoptric anamorphosis: images distorted in such a way that to reconstruct them easily, we need a mirror. Usually the mirror will be

Hans Troschel (1585–1628) engraving, "Satyrs Admiring the Anamorphosis of an Elephant," based on Simon Vouet's cylindrical anamorph. (Harris Brisbane Dick Fund, 1945, The Met, accession 45.97[78], The Met.)

cylindrical or conical. Diderot's *Encyclopedie* (1751) explained, "there are means of drawing on paper distorted objects which, seen through this kind of mirror, appear in their natural form."[102]

The earliest catoptric anamorphosis in Western art that has come to my attention is an engraving by Simon Vouet (1590–1649) made sometime between 1624 and 1627. It shows a large cylindrical mirror on a cloth-covered table. Upon the cloth is an anamorphic distortion of an elephant. The distortion is corrected by the mirror. Satyrs gather around the table, some pointing at the mirror, others at the cloth. Above the mirror is the inscription "FORMAT ET ILLUSTRAT."[103]

Jean-François Niceron (1613–1646) publicized various anamorphic transformations in his *La perspective curieuse* … (1638) and *Thaumaturgus opticus*… (1646). In 1638, Niceron produced a cylindrical anamorphosis of a painting of St. Francis of Paola. The illustration makes it clear exactly how the anamorphosis was constructed. In the center there is an undistorted sketch of the subject. Upon this Niceron superimposed a rectangular grid, divided by vertical and horizontal lines into square cells, and numbered the rows and columns. Around the original grid is the anamorphosed subject. The grid for it is polar rather than rectangular—that is, a center point was picked; rays were drawn from the center point with a constant angular separation, and then concentric circles were drawn around the point. Each vertical

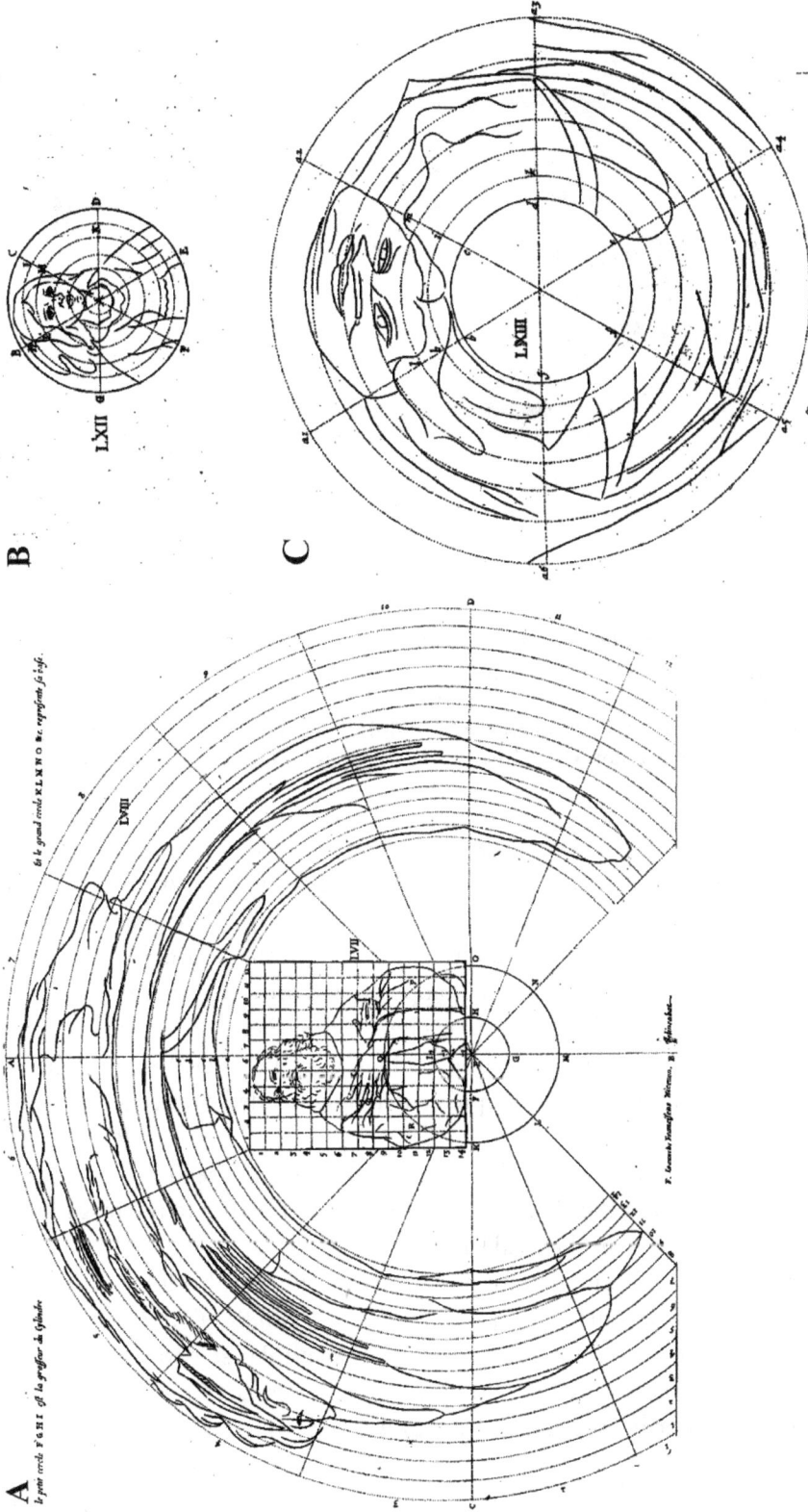

Cylindrical anamorphoses. A. Niceron's St. Francis. (Figs. LVII, LVIII, Plate 44 [Page Image 448/465].) B. Niceron's head and shoulders of a woman. Figs. LXII, LXIII, Plate 47 (Image 454/465). C. All from Niceron, Jean François, *La Perspective Curieuse...* (1663). (The Linda Hall Library of Science, Engineering & Technology. CC BY 4.0. https://catalog.lindahall.org/discovery/delivery/01LINDAHALL_INST:LHL.)

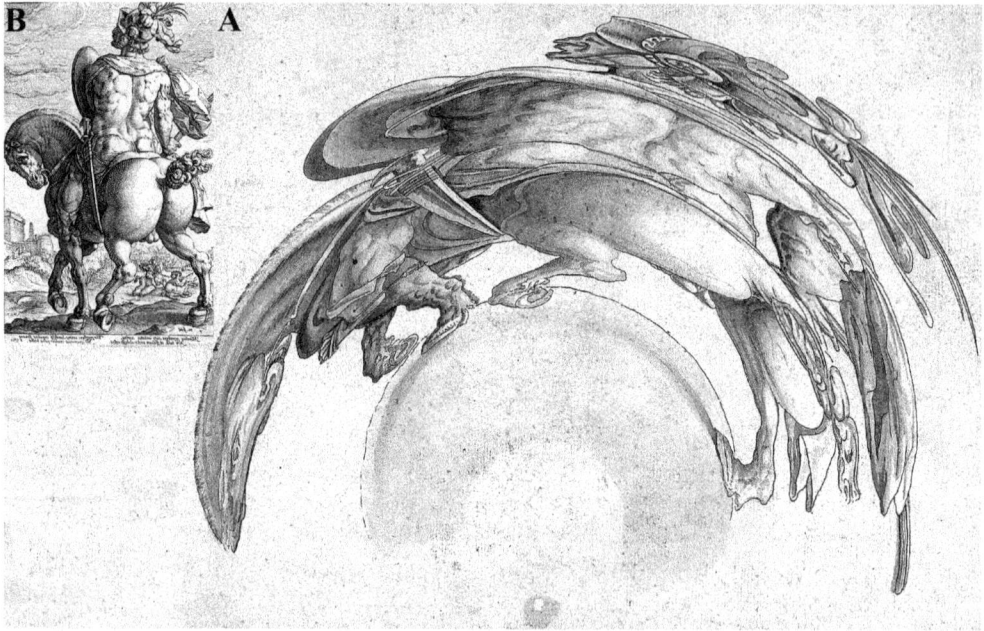

A. Jean François Niceron, "Soldier on Horseback in Catoptric Anamorphosis (after Hendrick Goltzius)" (ca. 1620–1640). This is a cylindrical anamorphosis based on B. (Purchase, Brooke Russell Astor Bequest, 2013, Accession 2013.203, The Met.) B (inset). Hendrick Goltzius, "Titus Manlius, from 'The Roman Heroes'" (1586). Image reversed to correspond to A. (The Elisha Whittelsey Collection, The Elisha Whittelsey Fund, 1951, Accession 51.501.39, The Met.)

in the original rectangular grid corresponds to a ray in the distorted grid; each horizontal corresponds to a concentric circle. These define distorted rows and columns which are numbered analogously to the original grid. To draw the anamorphosis, the artist copied (with appropriate distortion) each cell of the original grid to a cell of the target grid. The result is what would happen if you drew the original sketch on some stretchable material, pulled it out to each side, and then bent it into an arc.[104] The same approach can be seen in a cylindrical anamorphosis created by Athanasius Kircher (1602–1680) in 1646. This depicted a double-headed eagle, the Habsburg symbol.[105]

Since each square cell becomes a bowed rectangular cell, the artist must make adjustments in transferring the elements of the original drawing to the target grid. The finer the grid, the more mechanical the transfer is.

There are two basic parameters for a cylindrical anamorphosis: the radius of the center hole, and the angular size. In Niceron's *St. Francis of Paola*, the radius was equal to the height of the original sketch, and the angular size was 270 degrees. Kircher used the same radius, but an angular size of 252 degrees.

The rectifying mirror (anamorphoscope) was a cylindrical mirror with the same radius as the center hole, and it would be placed over that hole.

Further insight into the anamorphic transformation is provided by Mario

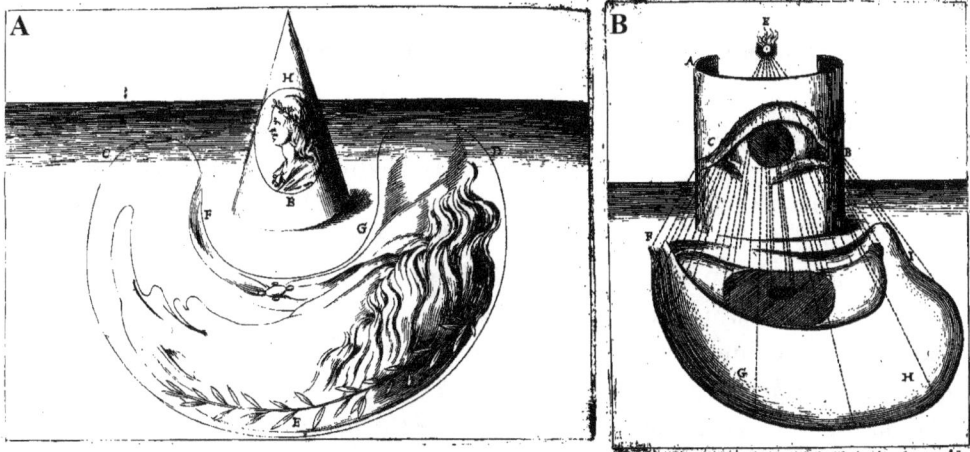

A. Conical anamorphosis of head-and-shoulders miniature. (Bettini, Mario, *Apiaria Vniversae Philosophiae Mathematicae,* Book V, Chapter 2 [1642]. Image 314/480.) B. Cylindrical anamorphosis of eye (Bettini, supra, Book V, Chapter 3. Image 315/480. Both courtesy of The Linda Hall Library of Science, Engineering & Technology. CC BY 4.0. https://catalog. lindahall.org/discovery/delivery/01LINDAHALL_INST:LHL/1289476080005961.)

Bettini's (1582–1657) anamorphosis of the eye of Cardinal Colonna, drawn in 1642.[106] The anamorph is in the foreground. Immediately behind it is the cylindrical mirror, with its reflection showing the rectified image. Behind the mirror is a viewpoint marked with an "E"; dashed lines emanate from it, going through the mirror and ending at the anamorphic representation. These are lines of projection. The true viewpoint, of course, is in front of the mirror, not behind it; it lies as much in front of the mirror as the represented "E" does behind it.

Niceron's and Kircher's cylindrical anamorphoses are only approximate; they would be correct only if the picture were viewed at an infinite distance. Since it is actually viewed at a finite distance, the vertical lines in the object grid, instead of becoming radial lines, are curved. (Following them outward from the center point, they bow toward the viewer. The radial line which starts directly toward the viewer is unaffected.)[107] The correct transformation was deduced by Jean-Louis Vaulezard and described in *Perspective cylindrique et conique* (1630). Niceron was certainly aware of Vaulezard's work but chose to use the simpler polar transformation.

Conical mirrors may also be used in conjunction with anamorphic paintings. In 1638, Niceron prepared a conical anamorphosis of Louis XIII. He superimposed a polar grid (one with radii and concentric circles) over the original sketch. The anamorphosed grid was a quadrant. It was as if the original sketch were on stretchable material, and one cut along a radius, then moved the two radial edges clockwise and counterclockwise away from each other, like closing a fan, until a quadrant shape was achieved.[108]

If you don't have the cone's dimensions or the original image, finding the correct cone to rectify the conical anamorph can be tricky. Cylindrical anamorphs are much easier to work with; the radius of the mirror is usually obvious from the hole in the anamorph, or from a circle drawn by the artist to mark the proper site of the mirror.

Drawing anamorphoses is laborious, and machines were soon invented to facilitate this endeavor. Machines for cylindrical and conical anamorphosis were described by Jacob Leupold in his *Acta eruditorum* (1712) and *Anamorphosis mecanica nova* (1715).[109] They appear to have worked somewhat like a pantograph. The input arm held a pointer or a pen of which the point touched a cylinder, and the output arm held a pen which rested on the flat drawing paper. The original sketch was mounted on the cylinder, and the artist then ran the input stylus over the appropriate lines and curves of that sketch, producing anamorphed output at the output pen. Presumably, there was a way to raise and lower the input stylus, and consequently the output pen, so discontinuous elements could be reproduced.

Nowadays, one can use a computer program to convert an ordinary picture into its mirror anamorph. One such program is Philip Kent's *Anamorph Me!* (2001); it supports conical, pyramidal, and cylindrical anamorphoses.[110]

At the other extreme, we have the Asian approach to catoptric metamorphosis. According to Baltrušaitis, the Chinese did not use grids or machines but drew their anamorphed pictures by eye. That is, they would paint on the canvas, while observing the resulting reflection in the mirror.[111]

Baltrušaitis believes that the Chinese invented the mirror anamorphosis. Based on the "licentious couples" shown in certain early Chinese anamorphs, he believes that they date to the Ming Dynasty reigns of Lung-Ch'ing (1567–1572) or Wan-Li (1573–1619). If that is true, then they certainly precede Vouet's work (~1625). However, he acknowledges that the "erotic cycle" in Chinese art lasted until 1644, when the Qing dynasty came into power, so these works could have been produced after Vouet's.

He also asserts that the first mirror anamorphoses of Western Europe were not the product of indigenous development (from the oblique metamorphoses of the sixteenth century), but rather were copied from Chinese examples which had made their way into the collection of the Sultan in Istanbul.[112] However, this assertion seems to rest largely on Vouet's choice of an elephant as the centerpiece of his cylindrical anamorph. While elephants were certainly known in Chinese art (Baltruišaitis provides a photograph of a Chinese cylindrical anamorph, *Personage mounted on elephant*, painted on silk, allegedly during the Wan-Li period[113]), they had been known to the West since Hellenistic times. Hanno, a white elephant, was given to the Pope by the King of Portugal. Hanno died in 1516, and in 1520 the Pope received another pachyderm, Leo, from a Knight of the Order of Rhodes.[114]

There are a considerable number of anamorphic paintings from the seventeenth and eighteenth centuries in art museums; they seem to have been something of a fad. However, not all artists were enamored of them. Samuel van Hoogstraten, in his *Introduction to the Advanced School of Painting* (1678), wrote:

I will pass over the manner in which, through reflected lines, malformed shapes may be restored to their correct aspect in reflecting globes, angled mirrors, and cylinders; for those are truly artifices rather than essential arts.[115]

Mirror Balls

A "myriad reflector" was patented by Louis Woeste in 1917. The claim recited that it comprised "mirrors mounted to form a polyhedron bounded by a convex system of plane faces in combination with means for suspending the device … so that it may be swung and rotated simultaneously to produce myriad reflections when light rays from an extraneous source are thrown thereon." The patent suggested that it might be used "for exhibition or theatrical purposes for the production of picturesque or scenic effects."[116]

Mirror balls became popular in the late 1960s. A spotlight would shine on a ball covered with small mirrors, flashing light off each mirror in turn as the ball rotated. Naturally, there were elaborations over the years: multiple beams, colored beams, and so on. What was the first American movie to feature a mirror ball? No, it wasn't *Saturday Night Fever*; there was a mirror ball in the 1942 film *Casablanca*.[117] But it appeared even earlier in a short silent film, *Die Sinfonie der Großstadt* [The Symphony of a Metropolis] (1927), just after the audience saw the feet of dancers doing a Charleston.[118]

Computer Games with Simulated Mirrors

A number of computer games have made use of simulated mirrors. One example is *Laser Chess*. This game was invented in 1987 and implemented on the Atari computer. The Basic Game had a single long-range attacking piece, the Laser, which destroyed enemy pieces if it hit them on an unprotected side. Several of the pieces had mirrored sides, which not only protected them from the Laser but also allowed them to redirect its attack against another piece—even the Laser itself! In the Basic Game, there were a Block and a Triangular Mirror (each with one mirrored side), and two reflective pieces that were immune to laser beams, the Straight Mirror and the Diagonal Mirror. In the Advanced Game, there was a One-Way Mirror (which could be used to protect the Laser from self-destruction if its light was reflected back upon it), a Stunner (which, after striking a nonreflective surface, nullified the reflective surfaces of its victim for a random length of time), the Triangular Mirror, and two better protected Blocks (one with protection on three of eight sides, the other with full protection). Both versions featured a Beam Splitter piece that could split a laser beam into two new ones.[119]

Another is *Aargon Deluxe*, a solitaire computer puzzle game. Your goal in each "level" is to cause all of the coins to spin. Some will spin when struck by light of any color, and others are color specific. You have at your disposal a variety of pieces, depending on the design of the level. These include a diagonal mirror that is reflective on one side, a diagonal mirror reflective on both sides, a double diagonal mirror with a slit so light can pass along the diagonal and through the slit, and various refractors, beam splitters, prisms, polarizers, color filters, and so forth. White or colored light is generated by a laser. A given piece, in a particular puzzle, may be rotatable, moveable, both, or neither.

Board Games with Real Mirrors

In *Mirror Mania* (Mattel, 1969), play began with eight "mirror blocks" placed in random positions and orientations on a three-by-three board. Mirrors were hidden inside the blocks, but the blocks were marked at the top to show how they directed light. On the four sides of the board, there were pawns worth a varying number of points. The mirror blocks were also worth points, depending on their color. On each turn, you could either turn a cube or move it to the empty spot. The goal was to set up high scoring paths by which a pawn could be seen via the mirror cubes from a viewing cube placed outside the grid.[120]

An unrelated product with the same name (Whitman, 1982) featured play sheets with various routes to be followed. The catch was that you had to move your pen along the route based on viewing it and the route via a reflected image.[121]

Khet: The Laser Game (Innovention Toys, 2005), originally named "Deflexion," is described by Boardgame Geek as "a chess-like board game that has two built in lasers and movable Egyptian-themed game pieces that have embedded mirrors that can be positioned to bounce the laser light around the board and hit opponent pieces."

> To play, players alternate moving their pieces around the board. Some pieces have mirrors and some do not. Bounding the board is a raised frame into which are built two low-power lasers, one for each player. The game pieces include a "pharaoh," obelisks, and pyramids with mirrors. After each move, a player must press the button on his/her laser. The beam bounces from mirror to mirror around the playing field. The challenge is to protect one's own pharaoh while maneuvering to "light up" the opposing player's pharaoh.[122]

This was succeeded by *Laser Chess* (2011):

> originally released as Khet 2.0.... To begin, players set up the pieces on the game board in one of several starting positions. Players then alternate turns, either moving a piece to an adjacent space, rotating a piece 90°, swap two adjacent pieces if one of them is a switch piece, or rotating their laser 90°; the pieces have two, one, or zero mirrored surfaces, and after moving or rotating a piece on your turn, you must then fire your laser, with the laser possibly being deflected by 90° whenever it hits a mirror on a piece (since the mirrors are angled at 45°). If the beam strikes a non-mirrored surface on almost any piece, that piece is removed from play. (One exception: The defenders have one surface immune from laser fire, which means you can use them to protect your king; they're vulnerable to laser fire from other directions, though.)
>
> When a king is hit by a laser, the other player wins the game! Friendly fire is possible, so try to make sure that you won't remove one of your one pieces when you shoot—unless you need to clear a shot to the enemy, that is![123]

There was also *Laser Battle* (MGA Entertainment, 2006). The pieces could be "rotated to sixteen different" orientations rather than four, the laser was "fired by a unit" rather than from the wall of the game board, and the game board was modular.[124]

Innovention had obtained a patent on its game,[125] and sued MGA for patent infringement; MGA raised an obviousness defense; the Federal Circuit held that computer games were "analogous art" and remanded the case to have the trial court consider them. On remand, the jury held that the claims were nonobvious and infringed.[126]

Mirror, Mirror (Eagle-Gryphon, 2011) was a two-player, turn-based game played on a checkerboard with pieces having mirrors on their backs. It is a bit reminiscent of Stratego, in that information about each piece is initially hidden from the opponent. Each player had nine pieces, of which eight were decoys. The decoys could be distinguished from the real piece by the color of the seal on the mirror (hidden) side. They were removed from play if you moved your piece on the same square and correctly guessed the color of its seal; if you were wrong, you lost your own piece. So you first used your pieces' mirrors to determine the color of the seal on an enemy piece, then targeted it. The pieces were of three types, each with a different move pattern as indicated by a card inserted into the mirror and seal holder.[127]

Mirror Garden (Gemblo, 2019) was a two- to four-player, real-time pattern-matching game: "you reveal a garden card from the top of the deck, then place it on the table. All players rush to make the garden shown on the card by using a foldable mirror and two semicircular garden boards. Use the boards to create a portion of the garden, then set the mirror to fill up the rest of the garden. Once your garden is complete, call out the code written on the boards. If correct, you claim that garden card. Whoever scores five points first wins!"

Board games with real mirrors are also described in the patent literature. For example, Marcel Mayas's U.S. Patent 4,314,191 (1994) describes "a reflective word game … in which letter tiles placed into a game plate by a player will simultaneously form both a real word and a reciprocal image word that is reflected in either one or two adjacent mirrors. A challenger player must make up a matching word by placing letter tiles into the game plate which will also be reflected in either one or two adjacent mirrors."

Commercializing Nature's Glitter

All that glitters isn't gold. Mica is one of the six most common minerals;[128] its name comes from Latin *micare*, "to shine." Mica flakes were occasionally used in early tapestries "to depict mirrors held by mermaids."[129]

Dry ground, it loses its distinctive sheen but acts as a moisture barrier, making it useful as a filler in drywall plaster. Wet ground, it adds sparkle to whatever product it is incorporated into: pearlescent paints, wallpaper, soap, toothpaste, and cosmetics, the latter including nail enamel (Revlon's "Foiled"), face powder (Chanel's "Lumiere Platine"), and body powder ("Pixie Dust").[130] "People notice that you're kind of glowing today," said Marie Condron, who markets the powder.

CHAPTER 7

Riding the (Light) Wave

The Nature of Light

Isaac Newton proposed, albeit somewhat hesitantly, that a ray of light was a stream of tiny projectiles ("corpuscles"), and quite a bit of the behavior of light could be explained by this "corpuscular theory." However, by the nineteenth century, it was apparent that it had its failings.

Newton had been aware of one problem: the concentric bright and dark rings formed when light was reflected off the combination of two glass surfaces, the upper one slightly convex down and the lower one flat, in contact at one point. These "Newton's rings" were actually first reported on by Robert Hooke, who explained them as a wave phenomenon. Newton described them, and attributed them, somewhat obscurely, to "fits" (some sort of internal vibration mode of the corpuscles).[1]

Thomas Young's double-slit experiment showed that light also exhibited wavelike properties. He set up a "homogeneous" (monochromatic?) light source, behind and equidistant from two narrow slits in a wall. Beyond the slit was a second, parallel wall. If the corpuscular theory were correct, then light was streaming in all directions from the light source but could pass the first wall at only two places. There would then be two spots of light on the second wall, each in line with the light source and one of the slits.

But that's not what Young observed. Instead he saw a pattern of dark and light bands. To better understand why this showed wavelike behavior, let's resort to a water wave analogy. Water waves are transverse waves, meaning that the change in level (vertical) is perpendicular to the direction in which they propagate (horizontal). The wave source would be a rhythmic "thumper." Each thump would send out a circular ripple; the repeated thumps would create concentric ripples. When they reached the slits, each slit would act like a new wave source and generate a semicircular pattern of concentric ripples. (The light waves are "diffracted," like water waves passing between two barrier islands.)

The secondary ripples from the two slits would crisscross. Where the crest of one wave met the trough of another, they would cancel each other out. This is called "destructive interference." And when crest met crest, or trough met trough, they would reinforce each other; this is "constructive interference."

Along the line of the second wall there would be cases of both constructive and destructive interference, as shown by changes in water level.

Young saw a light interference pattern, with the brightest spots indicating constructive interference and the darkest, destructive.

Later experiments showed that light wasn't merely a wave either and led to the development of quantum theory and the notion that light was made of photons having "wave–particle duality." Fortunately, for the "mirrors" we want to address in this chapter, we just need to understand interference.

Wave Terminology

The distance between successive peaks in a wave is called the wavelength; the time elapsed between the passage of successive peaks is the period; the number of peaks in a unit time period is the frequency.

The repeating unit of the wave is the cycle. At a particular place, the current point in the cycle is called the phase, and is measured in degrees (or radians), and a complete cycle is said to be 360 degrees. Peak and trough are thus 180 degrees (or a half-wavelength) apart.

The different colors of visible light are actually formed by light of different wavelengths (and frequencies), with red light being of longer wavelength (lower frequency) than blue light. Light of a single wavelength (and frequency) is said to be monochromatic. If all of the photons of a monochromatic light beam are also in phase with each other, the light is said to be coherent (a term that will be important when we consider lasers).

Interference and Phase Change

It follows that there is destructive interference when two waves at the same point in space and time are 180 degrees out of phase. A phase difference of 360 degrees is equivalent to no phase change at all, and thus to constructive interference.

Phase changes in two situations. First, there is a progressive change in phase with distance along the path of the light wave, 360 degrees per wavelength unit of distance. To determine the change of phase of a light wave over a particular optical path within a single medium, we need to know its optical path length. This is its geometric path length times the refractive index of the medium. The higher the refractive index, the slower is the speed of light in the medium. The geometric path length will depend on the angle at which the light crosses the medium. We take the optical path length, divide it by the wavelength of the light in a vacuum, and multiply by 360, to get the phase change in degrees.

(The reader may also encounter references to the "optical thickness" of a medium. Its optical thickness is the optical path length in the medium for a particular wavelength of incident light if the light enters the medium at zero angle of incidence.)

Second, there is a 180-degree change in phase when a light wave in a lower refractive index ("faster") medium is reflected from its boundary with a higher refractive index ("slower") medium—for example, when a light wave in the air is reflected off glass. We previously called this an "external reflection." The crest becomes a trough, and vice versa. There is no change of phase when the reflection is at a high/low boundary ("internal reflection"; e.g., glass/air).

Light Waves as Rulers: Interferometers

When monochromatic light from one source, initially in phase, travels on different paths to the same destination, the waves interfere, creating light and dark regions, called fringes. The center fringe, which is the brightest, is called the primary fringe; it is on the most direct light path in the interferometer. The primary fringe is flanked by secondary fringes; the distance between these fringes is related to the geometry of the interferometer and the wavelength of the light. The shorter the wavelength, the shorter the fringe separation.

Fresnel demonstrated that interference occurred even when light did not pass through a narrow slit. He allowed light to fall obliquely onto two adjacent mirrors that were almost, but not quite, in a straight line with each other. The virtual images of the light source, as seen in these mirrors, was at almost, but not quite, the same point, and in the region where the light waves forming these images overlapped, interference occurred.[2] Fresnel's device was the first interferometer.

In Lloyd's mirror interferometer, only one mirror was used. Part of the light stream struck the mirror at nearly a 90-degree angle of incidence (i.e., almost parallel to the mirror surface). This portion was almost completely reflected, falling upon a screen. The remainder of the initial light stream struck the screen directly. The path length for the reflected ray was of course longer than that for the direct ray. Thus, the light stream coming directly from the source was made to interfere with that reflected off the mirror.

The Fabry–Perot interferometer (1897) deserves special mention because of its relevance to laser design. In this interferometer, monochromatic light from a "broad" light source (as opposed to a point source) passes through a field lens, two partially silvered mirrors, and a telescope lens, and finally strikes a screen. Some of the light will just pass through unscathed; part will be reflected multiple times in the air gap between the mirrors before it finally escapes and reaches the screen. Interference will occur between the purely transmitted and the multiply reflected light.

Interferometers with a known geometry can be used to measure exactly the wavelength of a light source. Or a light source of known wavelength can be used in an interferometer to measure something which affects the geometry of the interferometer. For example, light can be split, with one part reflected off an object of known flatness, and the other off an object whose flatness is to be determined. The split beams are recombined, and the resulting interference fringes are examined.

The Michelson–Morley Experiment

An interferometer was used in an experiment designed to verify the existence of the "ether," the mysterious medium in which light waves were supposed to propagate. If the ether drifted past the Earth, the speed of light should depend on the velocity of the measurement apparatus, the interferometer, relative to the ether.

In 1887, Albert Michaelson and Edward Morley set up their interferometer on a stone block, 25 centimeters thick, floating on mercury. ("Horse drawn traffic in the street outside was enough to cause troublesome vibrations and many experimental runs were performed in the early hours of the morning to reduce these."[3])

Michelson's 1887 interferometer had a "cross" arrangement. Imagine that a monochromatic light source is in the west, mirror M1 in the south, mirror M2 in the east, and the observer (with a telescope) in the north. In the center there is a NW–SE oriented half-silvered mirror (beam splitter). Light from the source travels east, striking the beam splitter. Part is reflected south toward mirror M1, and part is transmitted east toward mirror M2. The light directed to M1 is reflected north and passes through the beam splitter, eventually reaching the observer. The light directed to M2 is reflected west, and then reflected again off the beam splitter, traveling north to the observer. (A glass compensator plate is placed in the beam-splitter-to-M2 light path so that this second beam travels through the same distance in glass as the first beam.) The effective length of the light path was about eleven meters.[4] While the interferometer used the "cross" pattern described above, the source and the telescope were close to the beam splitter, so that the interferometer looked like it had two arms, corresponding to the mirrors M1 and M2.

As Earth rotates about its axis, and revolves around the sun, there would be times when the M1 arm was parallel to the movement of the Earth through the ether, and the M2 arm perpendicular to it. At other times, their roles would be reversed. The interferometer could be rotated to hasten the change of roles.

If there was an ether wind, then the light should take less time to traverse the parallel arm than the perpendicular arm, as the light would be traveling "downwind" rather than "crosswind." The speed of light times the time difference is the optical path length difference. If this difference corresponded to an odd multiple of half the wavelength of the light, there would be destructive interference, resulting in a dark region. If it corresponded to an even multiple, there would be constructive interference, resulting in a light region. Light rays traveling off the optical axis would have a longer path length than the axial light rays, and hence there would be a pattern of light and dark bands visible in the telescope of the interferometer. These fringes should have shifted back and forth as the interferometer was rotated

360 degrees, and they should also have shifted slightly in response to the rotation of Earth about its axis and the revolution of the Earth about the sun.

Albert A. Michelson first tried to measure the effect of the ether in 1881. Using light of a wavelength of 600 nm, he expected to see a shift in the interference fringes corresponding to 0.04 times the normal distance between fringes as the interferometer was rotated 360 degrees. His apparatus was capable of detecting a shift of 0.02 fringes; he didn't see it. This implied that there was no "ether" in which light was propagating. Later experiments, with a more accurate interferometer, and one conducted on a high mountain, confirmed these results. Ultimately, he received the Nobel Prize in Physics (1907) for his "failed" experiment.[5]

Perfecting the Mirror

Polished metal is highly reflective, but even freshly deposited silver reflects just 89–97 percent (depending on wavelength) of visible light at normal incidence, and aluminum 91–92 percent under the same conditions. Paradoxically, one can do better with a special combination of nonmetallic materials.

The Distributed Bragg (Multilayer Dielectric) Reflector

In 1939, Turner and Cartwright described "a method of increasing the reflectivity of glass for a selected wavelength of light by depositing alternately thereon thin films of materials of respectively high and low indices of refraction and each having thicknesses of about one quarter wavelength of the light to be reflected."[6]

How does that work? Imagine a multilayer stack of transparent (dielectric) materials, since light absorption is our great foe. In the stack, we will alternate materials of high (H) and low (L) refractive index. Each layer will have a thickness such that the expected "optical path length" within it will be a quarter-wavelength[7] thick—this of course anticipates that the incident light is of a specific wavelength and striking the stack perpendicularly. The stack is actually a coating on glass; the glass provides rigidity.

Usually, since air has a low refractive index (1.0), the topmost layer will be an H layer (e.g., 2.32). Thus, the very first transition is L/H. The stack will have the form (HL)^{m}H; m is called the "period" of the stack. An L layer might be, say, 1.38. The final transition is from the last H layer to the glass substrate (1.52), if any, or air (1.0); this is effectively an H/L boundary.

Let us compare the phase of the reflection at a low index-to-high index boundary (L/H) to that of the returning reflection from the following high index-to-low index boundary (H/L). Let us say that, at the first L/H boundary, the incident light has a phase of 0. Then the light reflected off that boundary immediately has phase 180 (the result of the 180-degree phase change upon "external reflection"). The transmitted light, which is phase 0 at that L/H boundary, travels down to the first H/L

Spectral Reflectivity of Metals (Normal Incidence)

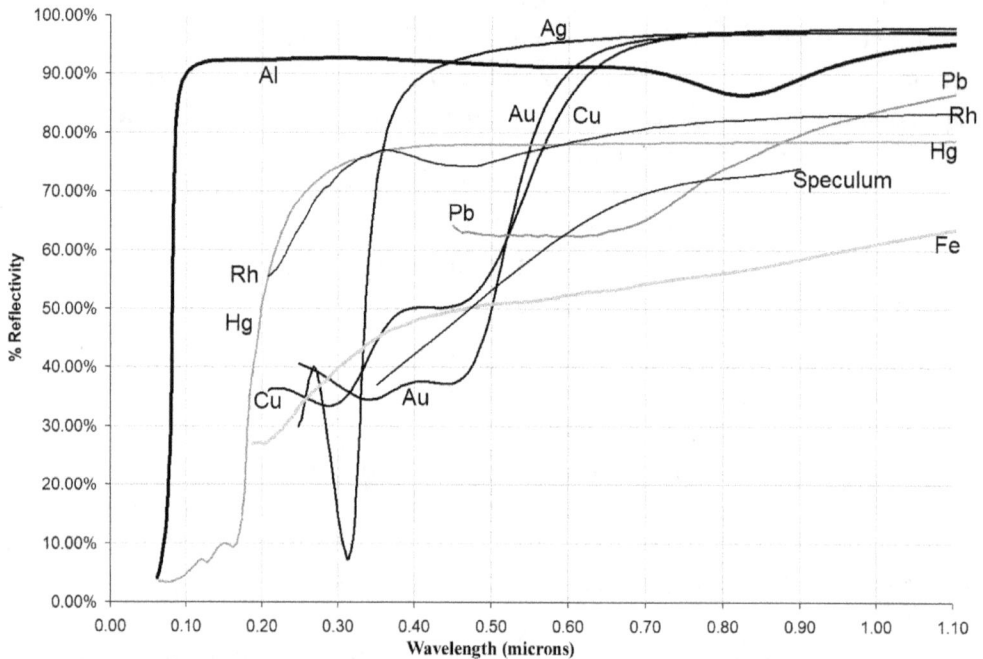

Spectral reflectivity of metals at normal incidence, calculated from complex refractive index data for aluminum (Al), copper (Cu), silver (Ag), gold (Au), iron (Fe), lead (Pb), tin (Sn), mercury (Hg), and beryllium (Be) from nk database of https://refractiveindex.info, and for rhodium (Rh) from Filmetrics, https://www.filmetrics.com/refractive-index-database/Rh/Rhodium. Reflectivity data for speculum alloy are from Nazimudeen, E.A., et al., "Early Compton Effect Experiments Revisited; Evidence for Outstanding Hard X-Ray Reflectivity of Speculum Metal," *ArXiv*, December 16, 2019, https://arxiv.org/abs/2001.02053. The visible range is roughly 0.4–0.7 nm. Less than 0.4 is ultraviolet, and greater than 0.7, infrared.

boundary, in the process increasing in phase to 90. There it is reflected without any change of phase. This second surface reflection travels back to the first L/H boundary, increasing in phase as it does so from 90 to 180. It is thus now in phase with the first surface reflection, so the two interfere constructively. The total reflection is the combination of the first and second surface reflections.

What about the third surface reflection, at the next L/H boundary? The original light has traveled two layers to reach it, so when it gets there, it is at phase 180. The reflection is phase 0. This light, returning to the original surface, again traverses two layers, changing phase to 180. So it is in phase with the first and second surface reflections. By similar arguments, we can show that the fourth, fifth, sixth, and so on surface reflections are all in phase with the first and second surface reflections. Thus, each reflection recaptures a portion of the originally transmitted light, enhancing the reflective power of the stack.

To increase the reflective power of the stack, one can increase the period m, or increase the ratio of the H index to the L index, or both.

A Constructive Interference from Quarter-Wave Stack of Hi and Lo Refractive Index Dielectrics

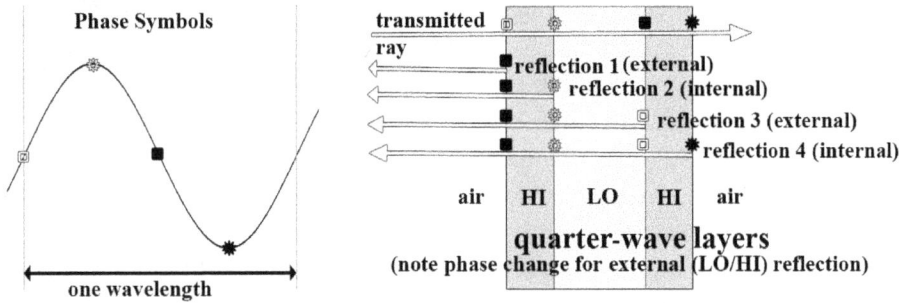

Phase Symbols

one wavelength

transmitted ray

reflection 1 (external)
reflection 2 (internal)
reflection 3 (external)
reflection 4 (internal)

air HI LO HI air

quarter-wave layers
(note phase change for external (LO/HI) reflection)

B Reflectivity, Quarter-Wave Stack (Hi 2.32, Lo 1.38)

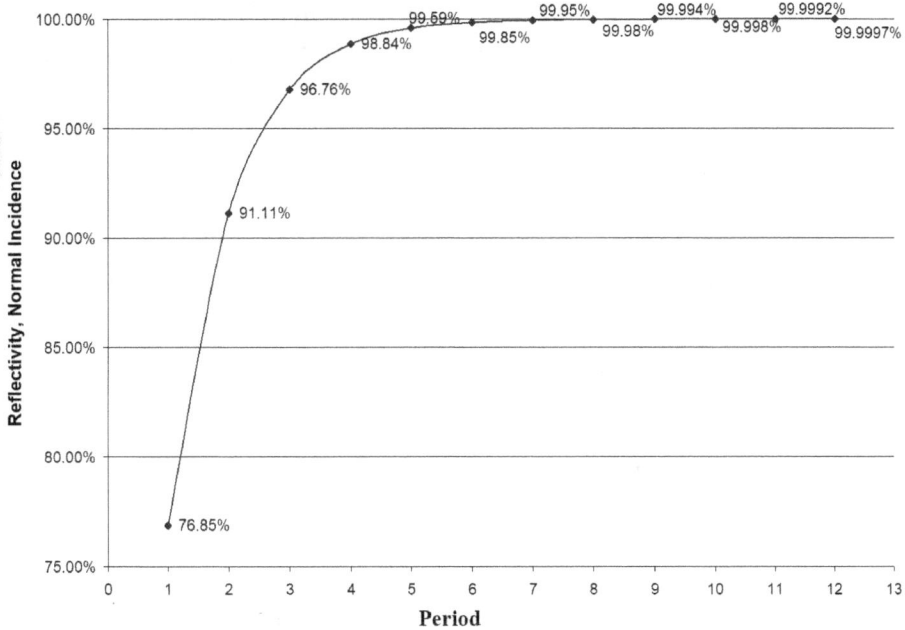

A. Schematic of constructive interference for light normal to multilayer quarter-wave stack. I. Left: Sine wave labeled to show symbols used on right for four phases: 0 (ascending node), white square; 90 (maximum), white star; 180 (descending node), black square; 270 (minimum), black star. Right: Incident/transmitted wave and reflected waves from all four surfaces in a period 1 (HLH) stack. Note that the same phase symbol (black square) appears at the front surface of the stack for all four reflected waves. (Ignore differences in size of arrowheads for reflected rays.) B. Reflectivity of monochromatic light at normal incidence as a function of the period of the stack, with H 2.32 and L 1.38, no substrate. A single H layer would have a reflectivity of 15.8 percent per surface, and a single L layer, 2.5 percent. Compare with the specular reflectivity of metals in the previous figure.

With nine "quarter-wave" layers (period 4), five high-index (2.32) zinc sulfide and four low-index (1.38) magnesium fluoride layers, the reflectivity is 98.84 percent. Increase that to seventeen layers (period 8), and it increases to 99.98 percent. If the low-index material were glass (1.52), the reflectivities would be 98.25 percent and 99.97 percent, respectively.[8]

So what's the catch? With these dielectric mirrors—unlike a metal mirror—the reflective power is very sensitive to the wavelength and angle of incidence of the incident light. They are optimized to reflect at the wavelength and angle of incidence for which the "optical path length" (geometric length times refractive index) through each layer is one-quarter wavelength long.

There will be a wavelength band for which the reflectivity is almost constantly high and typically over 90 percent. Outside that wavelength, there will be oscillation in reflectivity, a series of ever lower peaks. The mirror is said to be "tuned" to the high-reflectivity band.

The choice of the low and high index material will vary depending on the expected incident light wavelength, because, for a given material, both the refractive index and the absorption will vary with wavelength. In addition, the material must be one that can be deposited in a controlled thickness and have good durability (including resistance to the radiation).

Laser line dielectric mirrors are available commercially with advertised reflectivities ranging from 99 percent to 99.999 percent. It all depends on the laser wavelength you are interested in, and what you are willing to pay.

A dielectric mirror with a reflectivity of 99.9999 percent for infrared (1 micron) light was developed for use in a gravitational wave detector, which is essentially an interferometer. The H layer is tantalum pentoxide, and the L layer is silicon dioxide. The 36 layers were deposited by ion-beam sputtering, in controlled thicknesses.[9]

Dichroic Mirrors

A dichroic mirror is one designed to strongly reflect one wavelength band and to be essentially transparent to a second. This may be implemented with a multilayer dielectric.

Hot and cold mirrors are dichroic mirrors used for heat control. The hot mirror reflects infrared and transmits visible light. The cold mirror is the reverse.

Projectors often generate a lot of heat, and cold mirrors help to control this. For example, there may be a "cold mirror coating on the lamp reflector." The infrared radiation passes through the coating and is "absorbed in a heat sink."[10]

DIELECTRIC "SOLAR" SAILS

Because of their wavelength specificity, dielectric solar sails are best used with laser rather than solar propulsion. Each additional quarter-wave thickness layer increase reflectivity, but "the ratio of reflectance to mass is maximized for single layers of quarter wavelength material." Assuming 500-nanometer laser light, Landis proposed a single layer of diamond (figure of merit 10.3; reflectivity 50 percent), silicon carbide (17.4; 56%), or zirconia (34.7; 42%); the figure of merit reflects the maximum acceleration possible, relative to that of an aluminum sail (1).[11]

One complication is that the wavelength perceived by the sail will be Doppler-shifted, depending on its speed. This is an issue only for interstellar propulsion, where the speeds reached will ultimately be a significant fraction of the speed of light. By then, one may hope for the ability to vary the wavelength of the laser as the starship speeds up.[12]

Hybrid Dielectric–Metal Mirrors

Turner's 1950 patent (filed, 1946) proposed a hybrid mirror, with a "tuned" dielectric stack as a coating on a metal mirror. This both protected the metal and increased the mirror's reflectivity. For example, Turner "increased the white light reflectivity of a polished aluminum surface from about 92 percent to about 96 percent by applying a pair of quarter wave, low index-high index films formed, respectively, of cryolite and zinc sulphide. By applying a second superposed pair of these films the white light reflectivity is further increased to about 99 percent."[13]

Directionality

If a multilayer mirror designed for use at normal (zero) incidence is tilted, "the reflection features are shifted toward shorter wavelengths." This is because the phase difference at the surface of the stack between beams reflected from different layer interfaces is reduced as the angle of incidence is increased and this "can be compensated by using a shorter wavelength of light."[14]

In 1997, MIT researchers calculated that a Bragg reflector could be omnidirectionally reflective for a band of frequencies, and a prototype mirror was completed in February 1998.[15] "Yoel Fink, a graduate student in MIT's material science and engineering department, deposited a nine-layer stack of alternating layers [each about 1 micron thick] of polystyrene and tellurium on a sodium chloride substrate. Calculations predicted a reflectivity of nearly 100 percent for all angles from a wavelength of 10 to 13 μm, and experiments confirmed the prediction."[16] Ironically, the leader of the team, Dr. John D. Joannopoulos, had "proven" that an omnidirectional dielectric mirror was impossible in his 1995 textbook on photonic crystals.[17] While the MIT polystyrene–tellurium reflector was tuned to reflect infrared light, a visible/near-infrared light (600–800 nanometers) reflector has been made by the same group. It used 19 bilayers. Each bilayer was a 80-nanometer-thick layer of tin(IV) sulfide (refractive index 2.6) and a 115-nanometer-thick layer of silica (refractive index 1.46).[18] The mirror has a golden appearance to the naked eye.

Two-Dimensional Photonic Crystals

"A crystal is a periodic arrangement of atoms or molecules.… The optical analogy is the photonic crystal, in which the periodic 'potential' is due to a lattice of macroscopic dielectric media." The multilayer stack is a photonic crystal in which the periodic variation in refractive index occurs in only one dimension (down-stack).[19]

A two-dimensional photonic crystal can be created by making vertical holes in a horizontal dielectric layer, or in a multilayer dielectric stack. A laser incorporating such a crystal (an etched gallium arsenide) as a laser mirror was reported in 1996.[20]

Lasers

At night in the city, neon signs are everywhere. We may think about the message they tell us, but do we think about the medium which conveys it? Within the glass tubing are atoms which have been given a jolt. They stay at a higher energy level for a while, and then, at random times, drop back down, spontaneously emitting much of the energy as light. The light is emitted in all directions, and without any coordination of the peaks and troughs of the electromagnetic waves.

The purpose of the laser is to excite atoms in a controlled way, so that they emit light in one direction and in phase with each other. Inside the laser cavity is a material—ruby crystals were used in the earliest lasers—with atoms for which the electronic energy levels are such that they can be excited to emit light of the desired wavelength. These atoms are bombarded with photons of appropriate energy. The excited atoms emit photons of their own. This is what we call *stimulated emission of radiation*. The emitted photons strike other atoms and excite them in turn. The result is a chain reaction, an *amplification* of that emission.

In 1924, the Indian physicist S.N. Bose described light as a "gas" of photons. He predicted this gas to have peculiar properties; specifically, the photons tend to play "follow the leader." If one enters a particular state, others are likely to do the same. Albert Einstein speculated that if many atoms in a material are in an excited state—thus destined to re-emit the energy—then a passing photon can stimulate nearby atoms to emit their photons early. And these are emitted with the same wavelength (energy), phase, and direction of motion as the original photon. This meant that it was possible to have a cascade of emissions.[21]

Masers, the forerunners to lasers, used microwave radiation. Ammonia molecules were manipulated by electric fields so that only those in a higher energy state would pass into a cavity. Microwave radiation of the appropriate resonant frequency was beamed into the cavity, stimulating emission of the pent-up energy. The process was termed "Microwave Amplification by Stimulated Emission of Radiation," hence the name. The first maser was built in 1954. At the time, there was considerable doubt as to the cost-effectiveness of maser research, leading to the alternative acronym, "Means of Acquiring Support for Expensive Research."[22]

In 1957, Gordon Gould, then a graduate student at Columbia University, wrote in a notebook, "Some rough calculations on the feasibility of a LASER: Light Amplification by Stimulated Emission of Radiation." However, he did not apply for a patent until 1959.

In the meantime, in December 1958, Arthur L. Schawlow and Charles H. Townes published a paper titled "Infrared and Optical Masers." The first working laser, inspired by this article, was built by Theodore H. Maiman of Hughes Aircraft, and demonstrated in 1960.[23]

All of this activity resulted in a series of epic legal battles; Gould's patents took twenty years to issue.

Lasers usually have three components: the lasing medium, the energy source, and a resonator.

The lasing medium may be a solid, a liquid, or a gas; it must be possible to pump most of its atoms up to a higher energy state and to drop them back down. Usually the media have three or four energy levels.

The ideal elevated energy level is what is called "metastable." It is a state which the atom would drop down from spontaneously, given a long enough time, but which it can be kicked out of easily. A homely analogy would be a car stuck in a ditch on a hill. Once you get it over the lip of the ditch, it can roll down the hill easily.

The first lasing medium was a ruby. Its output wavelength was 694.1 nm, which is a deep red.

The energy source is usually an AC or DC electrical discharge, a high-energy xenon flash lamp, or a chemical reaction. Its purpose is to pump the atoms up to the higher energy state.

Finally, the resonator is a long cylindrical cavity, with mirrors at either end. Mirrors play an important part in the lasing process. The initially emitted photons are spontaneously emitted and hence are radiated out in all directions. The chance is extremely small that a photon emitted perpendicularly to the major axis of the cavity will strike an excited atom and thereby start a cascade. But what about a photon heading down the fairway, toward a mirror? If it doesn't strike an excited atom along the way, it gets a second chance after it is reflected. And a third chance when it is reflected off the second mirror. And so on.

Only a few photons head in the right direction initially. However, the photons which are repeatedly reflected are necessarily all traveling up or down the tube. Once enough of them stimulate parallel emissions, it is like starting an avalanche; you won't have to complain that the result is unimpressive.

Wait, you ask. If there are mirrors at both ends, how does the light escape? Usually, one of the mirrors is slightly more reflective than the other.

Around 1960, Ali Javan built the first continuously operating laser, with a helium–neon mixture as the lasing medium. Its plane mirrors each consisted of "13 layers of dielectric films. Reflectivity was 98.9 percent between 1.2 and 1.2μm [infrared wavelengths]."[24]

Modern "laser line" mirrors are usually multilayer dielectric mirrors. Such a mirror might have a reflectivity of 99.99 percent or better. Reflectivity is important, because on a single pass down the laser cavity, the increase in the number of photons might be just 5 percent. So you don't want to lose 5 to 10 percent each pass to absorption by a metal mirror.

The mirrors may both be perfectly flat, as in Javan's laser, or one or both may be concave with a typical radius (R) = 2 × focal length equal to the length of the cavity (L). The latter is a configuration called "confocal." Concave mirrors are easier to align. A resonant cavity terminated by curved mirrors was first used in a laser in 1962.[25]

The excited atoms have only a limited set of energy levels. Hence, the possible wavelengths of the emitted photons are limited to those corresponding to the energy difference between the current energy level and a lower energy level. The levels are not perfectly sharp, and hence the emitted light is not perfectly monochromatic (one color, hence one wavelength). The monochromaticity of the laser is enhanced if the mirrors are strongly reflective at the emission wavelength and not at other wavelengths. Then photons which are not of the right wavelength will be lost, even if they strike the mirrors.

It also helps that the cavity have the right length, and hence the right resonance. For it to resonate strongly, a standing wave pattern should be generated. That means that length of the cavity must correspond to an integral number of half-wavelengths (of the desired wavelength of emission).

The emerging beam is highly monochromatic (one wavelength) and coherent (in-phase) and usually collimated (parallel rays).[26]

To make a hologram, we need a laser and a partially reflecting mirror. The laser beam is split into two parts by the mirror. The reference beam is directed onto a photographic plate. The object beam is scattered off the object, illuminating the plate indirectly. The two beams create an interference pattern, which is recorded on the plate.

To reconstruct the image, we duplicate the reference beam and direct it onto the hologram. Light is scattered off the hologram and forms a view of the original object.[27]

Reflection of Polarized Light

I briefly referred to polarization in earlier chapters. Since light is a transverse wave, the direction of undulation is perpendicular to the direction of propagation. A light beam is made of many waves, and if the undulations are in every which direction, without any preference, the light is unpolarized. If they are all in the same direction, it is polarized.

Reflection can result in the partial or complete polarization of light. The light wave is an electromagnetic wave; there is a periodic fluctuation in the magnitude and direction of the electric and magnetic fields of force in time and space. (A field of force, by the way, is a shorthand description of the direction and magnitude of the force that would be felt by a susceptible particle, such as an electron, at each point in space.) The electric and magnetic fields have a directionality at right angles to each other as well as to the direction of propagation. By convention, the polarization is expressed as the direction of the electric force. If that direction is compared to the plane of incidence (the plane containing the incident and the reflected or refracted rays), the light is p-polarized if that direction is parallel to the plane, and s-polarized (from *senkrecht*, German for perpendicular) if perpendicular to it.

For light falling on a dielectric material (e.g., glass, water), the variation with angle of incidence of the reflection of p- and s-polarized light is quite substantial. As the angle is increased, the reflection of p-polarized light decreases, reaching zero at an angle called the Brewster angle. Beyond that angle, it increases sharply. In contrast, the reflection of s-polarized light increases with angle of incidence, first slowly and then more quickly. Metal mirrors are less sensitive to polarization state.

I previously alluded to von Helmholtz's choice of a particular angle for the plain glass reflectors in his ophthalmoscope. He recognized that there would be a reflection from the cornea that would tend to "wash out" the reflection from the central region of the retina. The reflection from the glass and from the cornea was specular and therefore polarized; the reflection from the retina was diffuse and therefore depolarized the light striking it. It was therefore possible to suppress the reflection from the cornea by inclining the plates so the angle of incidence would be Brewster's angle, hence only "s-polarized" light is reflected. For glass with a refractive index of 1.5, it is about 56 degrees. Hence, if the glass is at that angle, and the physician looks through a filter that only admits p-polarized light, only the diffuse light from the retina will be seen, not the s-polarized light from the cornea.[28]

With multilayer dielectric mirrors, as with other dielectric mirrors, if the angle of incidence is anything other than zero, p- and s-polarized light will be reflected to different degrees. If the incident light is unpolarized, this will result in a decrease in total reflectivity as the angle of incidence approaches Brewster's angle. "As a result, a multilayer interference mirror that is designed to have a 1 percent loss for reflection of p-polarized light (99% reflectivity) at normal incidence can have many times that loss at high incidence angles."[29]

Weber was able to overcome this problem using "highly birefringent polymers," namely, birefringent polyester (polyethylene naphthalate) and PMMA (polymethylmethacrylate). Birefringent materials have different refractive indexes depending on the direction in which light propagates through them. By appropriate orientation of the birefringence, his mirrors maintained near 100 percent reflectivity for visible p-polarized light as the incidence angle was increased from 0 to 45 degrees.

Multilayer stacks need not be simple repeated bilayer designs. Shen described a six-stack composite system, with each component stack tuned to a different visible light wavelength, which was "transparent (up to 98%) to p-polarized light at" an incidence angle of 55 degrees, but "behaves like a mirror at all other incident angles over the entire visible spectrum.... For s-polarized light, the sample behaves like a mirror at all angles."[30]

LCD and LCoS Digital Projectors

At the heart of a conventional LCD (liquid crystal display) projector lies a module consisting of a vertical polarizing filter, a negative electrode layer, a liquid crystal layer, a positive electrode layer, and a horizontal polarizing filter. A vertical polarizing filter only passes light vibrating in the vertical plane, and a horizontal one is analogously defined. A liquid crystal is capable of rotating the plane of polarization. If no voltage is applied to the electrodes, no light will pass through the module. At the highest voltage, essentially all the light will be transmitted. Thus, the LCD projector can display any grayscale value.[31]

In an LCoS (liquid crystal on silicon) module, light passes through a vertical polarizing filter and a liquid crystal layer and strikes a reflective coating. The reflected light traverses the liquid crystal layer and then a horizontal polarizing filter. Again, grayscale display is possible.[32]

Appendix

Technical Note on Archimedes and Burning Mirrors

This appendix is pitched at a higher technical level than the main text but is included for the benefit of readers who would appreciate a more sophisticated treatment.

To set the wood of a Roman warship on fire, Archimedes's mirror would need to deliver enough heat energy to the wood to raise its temperature to its ignition point and keep it there long enough for ignition to occur, despite the simultaneous dissipation of the heat by conduction and radiation.

Nominal Delivered Energy Flux

The intensity of solar radiation is dependent on time of year, time of day, latitude and cloud cover. The "maximum solar irradiance across a 1m² flat surface at ground level ... at an equatorial location on a clear day around solar noon ... is around 1000 watts."[1]

We do not know the time of year in which Marcellus launched his assault on Syracuse (latitude 37°N). I do not have instantaneous solar irradiance data for Syracuse, but I do for Athens (latitude 38°N). The average peak is in July around 2 p.m., about 900 watts per square meter. It's only about half that at 10:30 a.m. or 6:30 p.m. In December, the average peak is under 400 watts per square meter.[2]

On the other hand, the array is arranged to provide a concentration of the light. Let us define the rim angle of a dish mirror as the angle between the axis of symmetry of the mirror and a line connecting the focal point to the rim of the mirror. If the ratio of the radius to the focal length is small (i.e., the dish is shallow), the tangent of the rim angle approximates that ratio.[3]

The rim angle of a trough mirror is analogously defined.

For thermodynamic reasons, the maximum solar concentration obtainable with a concentrator having radial symmetry (3D, dish) is sin²(rim angle)/sin²(half solar angular size). The maximum concentration is at a rim angle of 90 degrees and since the angular size of the sun is about 0.5 degree, the concentration ratio is then 46,250. For a two-dimensional (2D, trough) concentrator, the maximum possible concentration is sin (rim angle)/sin (half solar angular size) and thus has a maximum value of just 215-fold.[4]

For a paraboloid dish, the maximum solar concentration is 11,600, one-quarter of its thermodynamic limit.[5] And for a parabolic trough, it is 108, one-half its thermodynamic limit.[6]

A spherical dish, receiving parallel rays, does not bring them to the same focus, and so it provides a lower concentration ratio. The theoretical limit is $\sin^2$ (rim angle/sin [solar angular radius]), and for a 90-degree rim angle, it is 115.[7] (I do not have a formula for the concentration ratio yielded by a circular trough, but it will certainly be less than for a parabolic trough.)

The distance of engagement is of critical importance. The focal length must equal the range, and the greater the focal length, the smaller the rim angle and thus the smaller the energy concentration. So how far away were the Roman ships when Archimedes attacked them? "Kircher took so much interest in the subject, that he went to Syracuse expressly to inquire into the probable position of Marcellus' fleet, and he arrived at the conclusion that it might have been within thirty yards of the walls."[8]

A 10-foot-diameter paraboloid dish mirror, with a focal length of 90 feet, would have a rim angle of about 6.37 degrees, and provide a solar concentration of about 36-fold, onto, ideally, a circular spot on the target.

A parabolic trough mirror with the same rim angle and focal length would only provide a solar concentration of sixfold, onto, ideally, a rectangular spot on the target.

These concentrations would be realized with a simple mirror only if the sun was directly behind the Roman ship; otherwise, the sunlight would not be coming in parallel to the mirror's optical axis and the solar image projected onto the enemy ship would be smeared out. With a mirror composed of independently oriented segments, an adjustment may be made for the misalignment. However, per Hunt, approximating a paraboloid dish or of a parabolic trough with planar segments would result in a reduced solar concentration.

Initial Energy Losses

1. The mirror must be aimed at a point halfway between the sun and the ship. Even if the sun were overhead, that would mean an angle of incidence slightly more than 45 degrees (since the ship is below the level of the mirror). The effective collecting area of the mirror is smaller (multiply the nominal area by cosine [theta], where theta is the angle of the solar rays to the axis). Cosine (45°) is about 0.71.

   The latitude of Syracuse, the city defended by Archimedes, and of Salamis, the site of the Sakkas experiment, is 37°N. That is above the Tropic of Cancer, so the sun is never overhead. The longitudes are 15°E and 23°E, respectively. Entering the Syracuse latitude and longitude into a solar position calculator for June 21 (when the sun would be closest to overhead in that region), I found the sun was at an elevation of 76.43 degrees. Doing the same for Salamis on November 6, 1973, I got a mere 35.79 degrees. With the target at 0 degrees, the mirror tilt from horizontal facing would be half the solar elevation.[9] The "cosine factor" (accounting for the spreading of the sun's rays over an oblique surface) would be reduced from 0.71 to 0.31. So the time to ignition would be more than double that attained with the sun overhead.

2. Archimedes would have used solid copper or bronze mirrors. The reflectivity of an ancient mirror would likely be less than that of a modern mirror.

3. In addition, the reflected rays probably aren't striking either the deck or the side of the ship perpendicularly. The hull is curved, and the mirrors are on the top of a wall, higher than the hull. So there would be a second correction for the spreading of the reflected light.

4. The wood would not absorb all of the light energy; some would be reflected. Light reflectance values vary widely, depending on the wood species: 4.1–41.9 percent for oak, and 47.8 percent for American pine.[10] In addition, reflectance will be altered by weathering and the presence of pitch, algal deposits, and water spray.

Ignition Temperature

For a burning mirror to be effective, it must concentrate sunlight sufficiently to raise the temperature of the target to one where it is ignited and maintain that temperature for a sufficient period of time. If the target cannot be immediately ignited when it comes into a fixed range, then it must further be possible to adjust the focal length of the mirror to keep the target in focus.

The final temperature must be the flash point of the target. As Hunt mentions, the flash point of paper is 451°F (233°C). However, it would be more relevant to look at the flash point for wood (572°F; 300°C).[11] Moreover, the wood of a sailing ship was treated with tar to make it waterproof. This treatment would have rendered it more flammable. The flash point of pine tar is variously reported as about 60–150°C;[12] a maritime history source says about 120°C.[13]

The burning mirror would be most effective in summer. The hottest month is August, with an average daily maximum of 29.8°C. The coldest is January, 9°C.[14] The wood in August would probably be a little cooler than the air, thanks to sea spray, but let's assume 30°C.

Heat Requirement

The absorption of heat raises the temperature of the wood. The specific heat capacity is the quantum of heat energy needed to raise the temperature of a unit mass of wood by one degree. For the range T = 0–100°C, with m = 0–27 percent moisture content, Volbehr (1896) proposed that it equaled 0.2590 + 0.000975m + 0.000605T + 0.000015mT (cal/g-°C). We multiply by 4.184 for kJ/kg-°C.[15] For Douglas fir, the heat capacity increases linearly up to about 275°C and then peaks at 300°C; red oak behaves similarly.[16] Hence, given all the other approximations necessary, I used Volbehr's equation to calculate the heat capacities of dry wood (the best case scenario for Archimedes) at 30°C and 300°C, and take their average value as the average heat capacity over that temperature range: 1.5 kJ/kg-°C.

Calculating the affected mass of wood is extremely difficult. The target is likely to be obliquely illuminated, so the irradiated area (if a flat surface) is elliptical. From there, the heat diffuses through the wood by conduction, and the heat distribution is nonuniform. Wood has a low heat conductivity, only about 0.04–0.12 W/m-K,[17] so conduction will be slow. If the wood has a high moisture content (conductivity 0.6), the heat will be transferred more quickly.

Let's say that the affected volume is a cylinder two feet in diameter and five inches deep; the volume is 0.001 m³. Wood density varies by species: Lebanese cedar (580 kg/m³); Corsican pine (510); English brown oak (740). If for convenience we take 500, the mass is 5 kg. And then it requires 2027 kJ (not counting heat losses) to take the indicated volume of dry wood from 30°C to 300°C.

Heat Losses

These reduce the temperature and thus delay or prevent ignition. Once the heat reached the submerged portion of the hull, there would be heat loss from the hull to the water by conduction.

The irradiated wood would also shed heat by radiation (to the air). The net rate of radiative heat loss by r is equal to the product of the Stefan–Boltzmann constant, the wood's emissivity (0.9 for "planed wood"[18]), the surface area, and the difference between the fourth powers of the wood temperature and the air temperature (K). For example, if the wood is at 300°C (573 K) and the air at 30°C (303 K), the radiative heat loss would be 5071 W/m². If the wood were at 120°C (393 K), it would instead be 787 W/m².

In this author's opinion, the calculated radiative heat loss is the nail in the coffin for the feasibility of an Archimedes-type incendiary beam at a target 90 feet away. It is unlikely that with a parabolic trough array to collect and direct the solar radiation, the irradiance of the enemy ship would ever reach 5071 W/m². Hence, the ignition temperature could not be maintained.

Chapter Notes

Chapter 1

1. Byron, George Gordon, "Childe Harold's Pilgrimage, Canto 4," Verse CLXXIII, http://web.archive.org/web/20051102234909/http://www.anglistyka.uw.edu.pl/literature/british/byron4.htm.

2. Frazer, The Golden Bough, "I. The King of the Wood," "1. Diana and Verbius" (1922) http://www.bartleby.com/196/pages/page1.html.

3. "Lake Nemi," https://www.themystica.com/-lake-nemi, citing Grimassi, Raven, Encyclopedia of Wicca & Witchcraft 213–14 (St. Paul, MN, Llewellyn Worldwide, 2000) {ISBN 1-56718-257-7}. For a photograph of the Moon in the lake water on Aug. 21, 2002, at about 4 a.m., see http://web.archive.org/web/20160128131711/http://www.digiter.it/albaen.htm.

4. For up-to-date information, see https://edits.nationalmap.gov/apps/gaz-domestic/public/search/names.

5. DeLaine, Linda S., "Creation of the National Mall," http://web.archive.org/web/20010620195912/http://dc.about.com/citiestowns/midlanticus/dc/library/weekly/aa051401b.htm.

6. "San Jacinto Monument and Museum," Handbook of Texas Online, http://web.archive.org/web/20050421192705/http://www.tsha.utexas.edu/handbook/online/articles/view/SS/lbs1.html.

7. Marvin Trachtenberg and Isabelle Hyman, Architecture: from Prehistory to Post-Modernism (H.N. Abrams, 1986), 223.

8. "Taj Mahal," Great Buildings Online, http://web.archive.org/web/20230715233406/http://www.greatbuildings.com/buildings/Taj_Mahal.html.

9. "Athens, Parthenon (Building)," Perseus Digital Library http://www.perseus.tufts.edu/hopper/artifact?name=Athens,+Parthenon&object=Building&redirect=true.

10. "Chinese Bronze Mirrors," http://web.archive.org/web/20131016150003/http://www.asiawind.com/antiques/mirrors.htm.

11. Li, Xueqin, The Wonder of Chinese Bronzes (Foreign Languages Press, Beijing, China, 1980), 20; for a side-view illustration, see page 19, Fig. 14(4).

12. Eaton, Elizabeth S., "A Group of Middle Kingdom Jewelry," Bull. Mus. Fine Arts, 32: 94, 98 (Dec. 1941).

13. Sinemoglu, Nermin, "The Passion Man Could Not Give Up for Thousands of Years: Mirrors and their Frames," Republic of Turkey, http://web.archive.org/web/20041024194705/http://www.mfa.gov.tr/grupc/cj/cja/mirrors.htm. See also "Catal Huyuk," http://web.archive.org/web/20021003214543/http://emuseum.mnsu.edu/offices/alpha/classes/book/farming/catalhuyuk.html.

14. Wilson, Hilary, "Mirrors in Ancient Egypt," The Past (2024); https://the-past.com/feature/-hilary-wilson-on-mirrors-in-ancient-egypt.

15. Caley, Earle R., and John F.C. Richardson, Theophrastus on Stones (Ohio State University, 1956), 52.

16. Caley and Richardson, 131.

17. Thoresen, Lisbet, "Archaeogemmology and Ancient Literary Sources on Gems and Their Origins," Gemstones in the First Millennium AD: Mines, Trade, Workshops and Symbolism (October 20–22, 2015, Mainz, Germany).

18. The Natural History of Pliny, Book XXXVI, Chap. 16, p. 409, Vol. VI, transl. John Bostock (Henry G. Bohn, 1857).

19. Engel, Frédéric André, An Ancient World Preserved: Relics and Records of Prehistory in the Andes (Crown Publishers, 1976), 105–6.

20. Lunazzi, Joseph, "Olmec Mirrors: An Example of Archaeological American Mirrors," arXiv (March 2007), https://www.researchgate.net/publication/2178322_Olmec_mirrors_an_example_of_archaeological_American_mirrors.

21. Lunazzi, 7; Miller, Mary, and Karl Taube, The Gods and Symbols of Ancient Mexico and the Maya (Thames and Hudson, 1993), 114.

22. International Gem Society, "Table of Refractive Indices and Double Refraction of Selected Gems" (2024), https://www.gemsociety.org/article/-table-refractive-index-double-refraction-gems.

23. Lunazzi, supra.

24. Corry, Charles E., "Physical Properties of Ore Minerals," Table 3: Hexagonal Crystal Class–Trigonal Subsystem, http://www.zonge.com/physical-property-lab-services/ore-minerals-physical-properties/prope_12.htm#HEADING12-0.

25. Healy, Paul F., and Marc G. Blainey, "Ancient Maya Mosaic Mirrors: Function, Symbolism, and Meaning," Ancient Mesoamerica, 22(2): 229–44 (2011).

26. Kerr Maya Vase Archives, Vase 625.00, photographed by Justin Kerr, http://research.mayavase.com/kerrmaya_list.php?_allSearch=&hold_search=&x=26&y=4&vase_number=625&date_added=&ms_number=&site=.

27. Kerr Maya Vase Archives, Vase 505.00, photographed by Justin Kerr, http://research.mayavase.com/kerrmaya_list.php?_allSearch=&hold_search=&x=26&y=4&vase_number=505&date_added=&ms_number=&site=; see also The Maya Vase Book Volume 1.

28. Miller, 125.

29. Pellant, Chris, *Rocks and Minerals,* Eyewitness Handbooks (Dorling Kindersley, 1992), 197; Busbey, Arthur III, et al., *Rocks and Fossils,* The Nature Company Guides (Time-Life Books, 1996), 140.

30. Oregon Department of Geology and Mineral Resources, "Gems and Minerals in Oregon: Obsidian," http://web.archive.org/web/20051118153453/http://www.oregongeology.com/learnmore/Obsidian.htm.

31. James, Peter, and Nick Thorpe, *Ancient Inventions* (Ballantine, 1994), 248.

32. Vedder, James, "Grinding It Out: Making a Mirror the Old-Fashioned Way," Online News, Archaeology, April 2, 2001, http://web.archive.org/web/20121119000724/http://www.archaeology.org/online/news/mirrors.html.

33. Miller, Mary, and Karl Taube, *The Gods and Symbols of Ancient Mexico and the Maya* (Thames and Hudson, 1993), 125.

34. Stocker, Terry, "A Technological Mystery Resolved," Corning Museum of Glass, http://web.archive.org/web/20040214234538/http://www.cmog.org/page.cfm?page=278.

35. "Obsidian Mirror Disk • A Selection from DUMA," *Duke Magazine* (September/October 2001), http://web.archive.org/web/20021129113621/http://www.adm.duke.edu/dukemag/issues/091001/depgall.html.

36. Campbell, Stuart, et al., "The Mirror, the Magus and More: Reflections on John Dee's Obsidian Mirror," *Antiquity*, 2021, https://doi.org/10.15184/aqy.2021.132.

37. Luke, Mary M., *Gloriana: The Years of Elizabeth I* (Coward, McCann & Geoghegan, 1973), 305–6.

38. Londesborough, (Baron) Albert Denison, *Miscellanea Graphica: Representations of Ancient, Medieval, and Renaissance Remains in the Possession of Lord Londesborough* (Chapman and Hall, 1857), 82.

39. Goldberg, Benjamin, *The Mirror and Man* (University of Virginia Press, 1985), 17–18. However, *Brewer's Dictionary of Phrase and Fable* says that it was a pink-tinted glass about the size of an orange. It is possible that Brewer is confusing the mirror with Dee's scrying globe of rock crystal. Both are now in the British Museum.

40. Goldberg, 18.

41. British Museum, "magical mirror; mirror-case," 1966.1001.1, https://www.britishmuseum.

org/collection/object/H_1966-1001-1; Campbell, supra.

42. Association of Belizian Archaeology, "Maya Archaeological Sites in Belize," http://www.ambergriscaye.com/pages/mayan/mayasites.html; Royal Ontario Museum, "Excavations of Altun Ha," http://web.archive.org/web/20050207231942/http://www.rom.on.ca/digs/belize/altun-ha.html.

43. NOVA Online, "Search for the Lost Cave People, Archaeologist Thomas Lee," http://www.pbs.org/wgbh/nova/laventa/lee.html.

44. Hellmuth, Nicholas M., "Tomb of the Jade Jaguar," http://web.archive.org/web/20040405121717/http://www.maya-art-books.org/html/jade_lecture.html.

45. For the latter, see Aguilar, Manuel, "Atlantes and Sculptured Columns," http://web.archive.org/web/20081203010332/http://instructional1.calstatela.edu/bevans/Art446-11-ToltecTula/WebPage-Info.00007.html.

46. Miller, 114.

47. Schuster, Angela M.H., "New Tomb Found at Teotihuacan," *Archaeology*, December 4, 1998; updated March 2, 1999; http://web.archive.org/web/20210422173725/https://archive.archaeology.org/online/features/mexico.

48. Valliant, George Clapp, *The Aztecs of Mexico*, 116, quoted in "Magic Mirrors, Time Cameras, and Catoptromancy: Part Three," http://web.archive.org/web/20091027165244/http://www.geocities.com/Athens/Olympus/6581/chronos10c.html. Marcasite is mentioned; it is a form of pyrite.

49. Sugiyama, Saburo, "Offerings at the Feathered Serpent Pyramid: Stone Disks" (August 20. 2001), Arizona State University, http://web.archive.org/web/20130607064347/http://archaeology.la.asu.edu/teo/fsp/Offer/OfStone.htm.

50. Aguilar, supra.

51. Gallaga, Emiliano, "How to Make a Pyrite Mirror: An Experimental Archaeology Project," in Gallaga, Emiliano, and Marc G. Blainey, *Manufactured Light: Mirrors in the Mesoamerican Realm* (University Press of Colorado, 2016), chapter 2.

52. Gallaga, supra.

53. To see how well hematite can form a reflected image, see McMeekin, Ann, "My Desk," The Mirror Project (July 2001), http://web.archive.org/web/20021102011054/http://www.mirrorproject.com/mirror/?id=1362. Or see the 5-centimeter hematite mirror from the Pedra Preto Pit, Brumado, Bahia, Brazil, in Alan Guisewhite's Mineral Collection Images: Miscellaneous Systematic Oxides Page, Row 11, Column 1, http://www-2.cs.cmu.edu/~adg/images/minerals/o/hematite5_sm.jpg.

54. Reed, Richard, review (August 2002) of Gibson, Eric, *The Nine Lords of the Night*, http://web.archive.org/web/20070813140319/http://www.ninelords.com/reviews.asp.

55. Hoopes, John W., "Chronological Periods, Ancient American Civilizations, Mesoamerica" (1998), http://web.archive.org/web/20060522175858/http://www.ku.edu/~hoopes/506/Chronology.htm.

56. "Structure 6F-3," Las Ruinas Antiguas de Yaxuná, http://web.archive.org/web/20020224121016/http://tesla.csuhayward.edu/sacredplaces/yaxuna/vr3.6f3.html.

57. Peabody Museum, "Technical Examination of a Mica Cutout" (1998), http://web.archive.org/web/20090326144255/http://www.peabody.harvard.edu/conservation/mica.html.

58. Riddle, James, "Prehistoric Heliographs?," http://web.archive.org/web/20020804093236/http://www.cableone.net/kd7aoi/the.htm.

59. "Gujurat Craft Traditions," India-Craft http://www.india-crafts.com/textile_products/gujaratculture.html.

60. Shorten, Judy, "Temari 'Faberge' Eggs," http://web.archive.org/web/20040201005523/http://www.geocities.com/jshorten_934/temari/2000/apr00-6.htm.

61. Freed, Rita, "The Nubian Collection of the Museum of Fine Arts, Boston," *Mitteilungen der Sudanarchaeologischen Gesellschaft zu Berlin E.V.* (July 1995), 32.

62. Education Development Center, Nubia, "C. The Emergence of the State: The A-Group and Pre-Kerma Periods: 3500–2500 B.C.," http://web.archive.org/web/20130528211511/http://www.nubianet.org/about/about_history3.html.

63. *The Natural History of Pliny*, Book XXXVI, Chap. 45, 369.

64. Id.

65. Purdue, "Petroleum and Coal," http://chemed.chem.purdue.edu/genchem/topicreview/bp/1organic/coal.html.

66. Arem, Joel E., *Gems and Jewelry* (Geoscience Press, 2nd ed., 1992), 95–96; Pellant, Chris, *Rocks and Minerals,* Eyewitness Handbooks (Dorling Kindersley, 1992), 245.

67. Whitby Museum, "FAQ About Jet," http://web.archive.org/web/20060311104702/http://www.durain.demon.co.uk/jet.

68. Engel, Frédéric André, *An Ancient World Preserved: Relics and Records of Prehistory in the Andes* (Crown, 1976), 33.

69. Rowe, John Howland, "Form and Meaning in Chavin Art," Berkeley (2001, March 16), http://web.archive.org/web/20020208200758/http://sunsite.berkeley.edu/Anthro/rowe/ch_art.html.

70. Malter Galleries, "Auction of Classical Antiquities and Pre-Columbian Art" (March 29, 1997), http://web.archive.org/web/20040615014427/http://www.maltergalleries.com/m29.html.

71. Burger, Richard L., *Chavin and the Origin of Andean Civilization* (Thames and Hudson, 1995), 120–21.

72. Tampere Art Museum, "The Cupinique Culture," http://web.archive.org/web/20050211003331/http://www.tampere.fi/tamu/peru/1024/cupinisque_en.htm. The writer dates the Chavin culture to 900 to 200 bce, making it younger than the Cupinisque culture. See "The Chavin Culture," http://web.archive.org/web/20031103012730/http://www.tampere.fi/tamu/peru/1024/2_en.htm.

73. Burger, Richard L., *Chavin and the Origin of Andean Civilization* (Thames & Hudson, 1995), 91.

74. Burger, 91.

75. Burger, 170.

76. Karen Albright Murchison Canadian Museum of Civilization, "What's a Whatsit? Interesting Artifacts from Canada's West Coast," http://web.archive.org/web/20041104071210/http://www.civilization.ca/educat/oracle/modules/kmurchison/page02_e.html.

77. Dante Aligheri, *Purgatorio*, Canto IX, Lines 94–96, The World of Dante, http://www.worldofdante.org/comedy/dante/purgatory.xml/2.9.

78. Gwilt translation, *De Architectura*, Book VII, Chap. 3, sec. 9, https://lexundria.com/vitr/7.3.9/ cf. See also sec. 10, https://lexundria.com/go?q=Vitr.+7.3.10&v=cf.

79. Aitken Spence Travels, "Sigiriya—'World Heritage Site,'" http://web.archive.org/web/20030629222818/http://www.aitkenspencetravels.com/pages/aboutsl/3_p07.htm; Holy Mountain Trading Company, "Pilgrimage to the Citadel of Sigiriya," http://web.archive.org/web/20170612051902/http://www.holymtn.com/SriLanka/pilgrimage.htm; Wiejesiriwardena, Dasith Sanjaka, "The Mirror Wall and the Sigiri Graffiti," http://web.archive.org/web/20040217063009/http://sigiriya.freeyellow.com/mirror.htm.

80. Tyler-Adam Corp., "The Mirror-Cut Warrior," Tidbits, http://web.archive.org/web/20051109221353/http://www.tyler-adam.com/282.html.

81. Bruton, Eric, *Legendary Gems, Or, Gems That Made History* (Chilton, 1956), 55–57.

82. https://collections.louvre.fr/en/ark:/53355/cl010103086. For a reflected image therein, see https://www.mediastorehouse.com.au/p/690/mirror-rock-crystal-enamel-gold-said-marie-22542240.jpg.webp.

83. Beale, S. Sophia, *The Louvre: A Complete and Concise Handbook* (Outlook Verlag, 2024 [1883]), 9.

84. For the discussion of metallurgy which follows, I am highly indebted to Lambert, Joseph B., *Traces of the Past* (Perseus Books, 1997), Chap. 7.

85. Albenda, Pauline, "Mirrors in the Ancient Near East," *Source: Notes in the History of Art*, 4(2/3): 2–9 (1985).

86. Albenda, 3.

87. "Historian Says Skeletons Are Jewish Princesses," *Archaeology News Digest* (Summer 2000), http://web.archive.org/web/20040501183900/http://www.msn.fullfeed.com/~scribe/digest2003.htm.

88. Gregory, Richard, *Mirrors in Mind* (W.H. Freeman/Spektrum, 1997), 48.

89. NMS Mummy Project, http://web.archive.org/web/20180123064702/http://www.akhet.co.uk/nmsmumm2.htm.

90. Thomas, Elizabeth, "Shining Light on Egyptian Mirrors: New Scientific Research into Their Metallurgy," *J. Archaeological Science: Reports*, 58: 104744 (2024), https://doi.org/10.1016/j.jasrep.2024.104744.

91. Wilkinson, Richard H., *Reading Egyptian Art* (Thames and Hudson, 1992), 177.

92. Wilkinson, 176.

93. Vin Callcutt, "Introduction to Copper: Mining and Extraction" (August 2001), https://www.copper.org/publications/newsletters/innovations/2001/08/intro_mae.html.

94. Davey, Christopher J., and W. Ian Edwards, "Crucibles from the Bronze Age of Egypt and Mesopotamia," *Proceedings Royal Society Victoria*, 120(1): 146–54 (2007).

95. Chwalkowski, Farrin, *Symbols in Arts, Religion and Culture: The Soul of Nature* (Cambridge Scholars, 2016), 114.

96. Gregory, 48.

97. Marie Parsons, "About Egyptian Pyramids," http://web.archive.org/web/20181205095106/http://www.touregypt.net/featurestories/kahun.htm.

98. Wilkinson, 121.

99. Eaton, Elizabeth S., "A Group of Middle Kingdom Jewelry," *Bulletin Museum Fine Arts*, 32: 94, 97, 98 (1941).

100. Egyptian Centre Canolfan Eifftaidd, "Cosmetics" (2010), http://web.archive.org/web/20120105104728/http://www.egypt.swansea.ac.uk/index.php/collection/141-cosmetics.

101. Newman, Jay Hartley, and Lee Scott, *The Mirror Book: Using Reflective Surfaces in Art, Craft, and Design* (Crown, 1978), 5.

102. Newman, 3.

103. Schiffer, Herbert F., *The Mirror Book: English, American and European* (Schiffer Publ. Ltd., 1983), 10.

104. Schorsch, Deborah. "Silver in Ancient Egypt," Heilbrunn Timeline of Art History, The Metropolitan Museum of Art (September 2018), https://www.metmuseum.org/toah/hd/silv/hd_silv.htm.

105. Freed, Rita, "The Nubian Collection of the Museum of Fine Arts, Boston," *Mitteilungen der Sudanarchaeologischen Gesellschaft zu Berlin E.V.* (July 1995), 33.

106. Wilson, supra.

107. Fragments, 384. Also translated as "Polished brass is the mirror of the body, and wine of the heart."

108. University of Pennsylvania Museum, "Ancient Greek World: Images of Women and Goddesses," entry for item L-64-23, Italic Low-Footed Red Figure Bowl with High Handles 4th century bce, on loan from Philadelphia Museum of Art, http://web.archive.org/web/20020602102510/http://www.museum.upenn.edu/Greek_World/Excerpts_Other/Women&Goddesses_excerpts.html.

109. Carpenter, T.H., *Art and Myth in Ancient Greece* (Thames & Hudson, 1991), Plate 160, Attic red figure *hydria* from Vulci, c. 430.

110. Panati, Charles, *Extraordinary Origins of Everyday Things* (Harper & Row, 1987), 229.

111. Calcutt, Vin, "Introduction to Copper: Mining and Extraction" (August 2001), https://www.copper.org/publications/newsletters/innovations/2001/08/intro_mae.html.

112. Bonati, Isabella, and Niccola Reggiani, "Mirrors of Women, Mirrors of Words: The Mirror in the Greek Papyri,' in Gerolemou, Maria, and Lilia Diamantopoulou, *Mirrors and Mirroring from Antiquity to the Early Modern Period* (Bloomsbury, 2020), 69.

113. Aristophanes, "The Clouds," Internet Classic Archives, http://classics.mit.edu/Aristophanes/clouds.html.

114. Bonati, 69.

115. For Etruscan mirrors, see generally Bonfante, Larissa, "Etruscan," in *Reading the Past: Ancient Writing from Cuneiform to the Alphabet* (University of California Press/British Museum, 1990), 347–54.

116. Pliny, *Natural History*, Book XXXIII, Chap. 45, Vol. VI, 127, and see note 1 therein.

117. Pliny, Book XXXIV, Chap. 48, Vol. VI, 124.

118. Gwilt translation, *De Architectura*, Book VII, Chap. 3, sec. 9, https://lexundria.com/vitr/7.3.9/cf.

119. Kowalski, "Tabula," Penn State, http://web.archive.org/web/20070212174356/http://www.personal.psu.edu/users/w/x/wxk116/roma/tabula.html.

120. Seneca, Epistle LXXXVI, from Lucius Annaeus Seneca, *Moral Epistles*, trans. Richard M. Gummere, The Loeb Classical Library (Harvard University Press, 1917–1925), Volume II, http://www.stoics.com/seneca_epistles_book_2.html. See also "The History of Plumbing: Pompeii and Herculaneum" (July 7, 1989), https://theplumber.com/pompeii-herculaneum/.

121. Li, Xueqin, 68–71.

122. Zhang Qian, "There Is More to a Mirror Than Your Reflection," *Shanghai Star* (December 8, 2000), http://web.archive.org/web/20030813142115/http://www.chinadaily.com.cn/star/2000/1208/cn4-2.html.

123. Kaplan, Edward, lecture, "AE12: Social and Economic Life in Early Imperial China" (December 1997), http://web.archive.org/web/20070701000000*/www.ac.wwu.edu/~kaplan/H371/ae12.pdf.

124. Dien, Albert E., *Six Dynasties Civilization* (Yale University Press, 2007), 265–66.

125. *Archaeology Magazine*, "Bronze Mirrors Made in Japan Earlier Than Previously Thought" (May 28, 2015), https://archaeology.org/news/2015/05/28/150528-japan-mirror-mold.

126. Gill, Robin D., *Topsy Turvy 1585* (Paraverse, 2004), 117.

127. Thompson, 10–11, 43.

128. Meek, James, "Russians Claim to Unearth Steppes' Ancient Amazons," *The Guardian* (November 23, 1998), http://web.archive.org/web/20050206201652/http://isd.usc.edu/~retter/amazons1.html.

129. Reticuli (pseud.), "The Doppelgänger," http://www.angelfire.com/or/reticuli/doppelganger.html.

130. Vanaeon, Elkin, ed., "Siberian Ice Maiden," http://web.archive.org/web/20080919165908/ http://members.aol.com/_ht_a/tammuz69/home/ Index/History/Ice_Man.html; see also Kato, Kyuzo, "Cultural Exchange on the Ancient Steppe Route: Some Observations on Pazyryk Heritage," *Senri Ethnological Studies*, 32: 5–20 (1992), at 13, https://doi.org/10.15021/00003094.

131. Joy, Jody, and Melanie Giles, "Mirrors in the British Iron Age: Performance, Revelation and Power," in Anderson, Miranda, ed., *The Book of the Mirror: An Interdisciplinary Collection Exploring the Cultural Story of the Mirror* (Cambridge Scholars, 2007), 16–31, 17.

132. Tsujita, Jun'ichiro, "The Chronology and Distribution of Iron Age Mirrors in Britain and Ireland" (2019), https://catalog.lib.kyushu-u.ac.jp/ opac_download_md/2230523/pa001.pdf.

133. Craddock, P.T., and J. Lang, "Crucible Steel—Bright Steel," *Historical Metallurgy*, 38(1): 35–46 (2004).

134. Goldberg, 141.

135. Man, John, *Gutenberg: How One Man Remade the World with Words* (MJF, 2002), 63–64.

136. Man, 68–69.

137. Childress, Diana, *Johannes Gutenberg and the Printing Press* (Lerner, 2008), 55.

138. Selin, Helaine, ed., *Encyclopaedia of the History of Science, Technology, and Medicine in Non-Western Cultures*, Vol. 1 (Springer, 2008), 1699.

139. Dungworth, David, "Roman Copper Alloys: Analysis of Artefacts from Northern Britain," *J. Archaeological Science*, 24: 901–10 (1997).

140. Smith, Cyril Stanley, and Martha Teach Gnudi, *The Pirotechnia of Vannoccio Biringuccio: The Classic Sixteenth-Century Treatise on Metals and Metallurgy* (Dover, 1990 [1959]), 385, 388; Pirotechnia, Book IX, Chap. 12. Surprisingly, he added that "nowadays most of the masters" reversed the copper-tin ratio.

141. Hall, A.R., "Sir Isaac Newton's Note-Book, 1661–65," *Cambridge Historical J.*, 9(2): 239–50 (1948), https://www.jstor.org/stable/3020622, citing Folio 111.

142. Mills, A.A., and P.J. Turvey, "Newton's Telescope. An Examination of the Reflecting Telescope Attributed to Sir Isaac Newton in the Possession of the Royal Society," *Notes & Records Royal Society London*, 33(2): 133–155 (1979), at 147, 152.

143. Mills, 139.

144. Seely, Oliver, "Composition and Physical Properties of Alloys," http://web.archive.org/ web/20020209215919/http://chemistry.csudh.edu/ oliver/chemdata/alloys.htm.

145. Mills, 147–48.

146. Id.

147. Vickers, 331–32.

148. Selin, supra.

149. Smith, 388–90 (*Pirotechnica*, Book IX, Chap. 12).

150. della Porta, Giambattista, *Seventeenth Book of Natural Magic*, Chapter 23 (1658 English edition),

http://web.archive.org/web/20030803063243/ http://members.tscnet.com/pages/omard1/ jportac17.html; facsimile, https://www.loc.gov/ item/09023451.

151. Howles, Matthew, "Looking at the Mirror: The Craftsmanship of Reflecting Telescopes" (May 2, 2020); https://blog.sciencemuseum.org. uk/looking-at-the-mirror.

Chapter 2

1. Anglicus, Bartholomew (transl. John Trevisa), *Mediaeval Lore*, https://www.gutenberg.org/ cache/epub/6493/pg6493-images.html.

2. Anglicus, supra.

3. Ellis, William S., *Glass: From the First Mirror to Fiber Optics, The Story of the Substance That Changed the World* (Avon Books, 1998), 3–6.

4. Ellis, 4.

5. Zerwick, Chloe, *A Short History of Glass* (Corning Museum of Glass, 1980), 12–14.

6. Seneca, *Episulae Morales* (Moral Epistles), Epistle LXXXVI (transl. Richard M. Gummere), Vol. II, 313 (The Loeb Classical Library, Harvard University Press, 1917-1925), http://www.stoics. com/seneca_epistles_book_2.html#'XC1.

7. Zerwick, 21–30.

8. Seneca, *Episulae Morales* (Moral Epistles), Epistle XC, Vol. II, 415, supra; http://www.stoics. com/seneca_epistles_book_2.html#'XC1.

9. See also Pliny, Natural History, XXXVI. xlv.160.

10. Butti, Ken, and John Perlin, *A Golden Thread: 2500 Years of Solar Architecture and Technology* (Van Nostrand Reinhold, 1980), 256–57, comment on p. 19 (which quoted Seneca).

11. PPG Glass, "History of Glass," http://web. archive.org/web/20030204150852/http://www. ppg.com/gls_ppgglass/architect/history.htm

12. Taylor, Mark, "No Pane, No Gain," (November 2003), http://www.theglassmakers.co.uk/ archiveromanglassmakers/articles.htm#Update. See also *Glass News* No. 9 (January 2001); "An Experiment in the Manufacture of Roman Window Glass," *ARA Bulletin* no. 13 (August 2002).

13. Glass Online, "A Brief History of Glass," http://web.archive.org/web/20040215053200/http:// www.glassonline.com/history.html. The effect is partly true decolorization (Lambert, 111–14), and partly masking with a complementary color, which reduces transparency. Maloney, F.J. Terence, *Glass in the Modern World: A Study in Materials Development* (Aldus Books, 1967), 55–56.

14. From a discussion of private houses in the Roman town of Caerwent, see Griffiths, J.J., "Private Houses," http://web.archive.org/ web/20040721105639/http://web.ukonline.co.uk/ jj.griffiths/1024/wc/caerwent/private.html.

15. Pliny, *Natural History*, XXXVI.66.

16. See, e.g., "Ancient Roman Glass," Ancient Touch, http://www.ancienttouch.com/roman_ glass.htm.

17. A photograph of the wall painting appears in Whitehouse, David, "Looking Through Roman Glass," *Archaeology* (September 8, 1997), http://web.archive.org/web/20040203124407/he.net/~archaeol/online/features/roman/roman.html.

18. Margaret Cool Root, "Wondrous Glass: Images and Allegories," University of Michigan, http://web.archive.org/web/20160829005930/http://www.umich.edu/~kelseydb/Exhibits/WondrousGlass/RomanGlass-Wondrous.html.

19. Pliny, *Natural History*, Book XXXIII, Chap. 45, from Bostock transl., Vol. VI, 127.

20. Id., Book XXXVI, Chap. 66, from Bostock transl., Vol. VI, 380.

21. Id., Bostock transl. Vol. VI, 380 note 85.

22. Ehrlich, Werner, *The History of Picture Frame Contents*, http://web.archive.org/web/20050907085430/http://www.iifs.org/ehlich/ehlichContents.html. "Two small 230 mm x 540 mm flat panes of glass were used in the ceiling window of the bathhouse in the plaza of Pompeii." See Glasstopia, "The Origin of Glass" (2000), http://web.archive.org/web/20051029094054/http://www.glasstopia.com/e_site/glassis/history/w_history/originofglass/originofglass.html.

23. MacFarlane, Alan, and Gerry Martin, *Glass: A World History* (University of Chicago Press, 2002), 19.

24. Gros-Galliner, Gabriella, *Glass: A Guide for Collectors* (Stein & Day, 1970), 14. See also Goldberg, 137.

25. The Met, Robert Lehman Collection, 1975, Accession 197.1.110, https://www.metmuseum.org/art/collection/search/459052.

26. Ehrlich, supra.

27. Goldberg, 108.

28. *FOCUS Online Magazine*, "Byzantian Glasses," (2005), http://web.archive.org/web/20080516070946/http://www.focusmm.com/civcty/glass/gla_byza.htm.

29. Tangram Technology, "Glass Timeline," http://web.archive.org/web/20030301174605/http://www.tangram.co.uk/TI-Glass_Timeline.html.

30. Gros-Galliner, 32–33.

31. Hawthorne, John G., and Cyril Stanley Smith, transl., *On Divers Arts: The Treatise of Theophilus*, Chapters 6 ("How to Make Sheets of Glass"), 54–55, and 9 ("Spreading Out the Glass Sheets"), 57 (University of Chicago Press, 1963).

32. Ellis, 23; Kunzig, Robert, "The Physics of Glass," *Discover* 20(10) (October 1999), https://www.discovermagazine.com/the-sciences/the-physics-of-glass.

33. Eliot, George, and John Walter Cross, *The Writings of George Eliot Romola* (Houghton Mifflin, 1909), 52.

34. Mehlman, Felice, *Phaidon Guide to Glass* (Prentice Hall, 1982), 53.

35. Dillon, Edward, *Glass* (Methuen and Company, 1907), 177.

36. Polak, Ada, *Glass: Its Tradition and Its Makers* (G.P. Putnam's Sons, 1975) 65.

37. Lambert, 109.

38. McCray, W. Patrick, *Glassmaking in Renaissance Venice* (Ashgate, 1999), 101–2; Macfarlane, 206.

39. Lambert, 104–5, 122, 124–26.

40. Lambert, 104, 113.

41. Ellis, *Glass* 19–20.

42. McCray, 102–5.

43. McCray, 105.

44. Melchior-Bonnet, Sabine, *The Mirror: A History* (Routledge, 2001), 20.

45. McCray, 106.

46. McCray, 112–13.

47. Melchior-Bonnet, 18. See also Gros-Galliner, 18; Goldberg, 140.

48. Schiffer, 6.

49. Goldberg, id.

50. Mehlman, 169.

51. Polak, 25.

52. Zerwick, 45–46; Polak, 24.

53. *Idem.*, p. 293.

54. Melchior-Bonnet 37.

55. Melchior-Bonnet 41, 44.

56. Gros-Galliner, 25.

57. Wills, Geoffrey, *English Looking Glasses: A Study of the Glass, Frames, and Makers, 1670–1820* (Country Life, 1965), 42.

58. Schiffer, 24.

59. Polak, 127.

60. Goldberg, 168.

61. Melchior-Bonnet 51.

62. Melchior-Bonnet, 33, 54–56.

63. Melchior-Bonnet, 57–58, 62.

64. Hunt, Robert, *Ure's Dictionary of Arts, Manufactures, and Mines*, Volume III (Longmans, Green, 6th ed., 1867), 473.

65. Hynd, W.C., "Flat Glass Manufacturing Processes," in Uhlmann, D.R., and N.J. Kreidl, eds., *Glass: Science and Technology*, Vol. 2, Processing (Academic Press, 1984), 49.

66. Pilkington, L.A.B., "Review Lecture. The Float Glass Process," *Proceedings Royal Society London*, Series A, 314 (1516): 1–25 (1969).

67. Pilkington, supra.

68. Pilkington, supra; Hynd, 97.

69. Pilkington, supra.

70. Schiffer, 6.

71. "Timeline: Technology," Facts on File, http://web.archive.org/web/20050314173830/http://www.fofweb.com/Subscription/Science/Timeline.asp?SID=2&Topic=Technology. I assume in *Perspectiva communis*, 1277, or in *Tractatus de Perspectiva*.

72. John Ciardi translation.

73. Pendergrast, Mark, *Mirror, Mirror* (Basic Books, 2003), 118.

74. In Venice, according to "Antique Mirrors—History," Antique Mirror Gallery, http://web.archive.org/web/20011213214708/http://www.antiquemirrorgallery.co.uk/history.htm.

75. della Porta, Book XVII, Chapter 22.

76. Quoted in Wills, 63.

77. Hunt, 190.

78. Hunt, 190–91.

79. Hunt, 191.

80. Bessemer, Henry, *Sir Henry Bessemer, F.R.S.: An Autobiography* (Offices of "Engineering," 1905), Chapter VIII, "Improvements in Glass Manufacture."

81. De Chavez, Kathleen Payne, "Historic Mercury Amalgam Mirrors: History, Safety and Preservation," *Art Conservator*, Spring: 23–26 (2010).

82. Schiffer, 7.

83. Mirror Mirror, "History of Mirrors" (2003), http://web.archive.org/web/20070814065059/http://www.mirrorx2.com/Non_frames/articles_hom.html.

84. Wills, 47–48.

85. Wills, 148.

86. Wills, 149.

87. Wills, 155.

88. Wills, 157–58.

89. Wilson, John Harold, *Nell Gwyn, Royal Mistress* (Pelligrini & Cudahy, 1952),159–61.

90. Edwards, Clive, "Eighteenth Century Mirrors," http://web.archive.org/web/20020615164914/http://www.antiquecc.com/articles/june96/jun1.html.

91. "Dining Room of the Hotel de Paris," Hotel de Paris Museum, http://web.archive.org/web/20091220034519/http://www.hoteldeparismuseum.org/dr.html.

92. Barnhouse, Mark, *Lost Denver* (Arcadia, 2015), 62.

93. Goodman, Bonnie K., *Silver Boom! The Rise and Decline of Leadville, Colorado as the United States Silver Capital, 1860-1896* (2008), 59, https://www.academia.edu/37978546/Silver_Boom_The_Rise_and_Decline_of_Leadville_Colorado_as_the_United_States_Silver_Capital_1860_1896.

94. "The Mansion," Ashland: The Henry Clay Estate, http://web.archive.org/web/20080622130142/http://www.henryclay.org/mansion.htm.

95. Lane, Joshua, and Rosie Grayburn, "Fade to Black: Reflections on the Aging Process of Mirror Glass" (2020), https://www.incollect.com/articles/fade-to-black-reflections-on-the-aging-process-of-mirror-glass.

96. Reat, Kay, and Gerry Munley, "Justus von Liebig: An Educational Paradox," http://web.archive.org/web/20040405053712/http://step.sdsc.edu/projects95/chem.in.history/essays/liebig.html.

97. Brock, William H., *Justus von Liebig: The Chemical Gatekeeper* (Cambridge University Press, 2002), 136–37.

98. Drayton, Thomas, "Improvement in Silvering Looking-Glasses," U.S. Patent [USP] 3,702 (August 12, 1844). See also British Patent 9,968 (1843).

99. Hunt, 192.

100. Cimeg, John, "Improvements in Silvering Glass," British Patent 619 (September 10, 1861).

101. Newman, 317, 322.

102. British Patent 12,358 (1848). Thomson and Varnish patented the use of this process for silvering double-walled vases: British Patent 12,905 (1849).

103. Bureau of Standards, "The Making of Mirrors by the Deposition of Metal on Glass," Circular 389 (GPO, 1931).

104. Newman, 15, 314.

105. Newman, 319.

106. Newman, 319–20.

107. De Chavez, supra.

108. James, Frank L., "The Deposition of Silver on Glass and Other Non-Metallic Surfaces," *Proceedings American Society Microscopists*, 6: 71–80 (1884).

109. Dodé, Edouard, British Patent 1520 (1864, June 18); "Chronicles of Science," *Quarterly J. Science*, 7: 497 (July 1865).

110. Mattox, Donald M., "Historical Timeline of Vacuum Coating and Vacuum Technology," Society of Vacuum Coaters (2003), https://www.svc.org/conferences/history-of-vacuum-coating.

111. Newman, 16.

112. Id.

113. Butti, 38.

114. Butti, 38–39.

115. Society of Plastic Engineers, "Plastics Injection Molding," http://web.archive.org/web/20030802000239/http://www.4spe.org/sections divisions/divisions/d23www.htm.

116. American Plastics Council, "History of Plastics," http://web.archive.org/web/20031202192355/http://www.americanplasticscouncil.org/benefits/about_plastics/history.html.

117. Newman, 81–82.

118. Goebel, Greg, "An Introduction to Plastics" (October 1, 2005), http://web.archive.org/web/20070810225820/http://www.vectorsite.net/ttplast.html.

119. Newman, 59.

120. Raymond, Robert, *Out of the Fiery Furnace: The Impact of Metals on the History of Mankind* (Pennsylvania State University Press, 1986), 30–31.

121. Id., 33.

122. Id., 45.

123. Pliny Natural History, Book XXXIII, Chaps. 49, 51, 52.

124. Newman, 6.

125. Gordon, J.E., *Structures: Or Why Things Don't Fall Down* (Cox & Wyman, 1978), 321–22, 386–87.

126. For acrylic, the figure of merit is 0.87 and for Mylar polyester film, 1.11. The metals that outperform glass are magnesium (2.00) and beryllium (3.56).

127. Howard-White, B., *Nickel: An Historical Review* (D. Van Nostrand, 1963), 104–5.

128. Pliny, *Natural History*, Book XXXIII, Chap. 20, Vol. VI, 99, and Chap. 32, 114.

129. Id.

130. Elite Cultural Tours to Russia, "St. Isaac's Cathedral—Construction of the Main Dome," http://web.archive.org/web/20030423123045/http://www.eliteculturaltours.com/isaaky/Construction_of_the_main_dome.html; "Isaac's Cathedral,"http://web.archive.org/web/20130425114

746/http://www.oksanas.net/isaacs.htm; Wacht, Michael, "Architecture: St. Isaac's Cathedral and the Architecture of Saint Petersburg, Russia, 1855–1881," http://web.archive.org/web/20031216 093807/http://webserver.rcds.rye.ny.us/id/Art/ artpageMAW.html.

131. Timbrell, John, *The Poison Paradox: Chemicals as Friends and Foes* (Oxford University Press, 2005), 166–67.

132. "Gilding," in "Glossary," "Russian Art from the Hulmer Collection," http://web.archive. org/web/20101204134134/http://hulmer.allegheny. edu/glossary.html.

133. Artisan Plating, "Electroplating History (and Electrodeposition of Metals)" (March 14, 2014), http://www.artisanplating.com/articles/ platinghistory.html The term "water gilding" is also applied to an unrelated technique which uses an adhesive.

134. Howard-White, 105.

135. Douglas Plating Limited, "Historical Perspective," excerpted from *MFSA Quality Metal Finishing Guide*, Vol. 1 No. 1-B, Decorative Precious Metal Plating, http://web.archive.org/ web/20030806011045/http://www.douglas-plating. co.uk/studentframe.html.

136. Howard-White, 107–8.

137. "Birmingham Jewellery Quarter Walk 5, Elkingtons' and the Assay Office," http://web. archive.org/web/20070928004810/http://jquarter. members.beeb.net/walk5.htm.

138. Peck, George Wilbur, *The Grocery Man and Peck's Bad Boy* (Belford, Clarke, 1883), Chapter XX, "Fourth of July Misadventures."

139. Lawrence, David Russell, *The Naturalist and His 'Beautiful Islands': Charles Morris Woodford in the Western Pacific* (Australian National University Press, 2014), 209, https://press-files. anu.edu.au/downloads/press/p298111/pdf/book. pdf.

140. Howard-White, 24–25.

141. Howard-White, 112; Bicycle Museum of America, "1810–1895," http://web.archive.org/ web/20051104011153/http://www.bicyclemuseum. com/Html/bike1.html.

142. "A Nickel-Plated Emperor," in *Marvelous Land of Oz*, http://web.archive.org/web/20041215 001603/http://www-2.cs.cmu.edu/People/rgs/ ozland-121.html.

143. Ley, Brian, "Diameter of a Human Hair" (1999), https://hypertextbook.com/facts/1999/ BrianLey.shtml.

144. Hashmi, M.S.J., ed., *Comprehensive Materials Processing* (Elsevier Science, 2014), 288–91. Cf. Simons, Eric, *Guide to Uncommon Metals* (Frederick Muller, 1967), 48–54.

145. INCO, "Finishing Facts: Nickel Plating 1843–1945," *Nickel Currents*, 7(1): 5, http:// web.archive.org/web/20031001000000*/http:// www.incoltd.com/Products/pdf/Newsletter/ NickelCurrents_439.pdf.

146. "GM Corporate History, 1920s," http:// web.archive.org/web/20070821013811/http://www.

gm.com/company/corp_info/history/gmhis1920. html.

147. Bicycle Museum of America, "1900–1934," http://web.archive.org/web/20050601074139/ http://www.bicyclemuseum.com/html/bike6. html.

148. "History of Electroplating Sterling Jewelry with Rhodium," Metal Finishing Q&As, http:// www.finishing.com/102/70.html.

149. Artisan Plating, "Electroplating of the Platinum Metals Group," http://web.archive.org/ web/20130404105033/http://www.artisanplating. com/articles/platingplatgrp.html.

150. Rubinstein, Richard, "Rhodium Mirrors," JedMed, http://web.archive.org/web/20090105 164012/http://www.jedmed.com/Dentistry/ rhodium_mirrors/rhodium_mirrors.html.

151. Miller, Mitchell D., et al., "The Development of the GCPCC Protein Crystallography Beamline at CAMD," in Duggan, J.L., and I.L. Morgan, *CP576, Application of Accelerators in Research and Industry—Sixteenth Int'l. Conf.*, https:// www.lsu.edu/camd/files/102-beamline.pdf.

152. JML Optical, "Reflective Coatings for Efficient Mirrors," http://web.archive.org/web/2002 0714032851/http://www.netacc.net/~jmlopt/ products/coatings/technical_reflective.html.

153. Photonics Spectra, "Liquid Metal Enables Switchable Mirrors" (June 16, 2021), https://www. photonics.com/Articles/Liquid_Metal_En ables_Switchable_Mirrors/a67085.

Chapter 3

1. Hecht, Jeff, *City of Light: The Story of Fiber Optics* (Oxford University Press, 1999), 39–51.

2. Waldman, Gary, *Introduction to Light: The Physics of Light, Vision, and Color* (Dover, 2002 [1983]), 58–61.

3. Shekhtman, Lonnie, "Laser Beams Reflected Between Earth and Moon Boost Science" (August 10, 2020), https://www.nasa.gov/missions/laser-beams-reflected-between-earth-and-moon-boost-science.

4. Potters, Rudolf H., "Method and Apparatus for Producing Glass Beads," USP 2,334,573 (issued November 16, 1943); cf. Viziglow, "History (Invention) of Retro Reflective Paint" (August 21, 2022), https://reflective-paints.com/retro-reflective-paint-articles/history-invention-of-retro-reflec tive-paint.

5. Id.

6. Hobbs, Philip, *Building Electro-Optical Systems: Making It All Work* (Wiley, 2000), 224.

7. Southall, James Powell Cocke, *Mirrors, Prisms and Lenses: A Text-Book of Geometrical Optics* (Macmillan, 1918), 113.

8. Malacara, Daniel, *Geometrical and Instrumental Optics* (Academic Press, 2014), 67.

9. Yoder, Paul, *Design and Mounting of Prisms and Small Mirrors in Optical Instruments* (SPIE Optical Engineering Press, 1998), 136.

10. Electrical4U, "Reflectance, Reflectivity,

and the Solar Reflectance Index" (May 29, 2024), https://www.electrical4u.com/what-is-reflectance.

11. This assumes unpolarized incident light, no absorption, glass of refractive index 1.52, and no multiple reflection.

12. Friedman, Joseph Solomon, *History of Color Photography* (American Photographic Publishing, 1947), 43–44.

13. Friedman, 46.

14. Friedman, 47.

15. James, 249–50, quoting Michael Sullivan, *The Arts of China* (University California Press, 1961), 171–72.

16. Wu, Ling-An, et al., "Optics in Ancient China" (February 18, 2015), https://light2015blog dotorg.wordpress.com/2015/02/18/optics-in-ancient-china.

17. Gregory, 89.

18. Cohen, Morris R., and I.E. Drabkin, *A Source Book in Greek Science* (Harvard University Press, 1948), 262.

19. Cohen, 267–68.

20. della Porta, *Natural Magic*, Book XVII, Chapter 23.

21. Id.

22. Id.

23. Feist, Ulrike, "The Reflection Sundial at Palazzo Spada in Rome: The Mirror as Instrument, Symbol and Metaphor," in Frelick, Nancy M., ed., *The Mirror in Medieval and Early Modern Culture: Specular Reflections* (Brepols, 2016), 271–86, at 275.

24. Small World Productions, "Denmark: Copenhagen and Aero" (1997), "snooping periscopes" at [47], https://web.archive.org/web/2002 0217033508/http://www.travelsmallworld.com/travels/travels_files/scandinavia/scandexc.html.

25. Davis, Burke, *The Civil War: Strange and Fascinating Facts* (Random House, 1960), 30.

26. https://web.archive.org/web/20050216 001153/http://home.europa.com/~telscope/trsg26.txt; quoting Hans Seeger, *Militaerische Fernglaeser und Fernrohre*, 2.6, 85, Scissor Telescopes. Hans Seeger, Hamburg, and Alfred Koenig, Herborn.

27. Youlten, William, "Sighting Device for Firearms," USP 694,904 (issued March 4, 1902).

28. U.K. Historic Arms Resource Centre, "The Youlten Hyposcope," https://www.rifleman.org.uk/Youlten_Hyposcope.html.

29. Id.

30. U.K. Historic Arms Resource Centre, "A Trench Periscope Adaptation of the Lee-Enfield Rifle," https://www.rifleman.org.uk/Enfield trench_periscope_adaptation.html.

31. See post by "Centurion," "Invention of the Periscope Rifle" (2012, March 29), Great War Forum, which includes a photo purporting to be the periscope rifle used by Corporal Kent at Ypres in early 1915, and referring to other "early adopters," too: https://www.greatwarforum.org/topic/177884-invention-of-the-periscope-rifle.

32. Wahlert, Glenn, and Russell Linwood, *One Shot Kills: A History of Australian Army Sniping* (Big Sky Publishing, 2014), chapter 3.

33. Laemlein, Tom, "Sniping from Below: Periscope Rifles in World War I," American Rifleman (April 5, 2018), https://www.americanrifleman.org/content/sniping-from-below-periscope-rifles-in-world-war-i.

34. The Armourer's Bench, "Periscope Rifles in Ukraine," https://armourersbench.com/2024/03/22/periscope-rifles-in-ukraine.

35. Submarine Periscope Manual, supra, chapter 1.

36. Multieducator, History Central, "Civil War Naval History, April 12, 1864," https://web.archive.org/web/20100214073507/http://www.multied.com/navy/cwnavalhistory/April1864.html; see also Navy, "History of USS Lexington," Dictionary of American Naval Fighting Ships (DANFS), https://web.archive.org/web/20070427014051/http://www.history.navy.mil/danfs/l/lexington.htm.

37. Flatow, Ira, *They All Laughed ...* (Harper-Perennial, 1992), 189.

38. Merrill, John, "Looking Around: A Short History of Submarine Periscopes Part 1" (January 2002), https://archive.navalsubleague.org/2002/looking-around-a-short-history-of-submarine-periscopes-part-1.

39. See, e.g., "About Submarines and a Little History," http://web.archive.org/web/2004 0510161223/http://www.boomersailors.net/subclass/subclass.html.

40. "Submarine Periscope Manual," Chap. 1, 1, supra.

41. Goebel, Greg, "The Invention of the Submarine," http://web.archive.org/web/20041209 093257/http://www.vectorsite.net/twsub1.html.

42. "Submarine Centennial," United States Navy, http://web.archive.org/web/20060927230 507/http://www.chinfo.navy.mil/navpalib/ships/submarines/centennial/subhistory.html.

43. Submarine Periscope Manual, supra.

44. Sueter, Murray Fraser, *The Evolution of the Submarine Boat, Mine and Torpedo, from the Sixteenth Century to the Present Time* (J. Griffin, 1908), 238.

45. "How It Works: VI—The Periscope," *Illustrated War News*, February 24, 1915, 40; Sueter, supra.

46. Sueter, supra.

47. "Submarine Periscope Manual," Navpers 16165 (June 1946), Chap. 1, p. 3, reprinted at https://maritime.org/doc/fleetsub/pscope/index.php.

48. Hegde, P.D., *A Brief History of Great Inventions* (KK, 2021), 91–92.

49. "Periscope Aids Study of Underwater Life," *Popular Science* (December 1937), 36.

50. Whittell, Giles, *Bridge of Spies* (Simon & Schuster UK, 2011).

51. Swenson, Loyd S., Jr., et al., *This New Ocean: A History of Project Mercury* (National Aeronautics and Space Administration, 1998), 355.

52. Newton, Isaac, *Opticks ...* (William Innys, 4th ed., corrected, 1730), 83.

53. Newton, 91.

54. Wilson, R.N., *Reflecting Telescope Optics*, vol. I (Springer-Verlag, 1996), 2.

55. Rybski, Paul M., "Important Astronomers, Their Instruments and Discoveries, Part 3: The Story of the Earliest Reflecting Telescopes," http://web.archive.org/web/20040803144226/http://www.seds.org/billa/psc/hist3.html; and Hong, Sungook, "Optics and Light," University of Toronto, http://web.archive.org/web/20020818153706/http://www.chass.utoronto.ca/~sungook/hps311/optics.htm.

56. Zoom Inventors and Inventions, Inventors and Inventions from the 1600s—Seventeenth Century, "Gregory, James," http://www.enchantedlearning.com/inventors/1600.shtml.

57. Zebrowski, Ernest, *A History of the Circle* (Free Association Books, 1999), 113.

58. White, Michael, *Newton: The Last Sorcerer* (Perseus Books, 1997), 169.

59. Newton, 91.

60. Newton, 91.

61. Cybrations, "Reflections of the Heavens," http://web.archive.org/web/20160514172058/http://www.cybrations.com/Clips/Voyage.htm.

62. White, 168.

63. White, 168.

64. White, 169–71; See also Van Helden, Albert, "The Telescope," Rice University (1995), http://web.archive.org/web/20011213235248/http://es.rice.edu/ES/humsoc/Galileo/Things/telescope.html.

65. Newton, 71.

66. Newton, 84.

67. Cybrations, supra.

68. "Astronomical Instruments," http://web.archive.org/web/20060216181442/http://members.tripod.com/~worldsite/astronomy/astroinst.html.

69. Amateur Astronomy Foundation, "Tools of the Trade, Lecture 2, The Telescope, Part 2," http://web.archive.org/web/20050927184900/http://members.ozemail.com.au/~swadhwa/m2c2lect2.html.

70. Newton, 94.

71. CapellaSoft, "Scenario: Georgium Sidus," Skyhound, http://web.archive.org/web/20020606220720/http://www.skyhound.com/george.htm.

72. Gregory, 163–64.

73. Zebrowski, 114.

74. Zoom Inventors and Inventions, Inventors and Inventions from the 1600s—Seventeenth Century, "Cassegrain Telescope," http://www.enchantedlearning.com/inventors/1600.shtml.

75. Gills, John F., "From James Gregory to John Gregory: The 300 Year Evolution of the Maksutov-Cassegrain Telescope" (1998), http://www.weasner.com/etx/guests/mak/MAKSTO.HTM.

76. Anderson, Kevin J., "Reflections on Mirror Coating Materials," *MRS Bulletin*, April 1989, 68.

77. Pendergrast, 161.

78. Anderson, supra.

79. King, Henry C., *The History of the Telescope* (Dover, 2003 [1955]), 262.

80. Steinheil, Karl, "On the Advantages to be Derived from the Use of Silver Mirrors for Reflecting Telescopes, etc.," *Monthly Notices Royal Astronomical Society*, 19(1): 57, 58 (1858).

81. Tobin, William, "Evolution of the Foucault-Secretan Reflecting Telescope," *J. Astronomical History Heritage*, 19(2): 106–84 (2016).

82. Tobin, supra; King, supra.

83. Marion, Fulgence, *The Wonders of Optics* (Scribner, 1871), 169.

84. Learner, Richard, *Astronomy Through the Telescope* (Van Nostrand Reinhold, 1981), 107–9.

85. Ellerman, Ferdinand, "Silvering the 100-Inch Hooker Telescope," Leaflet 52 (May 1933), https://adsabs.harvard.edu/full/1933ASPL....2....5E.

86. Learner, 110.

87. Kramer, Jack, "A Short History of the Telescope," Lake County Astronomical Society (1998), http://web.archive.org/web/20010506224440/http://homepage.interaccess.com/~purcellm/lcas/Articles/telehist.htm.

88. Ellerman, 5–6.

89. Ellerman, 6–8.

90. "Amateur Telescope Making," http://web.archive.org/web/20061230151255/http://www.willbell.com/tm/tm7.htm.

91. Teare, Scott W., "Aluminization of Large Telescope Mirrors using the 108" Bell Jar at Mount Wilson Observatory" (July 17, 2001), http://web.archive.org/web/20150909172434/http://www.ee.nmt.edu/~teare/mwo_al_main.htm.

92. Wilson, Ray N., *Reflecting Telescope Optics II: Manufacture, Testing, Alignment, Modern Techniques* (Springer, 1996), 216–20.

93. Angel, J.R.P., and J.M. Hill, "Honeycomb Mirrors of Borosilicate Glass," Proceedings ESO Conference, "Scientific Importance of High Angular Resolution at Infrared and Optical Wavelengths" (Garching, March 24–27, 1981), https://adsabs.harvard.edu/pdf/1981siha.conf...61A.

94. Mailly, Ed, *De L'Astronomie dans l'Académie Royale de Belgique: Rapport Séculaire (1772–1872)* (F. Hayez, 1872), 99–100 (transl. with DeepL free version).

95. Koupelis, Theo, *In Quest of the Universe* (Jones & Bartlett, 6th ed., 2010), 139.

96. Ruebush, Jim, "Big Telescope Mirrors: Spin Casting" (April 4, 2014), https://jarphys.wordpress.com/2014/04/04/big-telescope-mirrors-spin-casting.

97. Davis, Joel, *Alternate Realities: How Science Shapes Our Vision of the World* (Plenum Publishing, 1997), 87.

98. Davis, 82–90.

99. Osterbrock, Donald E., *Yerkes Observatory, 1892-1950* (University of Chicago Press, 1997), 189–90.

100. NASA, "Webb's Mirrors," https://science.nasa.gov/mission/webb/webbs-mirrors; "JWST Telescope," https://jwst-docs.stsci.edu/jwst-observatory-hardware/jwst-telescope#gsc.tab=0; McElwain, Michael W., et al., "The James Webb Space

Telescope Mission: Optical Telescope Element Design, Development, and Performance," *Publications Astronomical Society Pacific*, 135 (1047): 058001 (2023), doi:10.1088/1538-3873/acada0.

101. Large Binocular Telescope Observatory, "The First of the Extremely Large Telescopes" (accessed September 6, 2024), https://www.lbto.org.

102. Wikipedia, "List of Largest Optical Reflecting Telescopes," https://en.wikipedia.org/wiki/List_of_largest_optical_reflecting_telescopes.

103. SALT, "Primary Mirror," in "Inside SALT," in "About the Telescope," https://www.salt.ac.za/telescope; and Giant Magellan Telescope, "Primary Mirrors," https://giantmagellan.org/explore-the-design.

104. NASA Goddard Space Flight Center, Laboratory for High Energy Astrophysics, "Imagine the Universe," various webpages, accessible from http://imagine.gsfc.nasa.gov/docs/science/advanced_science.html.

105. The analogy is from "Overview: MCP Optics," University of Leicester, http://web.archive.org/web/20020905023402/http://www.src.le.ac.uk/lobster/ov_optics.htm.

106. See Mattson, Barbara, "X-Ray Imaging Systems," NASA Goddard Space Flight Center, Laboratory for High Energy Astrophysics, http://web.archive.org/web/20120724081110/http://imagine.gsfc.nasa.gov/docs/science/how_l2/xtelescopes_systems.html.

107. Hudec, Rene, and Charly Feldman, "Lobster Eye X-Ray Optics" (August 15, 2022), https://arxiv.org/abs/2208.07149, arXiv:2208.07149v1 [astro-ph.IM].

108. Id.

109. Id.

110. Chinese Academy of Sciences, "Follow-Up X-Ray Telescope" (2024), https://ep.bao.ac.cn/ep/cms/article/view?id=25.

111. Borra, Ermanno, "Liquid Mirrors: A Review," October 1994, https://arxiv.org/abs/astro-ph/9410008.

112. Mailly, Ed, *De L'Astronomie dans l'Académie Royale de Belgique: Rapport Séculaire (1772–1872)* (F. Hayez, 1872), 99 (transl. with DeepL free version).

113. Gibson, Brad K., "Liquid Mirror Telescopes: History," *J. Royal Astronomical Society Canada*, 85(4): 158, 162 (1891).

114. Gibson, 164; Borra, "Review," supra.

115. Shanko, Barry, "The Liquid Lens: Telescope Technology Takes a Leap," SPAce.com (September 24, 2000), http://web.archive.org/web/20100612122113/http://www.space.com/scienceastronomy/astronomy/liquid_mirror_000924.html.

116. Pospieszalska, Anna, "First Light May/June 2022," http://www.ilmt.ulg.ac.be/first-light-june-3rd-2022.

117. Borra, Ermanno, et al., "Gallium Liquid Mirrors: Basic Technology, Optical-Shop Tests and Observations," *Publication Astronomical Society Pacific*, 109: 319–25 (1997).

118. Borra, E.F., et al., "Large Magnetic Liquid

Mirrors," *Astronomy & Astrophysics*, 446: 389–93 (2006).

119. Hickson, Paul, "Liquid-Mirror Telescopes," *American Scientist*, https://www.americanscientist.org/article/liquid-mirror-telescopes.

120. Borra, "Review," supra.

121. Borra, "Review," supra.

122. Shuter, William, and Lorne Whitehead, "A Wide Sky Coverage Ferrofluid Mercury Telescope," *Astrophysical J.*, 424: L139–L141 (1994).

123. Borra, Ermanno, et al., "Floating Mirrors," *Astrophysical J.*, 516: L115–L118 (1999).

124. Hickson, supra.

125. Thibault, Simon, and Ermanno Borra, "Liquid Mirrors: A New Technology for Optical Designers," *Proceedings SPIE (International Society Optics Photonics)*, 3482: 519–27 (1998).

126. Jones, Thomas E., *History of the Light Microscope*, Chapter 4, "Mechanical Improvements of the Eighteenth Century" (University of Tennessee, Memphis, 1995, 1997), http://web.archive.org/web/20021015005107/http://www.utmem.edu/~thjones/hist/c4.htm.

127. Ford, Brian J., *Single Lens: The Story of the Simple Microscope* (Harper & Row, 1985), 95–96.

128. Ells, William, "Incident Lighting and the Lieberkühn Speculum," http://www.microscopy-uk.org.uk/mag/indexmag.html?http://www.microscopy-uk.org.uk/mag/artsep98/beincid.html.

129. "Compound Microscope as Rendered by Artist in Descartes 'La Dioptrique,'" https://www.loc.gov/resource/cph.3c10450/.

130. Spitta, Edmund Johnson, *Microscopy; the Construction, Theory and Use of the Microscope* (Murray, 1909), 206.

131. Sobel, Barry J., and Jurriaan de Groot, "Vertical Illuminator with Variable Apertures" (2015), https://microscope-antiques.com/watsonvertillA.html.

132. Spitta, 207; Ells, supra.

133. Croft, William J., *Under the Microscope: A Brief History of Microscopy* (World Scientific, 2006), 47–48.

134. Bradbury, S., *The Evolution of the Microscope* (Pergamon, 2014 [1967]), 153–54.

135. Morawetz, Leopold, and Charles Volkmar, "Automatic Heliotropes," USP 55,523 (June 12, 1866).

136. Newton, Isaac, Feb. 6, 1672, Letter to Henry Oldenberg (secretary of the Royal Society of London), *Philosophical Transactions Royal Society London*, 6 (1671–1672), reprinted at http://web.archive.org/web/20090214024213/http://www.ugcs.caltech.edu/~plavchan/ses158/library.htm.

137. Barker, Robert, "A Catoptric Microscope," *Philosophical Transactions Royal Society London*, 39 (442): 259 (1735–1736).

138. Jones, Thomas E., "History of the Light Microscope, Chapter 5, "The Achromatic Lens Dispute, and other Optic Improvements," University of Tennessee, Memphis (1995, 1997), http://web.archive.org/web/20021015004634/http://www.utmem.edu/~thjones/hist/c5.htm.

139. Davidson, Michael W., "Jecker Horizontal Catoptric Microscope" (2022), https://micro.magnet.fsu.edu/primer/museum/jeckercatoptric.html.

140. "Sir Charles Wheatstone," Hebrew University of Jerusalem, Institute of Chemistry, http://web.archive.org/web/20061002065646/http://chem.ch.huji.ac.il/~eugeniik/history/wheatstone.html; and Naughton, Russell, "Adventures in CyberSound: Charles Wheatstone," Australian Centre Moving Image, http://web.archive.org/web/20111120052618/http://www.acmi.net.au/AIC/WHEATSTONE_BIO.html.

141. Naughton, Russell, "Adventures in CyberSound: Stereoscope—1838," Australian Centre Moving Image, http://web.archive.org/web/20110605030303/http://www.acmi.net.au/aic/magic_machines_2.html.

142. Calvert, J.B., 'Steropsis," University of Denver (2000), http://web.archive.org/web/20070701210416/http://www.du.edu/~jcalvert/optics/stereops.htm.

143. Layer, Harold A., "Stereoscopy: Where Did It Come From? Where Will It Lead?," *Exposure*, 17(3): 34–48 (1979), http://web.archive.org/web/20140506185547/http://online.sfsu.edu/hl/stereo.html.

144. Manekshaw, Bob, "The Art of Stereo Photography," Photostuff, http://www.photostuff.co.uk/stereo.htm.

145. Baird, Keith, et al., eds., "The View-Master Database" (2024), https://viewmasterinfo.com.

146. Schwartz, John, "The Body in Depth," *New York Times* (April 22, 2008), http://web.archive.org/web/20210308140251/https://www.nytimes.com/2008/04/22/science/22bass.html?pagewanted=1&ei=5087&em&en=4a6e8113b82be3ac&ex=1209096000.

147. Anonymous, "The Earliest Packets," https://viewmasterinfo.com/articles/first_packets.

148. Wikipedia, "View-Master," https://en.wikipedia.org/wiki/View-Master#cite_note-27.

149. Chiesa, Gabriele, and Paolo Gosio, *Daguerrotype Hallmarks* (Gabriele Chiesa, 2020), 39.

150. Wolcott, Alexander S., "Means of Taking Likeness by Means of a Concave Reflector…," USP 1,582 (1840, May 8).

151. McCardle, James, "February 4: Capital" (2018), https://onthisdateinphotography.com/2018/02/04/february-4-capital.

152. Manekshaw, supra.

153. Spira, S.F., "The Deceptive Angle Graphic," *Graflex Historic Quarterly*, 1(2): 2–4 (1996). See also "The Deceptive Angle Graphic," *American Amateur Photographer* (July 1901) https://digital.library.yale.edu/catalog/2112292.

154. Photo-Miniature, supra, 126–27.

155. "Spiratone Photo & Video Accessory Catalog," No. 852 (Fall/Winter 1985), 44, https://docs.google.com/file/d/0B3pfbneT7Br2elpiUjFZaE9CQTg/view?resourcekey=0-W9baFJX4qJEub5-7098Bmw.

156. https://fotodioxpro.com/products/fltr-spy.

157. Im, Ah Hyeon, et al., "Reflecting Module for Optical Image Stabilization (OIS) and Camera Module Including the Same," US20240295746 (published September 5, 2024).

158. Stereoscopic Company, "Hand Cameras for 1894–5" (advertisement), in Sturmey, Henry, *Photography Annual for 1894* (Iliffe & Son, 1894), I.

159. Rolleiclub, "The First Rolleiflex Cameras," http://www.rolleiclub.com/cameras/tlr/info/early_tlr.shtml.

160. Sutton, Thomas, "Photographic Camera," British Patent 2073 (August 20, 1861).

161. Smith, C.R., "Photographic Camera," USP 301,400 (July 1, 1884).

162. Lothrop, Eaton S., Jr., and Jason Schneider, "The SLR," Part 1, *Popular Photography* (April 1994), 42–44.

163. Herrick, Francis Hobart, *The Home Life of Wild Birds: A New Method of the Study and Photography of Birds* (G.P. Putnam & Sons, 1901), 32.

164. "Hand Cameras," *The Photographic Dealer* (March 1898), 74.

165. Lim, Khen, "A Historic Timeline of the Single Lens Reflex (SLR) Camera 1676–2010…" (2013), https://www.ayton.id.au/wiki/doku.php?id=photo:kl:slr_timeline.

166. Anonymous, "Reflex Cameras," *The Photo-Miniature*, 9(99), 99–144 (1909), at 116–17.

167. "Instant-Return Mirror," http://camera-wiki.org/wiki/Instant-return_mirror.

168. Goldberg, Norman, "Shoptalk," *Popular Photography* (1983, December), 10.

169. Lim, supra.

170. *Photo-Miniature* supra, 127–28.

171. Staudinger, Kurt, "Vorrichtung fuer Reflexkameras," DE 556783 (January 28, 1933), https://patents.google.com/patent/DE556783C/en?oq=DE556783.

172. Wade, John, "c. 1957: Wrayflex II Prototype," https://www.johnwade.org/wray-prototype.

173. "Who Had the First 35mm SLR with a Pentaprism?" (April 9, 2024), https://pixelcraft.photo.blog/category/analogue-photography.

174. Saxby, Graham, *The Science of Imaging* (CRC Press, 2016), 136.

175. Miller, Cearcy D., "The NACA High-Speed Motion-Picture Camera: Optical Compensation at 40,000 Photographs Per Second," National Advisory Committee for Aeronautics Report 856 (GPO, 1946), https://ntrs.nasa.gov/citations/19930091928.

176. Saxby, supra.

177. Saxby, 137–38.

178. Saxby, 138.

179. Miller, C.D., "High-Speed Motion-Picture Camera," USP 2,400,887 (issued May 28, 1946; application filed December 28, 1940).

180. Brixner, Berlyn, "A High Speed Rotating Mirror Frame Camera" (declassified July 15, 1952), 15, https://digital.library.unt.edu/ark:/67531/metadc172736/m1/1.

181. Krehl, Peter O.K., *History of Shock Waves, Explosions and Impact: A Chronological and Bio-*

graphical Reference (Springer Berlin Heidelberg, 2008), 556.

182. Cordin Scientific Imaging, Model 510, http://www.cordin.com/pdfs/Cordin510.pdf; Uhring, Wilfried, "High Speed Imaging," Netware 2015, https://www.iaria.org/conferences2015/filesFASSI15/WilefriedUhring_ HighSpeed_Imaging.pdf.

183. Critchfield, Robert R., et al., "Synchro-Ballistic Recording of Detonation Phenomena" (Los Alamos National Laboratory, 1997), https://digital.library.unt.edu/ark:/67531/metadc690865.

184. Groueff, Stephane, *Manhattan Project: The Untold Story of the Making of the Atomic Bomb* (Plunkett Lake Press, 2023 [1967]).

185. Krehl, 374, 397, 1017.

186. Krehl, 416.

187. Atomic Heritage Foundation, "High-Speed Photography," https://ahf.nuclearmuseum.org/ahf/history/high-speed-photography.

188. "Berlyn Brixner's Interview," *Voice of the Manhattan Project* (1992 interview), https://ahf.nuclearmuseum.org/voices/oral-histories/berlyn-brixners-interview.

189. Ray, Sidney, *Applied Photographic Optics* (Taylor & Francis, 2002), 60.1.2.

190. Meeker-O'Connell, Ann, "How Photocopiers Work," https://home.howstuffworks.com/photocopier.htm.

191. Klein, Stephen B., *Biological Psychology* (Worth, 2006), 218.

192. Maxwell, James Clerk, "Experiments on Colour, as Perceived by the Eye...," *Transactions Royal Society Edinburgh*, 21(Part II): 275–98 (1855), at 284. The same process was patented by Frederic Ives, "Composite Heliochromy," USP 432,530 (July 22, 1890)!

193. Wall, Edward John, *The History of Three-Color Photography, Part 1* (American Photographic Publishing, 1925), 105–7.

194. Cros, Charles (transl. W.R. Harrison), "Note Upon the Classification of Colours, and Upon the Reproduction of Them by Means of Photography, *British J. Photography*, 26: 29 (1879).

195. Wing, Paul, "The Ives Kromskop" (March 19, 2023), https://stereosite.com/collecting/the-ives-kromskop.

196. Ives, Frederic, "Photochromoscope and Photochromoscope Camera," USP 531,040 (December 18, 1894).

197. Wing, supra.

198. Hannavy, John, *Encyclopedia of Nineteenth-Century Photography* (Routledge, 2008), 762.

199. Porter, Albert B., "Kromskop Color-Photography," Scientific Shops, Circular 348 (2nd ed., May 1907), https://www.sil.si.edu/DigitalCollections/trade-literature/scientific-instruments/pdf/sil14-52573.pdf.

200. "1861–1935: Finding Colour," https://colour.photography/history/1861.

201. Wing, supra.

202. Porter, supra.

203. Hamilton, Frederick William, *A Brief History of Printing in America* (United Typothetae of America, 1918), 84.

204. Wall, 107–8.

205. Ives, Frederic, "Camera," USP 475,084 (May 17, 1892).

206. Ives, Frederic, *Krōmskōp Color Photography* (Photochromoscope Syndicate Ltd., 1898), 30–33.

207. Porter, supra.

208. Snodgrass, Lloyd, *The Science and Practice of Photographic Printing* (Worman Printery, 1923), 268.

209. National Photocolor Corporation, "Action in Color," *Popular Photography* (November 1939), 121.

210. Hiscox, Gardner Dexter, and Thomas O'Conor Sloane, *Fortunes in Formulas, for Home, Farm, and Workshop* (Books, Incorporated, 1939), 840.

211. National Photocolor Corporation, "The Color Shot of the Century... with a National Photocolor Camera," *Popular Photography* (November 1945), 115.

212. Friedman, 82.

213. Friedman, 37.

214. "Professor Dr. Miethe's Dreifarben-Camera" (December 20, 2020), http://www.vintage photo.tv/mb.shtml.

215. Lipton, Norman C., "Illusion," *Popular Photography*, 1949, August; 34–35, 114–15.

216. Lipton, supra.

217. Shastid, Thomas Hall, "An Outline History of Ophthalmology," *American J. Physiological Optics*, 7: 568, 592 (1926).

218. Ings, Simon, *A Natural History of Seeing: The Art and Science of Vision* (W.W. Norton, 2008), 186.

219. Shastid, supra.

220. von Helmholtz, Hermann (transl. Thomas Hall Shastid), *The Description of an Ophthalmoscope...* (Cleveland Press, 1916 [German original A. Förstner, 1851), 11–12 and Fig. 1.

221. von Helmholtz, 14.

222. von Helmholtz, 14–16.

223. von Helmholtz, 17.

224. Keeler, C. Richard, "The Ophthalmoscope in the Lifetime of Hermann von Helmholtz," *Archives Ophthalmology*, 120: 194, 198 (2002).

225. Zander, Adolf (transl. Robert Brudenell Carter), *The Ophthalmoscope: Its Varieties and Its Use* (Robert Hardwicke, 1864), 18–24.

226. Ravin, James G., "Sesquicentennial of the Ophthalmoscope," *Archives Ophthalmology*, 117: 1634–1638 (1999).

227. Armour, Roger H., "Manufacture and Use of Home Made Ophthalmoscopes: A 150th Anniversary Tribute to Helmholtz," *British Medical J.*, 321(7276): 1557–59 (2000).

228. Alkatout, Ibrahim, et al., "The Development of Laparoscopy—A Historical Overview," *Frontiers Surgery*, 8: 799442 (2021), doi:10.3389/fsurg.2021.799442.

229. Ramai, Daryl, et al., "Philipp Bozzini (1773–1809): The Earliest Description of Endoscopy," *J. Medical Biography* (2018), doi:10.1177/0967772018755587.

230. Engel, Rainer M.E., "Philipp Bozzini—The Father of Endoscopy," *J. Endourology*, 17(10): 859–62 (2003).

231. Sircus, W., "Milestones in the Evolution of Endoscopy: A Short History," *J. Royal College Physicians Edinburgh*, 33: 124–34 (2003).

232. Ramai, supra.

233. Ellison, Sarah, "The Historical Evolution of Endoscopy," Honors Thesis 2571 (Western Michigan University, April 24, 2015), https://scholarworks.wmich.edu/honors_theses/2571.

234. Sircus, supra.

235. Ramai, supra.

236. Alkatout, supra.

237. Sircus, supra.

238. Alkatout, supra.

239. Sircus, supra.

240. Sircus, supra.

241. Sircus, supra.

242. Beaty, William, "Question #293: What Is a One-Way Mirror and How Does It Work?," PhysLink (2022), http://web.archive.org/web/20221208225720/http://www.physlink.com/Education/AskExperts/ae293.cfm.

243. Wilson, Richard, "Transparent Mirror," USP 728,063 (September 8, 1902).

244. Bloch, Emil, "Transparent Mirror," USP 720,877 (February 17, 1903).

245. Naughton, Pepper, supra.

246. Friend, John Newton, *Iron in Antiquity* (Griffin, 1926), 116.

247. Tsujita, supra, 1–2.

248. Jones, Alexander, "Pseudo-Ptolemy De Speculis," *SCIAMVS*, 2: 145–86, 107 (2001).

249. "Indy 500 History," Pagewise (2002), http://web.archive.org/web/20060519053938/http://wiwi.essortment.com/indyhistory_rgur.htm.

250. Carparts.com, "A Brief History of Automotive Mirrors," https://www.carparts.com/blog/a-brief-history-of-automotive-mirrors.

251. Id.

252. Leavitt, Dorothy, *The Woman and the Car: A Chatty Little Handbook for All Women who Motor Or who Want to Motor* (John Lane, 1909), 29; Kelly, Kate, "First Rearview Mirror Marketed as 'Cop-spotter,'" https://americacomesalive.com/-first-rearview-mirror-marketed-as-cop-spotter.

253. Weed, Chester, "Mirror Attachment for Automobiles," USP 1,114,559 (October 20, 1914).

254. Advertisement, *Hardware World*, 15(8): 176 (2020).

255. Cutnell, John D., et al., *Physics, Volume 2* (Wiley, 2009), 723.

256. Bell, Walter, "Rear Vision Mirror," USP 1,699,043 (January 15, 1929); Id., "Transparent, Prismoidal Mirror," USP 1,949,138 (February 27, 1934).

257. Colbert, William, "Rear View Mirror," USP 2,397,947 (April 9, 1946).

258. "Sunglasses Have Small Mirrors to Provide a Backward View," *Popular Mechanics,* April 1953, 121.

Chapter 4

1. "Siege of Syracuse, Introduction," http://web.archive.org/web/20160408103940/http://www.mcs.drexel.edu/~crorres/Archimedes/Siege/Summary.html; has pointers to excerpts from these works.

2. Gibbon, Edward, *Decline and Fall of the Roman Empire*, Vol. 4, "Chapter XL: Reign of Justinian," Part V, note 95. For electronic text, see, e.g., http://web.archive.org/web/20070208160501/http://www.ccel.org/g/gibbon/decline/decline4.txt.

3. "Siege of Syracuse, Dio Cassius," *Dio's Roman History* (Volume II: Fragments of Books XII—XXV), transl. E. Cary (Harvard University Press, 1914), http://web.archive.org/web/20150920014232/http://www.mcs.drexel.edu/~crorres/Archimedes/Siege/DioCassius.html.

4. Id., *Greek Mathematical Works*, Volume II, transl. Ivor Thomas (Harvard University Press, 1941), 19; same URL.

5. Rorres, Chris, "Burning Mirrors: Refuting the Legend," https://math.nyu.edu/Archimedes/Mirrors/legend/legend.html.

6. Buffon, *Oeuvres completes de Buffon avec les supplémens,* Suppl., Vol. I, Sixième tome, 424 (1835).

7. Rorres, supra.

8. Kayserstuhl Reenactment Gear, "Roman, Celtic & Greek Shields," https://kayserstuhl.com/collections/roman-celtic-greek-shields/-hellenistic-period.

9. This was not a new idea; see Tzetzes's description of Archimedes in action, quoted earlier.

10. Hunt, J.L, "A Burning Question" (1976), http://web.archive.org/web/20051023113659/http://quiz.thphy.uni-duesseldorf.de/96/brs.txt.

11. Sunflower (pseud.), Aug. 6, 1997, email, thread "Re: [SOLAR] Heliostats as instruments of War," http://web.archive.org/web/20031121075039/http://www.cichlid.com/solar-concentrator-archive.1997/msg00091.html.

12. Beaty, William, "An Infinitely Large Solar Furnace" (1996), http://www.eskimo.com/~billb/amateur/mirror.html.

13. Claus, Albert C., "On Archimedes' Burning Glass," *Applied Optics* 12: A14 (1973). See also "Making a Signal Mirror," Boy Scouts of America, http://web.archive.org/web/20030320121136/http://www.ontargetbsa.org/make_mirror.htm.

14. Buffon, *Oeuvres completes de Buffon avec les supplémens,* Suppl., Vol. I, Sixième tome, (1835), 428 says that the mirrors are "six pouces sur huit pouces"; a "pouce" is an inch. However, Mielenz, Klauz D., "Eureka!," *Applied Optics* 13: A14 (1974) citing Buffon, *Mémoires de l'Académie Royale des Sciences pour 1747* (1752), at pp. 82–101, says that

the mirrors are 8 by 10 inches in size. He was clearly describing the same experiments, so I am puzzled by the discrepancy.

15. Gibbon, Edward, *Decline and Fall of the Roman Empire*, Vol. IV (Methuen, 1898), 243 note 99.

16. "Dr. Skyskull," "Mythbusters Were Scooped—by 130 Years! (Archimedes Death Ray)," (February 7, 2010), https://skullsinthestars.com/2010/02/07/-mythbusters-were-scooped-by-130-years-archimedes-death-ray.

17. My sources include Walker, Jearl, *The Flying Circus of Physics, with Answers*, answer to problem 3.76 (John Wiley and Sons, 1977); Hunt, supra; Technology Museum of the Thessaloniki, "Ancient Greek Scientists, Archimedes," http://web.archive.org/web/20091111044636/http://www.tmth.edu.gr/en/aet/1/13.html. Walker cites "Archimedes' Weapon," *Time*, 102, 60 (November 26, 1973); "Re-enacting History—With Mirrors," *Newsweek*, 82, 64 (November 26, 1973); Hunt also cites *The Times*, November 11, 1973; *The New York Times*, November 11, 1973; *International Herald Tribune*, November 8, 1973.

18. Anonymous, "Archimedes' Weapon," *Time*, 102: 60 (November 26, 1973).

19. "MythBusters, Season 2, Episode 8," https://watch.plex.tv/watch/show/mythbusters/season/2/episode/8?uri=provider%3A%2F%2Ftv.plex.provider.vod%2Flibrary%2Fmetadata%2F5fbd8106dcbaf9002f97d37b.

20. Massachusetts Institute of Technology, "Archimedes Death Ray: Idea Feasibility Testing," Product Engineering Process Gallery, https://web.mit.edu/2.009_gallery/www/2005_other/archimedes/10_ArchimedesResult.html.

21. Massachusetts Institute of Technology, 'FAQ About Archimedes Death Ray," https://web.mit.edu/2.009_gallery/www/2005_other/archimedes/10_ArchimedesFAQ.html#FAQiv.

22. Massachusetts Institute of Technology "2.009 Archimedes Death Ray: Testing with MythBusters," https://web.mit.edu/2.009_gallery/www/2005_other/archimedes/10_Mythbusters.html#details.

23. Wikipedia, MythBusters (2010 Season), https://en.wikipedia.org/wiki/MythBusters_(2010_season)#Episode_157_%E2%80%93_%22President's_Challenge%22.

24. Bader, Paul, "Archimedes Burning Mirror Demo" (uploaded February 11, 2011), https://vimeo.com/19836043.

25. See Fig. 16, Map of Ancient Syracuse, Drexel University, http://web.archive.org/web/20000919010217/http://www.mcs.drexel.edu/~crorres/bbc_archive/MapSyracuse3D.jpg.

26. Peddie, John, *The Roman War Machine* (Combined Books, 1994), 137.

27. Id.

28. Delbruck, Hans, *Warfare in Antiquity* (trans. Walter J. Renfroe, University of Nebraska Press, 1975), 89–90, see Vegetius' estimate on p. 90.

29. Museo Galileo, "Architronito" (2013), https://exhibits.museogalileo.it/archimedes/object/Architronito.html.

30. Merchant, Jo, "Reconstructed: Archimedes's Flaming Steam Cannon," *New Scientist* (July 13, 2010), https://www.newscientist.com/article/dn19170-reconstructed-archimedess-flaming-steam-cannon.

31. MythBusters, Episode 55, "Steam Cannon" (July 19, 2006).

32. Prenderghast, Gerald, *Repeating and Multi-Fire Weapons: A History from the Zhuge Crossbow Through the AK-47* (McFarland, 2018).

33. Joslin, William, "Machine Gun," USP 24,031 (issued May 17, 1858); Dickinson, Charles, "Machine Gun," USP 24,997 (issued August 9, 1859).

34. MIT, "Archimedes's Steam Cannon," http://designed.mit.edu/gallery/data/2011/homepage/experiments/steamCannon/ArchimedesSteamCannon.html.

35. Rorres, supra.

36. Gibbon, supra, Vol. 4, Chap. 39, text at footnotes 96–97, citing Zonaras (l. xi. c. p. 55). McDaniels, David, *The Sun: Our Future Energy Source* (John Wiley & Sons, 1979), 68, gives the inventor's name as "Procleus," the year of the battle as 626 ce, and the enemy commander's name as "Vitellius."

37. Id.; see also Beck, Sanderson, "Goths, Franks, and Justinian's Empire 476–610," https://san.beck.org/AB12-GothsFranksJustinian.html; chapter 8 of his *Roman Empire: 30 BC to 610*. ("Marinus commanded the imperial forces that defeated the rebels in the naval battle of the Golden Horn. A chemical compound invented by an Athenian that set fire to ships greatly aided the imperial victory.")

38. Butti, 30–31.

39. Butti, supra.

40. Book IX, Chap. 12, transl. Smith, supra, 387.

41. Rossen, Eric, "Heliostats as Death Rays" (2000), http://web.archive.org/web/20011006012136/http://www.multimania.com/rossen/solar/deathray.html.

42. Florida Solar Energy Center, "Solar Energy Timeline," Solar Matters (1999), http://web.archive.org/web/20051216125826/http://www.fsec.ucf.edu/ed/sm/Ch1-General/Timeline.htm.

43. Butti, 29, citing Needham, Joseph, *Science and Civilization in China* (Cambridge University Press, 1954), 87–89.

44. *Plutarch's Lives*, edited by A.H. Clough, Project Gutenberg Etext #674, https://www.gutenberg.org/files/674/674-h/674-h.htm.

45. Lunazzi, Jose, "On the Quality of the Olmec Mirrors and Its Utilization," 4; pr-print version of the one published in the Proceedings of the "II Reunión Iberoamericana de Óptica," Guanajuato—GTO—Mexico, September 18–22, 1995, SPIE V 2730, 2–7, https://www.researchgate.net/publication/260457130_Quality_of_the_Olmec_mirrors_and_their_utilization.

46. Lunazzi, supra.

47. Westfall, Richard S., "Magini, Giovanni Antonio," Rice University (1995), http://web.archive.

org/web/20040811185724/http://es.rice.edu/ES/humsoc/Galileo/Catalog/Files/magini.html.

48. Butti, 34.

49. Wesley, John, *Compendium of Natural Philosophy*, Part IV, Chap. 2, sec. 29 (1810), http://web.archive.org/web/20021129062400/http://wesley.nnu.edu/Wesley_Natural_Philosophy/natural_p4_ch02b.htm. Butti 259 (note on p. 37), says that Villette's work is recorded at *Philosophical Transactions Royal Society Londo*n 1(6)(1665), 4(47) (1669), and 30(360)(1719); http://www.jstor.org/journals/03702316.html.

50. Butti, 33.

51. "A Brief History of Solar Energy" (April 5, 1999), University of Colorado, Colorado Springs, http://web.archive.org/web/20080905121931/http://www.uccs.edu/~energy/courses/160lectures/solhist.htm.

52. "Leonardo da Vinci," European Educational Project, http://web.archive.org/web/20040120000547/http://www.liceomarconi.it/Progetti/PEE-1999_2000/engineers2e.html.

53. Butti, supra, referring to the writings of Adam Lonicer, a botanist.

54. Newcomb, Sally, *The World in a Crucible: Laboratory Practice and Geological Theory at the Beginning of Geology* (Geological Society of America, 2009), 40–42.

55. Butti, 38, quoting *Philosophical Transactions Royal Society London*, 16(188) (1687).

56. Gleeson, Janet, *The Arcanum: The Extraordinary True Story* (Warner Books, 1998), 48. For the value of a thaler, see p. 11.

57. Rujivacharakul, Vimalin, ed., *Collecting China: The World, China, and a History of Collecting* (University of Delaware Press, 2011), 57; see also Gleeson, 36–37, 52, 66.

58. Butti, 66.

59. "Energy-Conserving Devices: Solar Cooker," 21Design http://web.archive.org/web/20030602160357/http://www.21design.com/prodinfo/devices.html.

60. Butti, 66–68.

61. Id., 74.

62. Public Affairs Office, White Sands Missile Range, "White Sands Solar Furnace," http://web.archive.org/web/20040510185505/http://www.wsmr.army.mil/paopage/Pages/solar.htm.

63. Jones, Steven E., "The Solar Funnel Cooker: How to Make and Use the BYU Solar Cooker/Cooler," Brigham Young University, http://solarcooking.org/funnel.htm and related pages.

64. Jones, Stephen E., "Basic Principles of Solar Cooking, and Introducing the Foil-ware Solar Cooker," Brigham Young University (July 25, 2001), http://web.archive.org/web/20050308160946/http://physics1.byu.edu/jones/rel491/solarbowl.htm.

65. DigtheHeat, "Light Shelves: How to Maximize Daylight Penetration," http://www.digtheheat.com/Solar/lightshelves.html.

66. Wu, Yampeng, et al., "Integrated Systems of Light Pipes in Buildings: A State-of-the-Art Review," *Buildings*, 14(2): 425 (2024), https://doi.org/10.3390/buildings14020425.

67. Cohen, 262.

68. Flandrau Science Center, "Sundial Artist Donates Heliochronometer to Flandrau Science Center," University of Arizona, http://web.archive.org/web/20030408032907/http://www.flandrau.org/exhibits/heliochronometer.htm.

69. Waugh, Albert, *Sundials: Their Theory and Construction* (Dover, 2012 [1973]), 116.

70. Kriegler, Reinhold, "The First Mirror Sundial in Bremen," Planetarium Bremen, http://web.archive.org/web/20011120220026/http://www.hs-bremen.de/planetarium/astroinfo/sonnenuhren/kriegler/r1e.htm.

71. Feist, 271–75.

72. Feist, 281–82.

73. Museo Galileo, "L'astrolabio catottrico gnomonico di Emmanuel Maignan a Palazzo Spada," https://www.youtube.com/watch?v=gPQNDe8YNvo, ~1:28.

74. Feist, 276.

75. Feist, 275.

76. Pagliano, Alessandra, et al., "Geometry and the Restoration of Ancient Sundials: Camera Obscura Sundials in Cava de' Tirreni and Pizzofalcone," *Nexus Network J.*, 19: 121–43 (2017); online 2006, https://link.springer.com/article/10.1007/s00004-016-0318-4.

77. Middleton, W.E. Knowles, "Giovanni Alfonso Borelli and the Invention of the Heliostat," *Archives History Exact Sciences*, 10: 329–41 (1973).

78. Bud, Robert, and Deborah Jean Warner, eds., *Instruments of Science: An Historical Encyclopedia* (Taylor & Francis, 1998), 306; Desaguliers, J.T., transl., *Mathematical Elements of Natural Philosophy*, ... Vol. II (Innys, Longman etc., 1747), 107–15.

79. Learner, 122.

80. Learner, 178–180.

81. "One Central Park—Heliostat and Reflector System," https://good-design.org/projects/one-central-park-heliostat-and-reflector-system.

82. BBC News, "Italy Village Gets 'Sun Mirror,'" (December 18, 2006), http://news.bbc.co.uk/2/hi/europe/6189371.stm.

83. BBC News, "Mirrors Finally Bring Winter Sun to Rjukan in Norway" (October 30, 2013), https://www.bbc.com/news/world-europe-24747720.

84. Dunnington, G. Waldo, "The Sesquicentennial of the Birth of Gauss," *The Scientific Monthly*, 24: 402–414 (1927); see also Reid, Frank, "The Mathematician on the Banknote: Carl Friedrich Gauss," *Parabola*, 36(2): 2, 4 (2000), reprinted at http://web.archive.org/web/20050615000000*/www.maths.unsw.edu.au/Parabola/vol36/no1.pdf.

85. Cung, Nelly, "Gauss' Four Main Inventions," http://web.archive.org/web/20091027155023/http://www.geocities.com/RainForest/Vines/2977/gauss/commentary/110100.html.

86. "Years of Struggle," in Theberge, Albert E., *The Coast Survey 1807–1867*, Vol. I (NOAA, 2001),

http://web.archive.org/web/20070824161157/ http://www.lib.noaa.gov/edocs/HASSLER2.htm.

87. From "The Heliotrope: The Shy Flower of Surveying," by Harold Nelson, quoted in Roeder, Fred, "Carl Friedrich Gauss," *Backsights Magazine,* Surveyors Historical Society (1993), http://web. archive.org/web/20040502180500/http://www. surveyhistory.org/carl_friedrich_gauss.htm.

88. Dracup, Joseph F., "Geodetic Surveys in the United States: The Beginning and the Next One Hundred Years: 1807–1940," NOAA, http://web. archive.org/web/20081026041335/http://www.ngs. noaa.gov/PUBS_LIB/geodetic_surveying_1807. html.

89. In a letter to Olbers. See Darling, David, "Gauss, Karl Friedrich (1777–1855)," The Encyclopedia of Astrobiology, Astronomy, and Spaceflight, http://web.archive.org/web/20030821131205/ http://www.angelfire.com/on2/daviddarling/ Gauss.htm.

90. Herodotus, *The Persian Wars,* Book 6, chapter 115, http://www.perseus.tufts.edu/hopper/text? doc=Perseus%3Atext%3A1999.01.0126%3Abook% 3D6%3Achapter%3D115%3Asection%3D1.

91. Xenophon, *Hellenica,* Book II, 1.27. Holzmann, Gerald J., and Björn Pehrson, "The Early History of Data Networks: Mirrors and Flags," http://web.archive.org/web/20020402051025/ http://www.it.kth.se/docs/early_net/ch-2-1.2.html.

92. Internet Classic Archives, *Plutarch's Lysander* (75 ce), transl. John Dryden, http://classics. mit.edu/Plutarch/lysander.html.

93. James, 533–36.

94. Holzmann, supra.

95. Holzmann, supra.

96. Plum, William Rattle, *The Military Telegraph During the Civil War in the United States...* (Jansen, McClurg & Co., 1882), 29–30.

97. Ciolek, T. Matthew, "Global Networking: A Timeline 1800–1899" (March 9, 2002), citing Coe, Lewis, *The Telegraph: A History of Morse's Invention and Its Predecessors in the United States* (McFarland and Company, 1993), 8.

98. Plum, 30.

99. Wrixon, Fred B., *Codes Ciphers & Other Cryptic & Clandestine Communication* (Black Dog & Leventhal, 1998), 433.

100. Kipling, Rudyard, "A Code of Morals," in *Departmental Ditties and Other Verses* (1886), http: //web.archive.org/web/20080126090859/http:// whitewolf.newcastle.edu.au/words/authors/K/ KiplingRudyard/verse/p2/codeofmorals.html.

101. *The War of the Worlds,* Chapter 13, http:// web.archive.org/web/20220808050834/http:// www.fourmilab.ch/etexts/www/warworlds/b1c13. html.

102. Wrixon, 435.

103. "Heliograph Peak," Arizona State University, http://web.archive.org/web/20080706143216/ http://www.public.asu.edu/~bvogt/20-20/ heliograph/in-heliograph.html. It is the sixth tallest peak in Arizona.

104. Holzmann, Gerald J., "MEMS the Word," *Inc magazine* (November 15, 2000), http://spinroot. com/gerard/pdf/gjh_cv.pdf.

105. "Personal View of C.F. von Hermann," in Grice, Gary K., *A National Weather Service Publication in Support of the Celebration of the American Weather Services ... Past, Present and Future* (National Oceanic and Atmospheric Administration, 1991), https://repository.library.noaa.gov/ view/noaa/6343.

106. Rolak, Bruno J. "General Miles' Mirrors: The Heliograph in the Geronimo Campaign of 1886," http://web.archive.org/web/20040826210 824/http://138.27.35.32/history/html/Rolak.html.

107. Wrixton, 433.

108. Harris, J.D., "Wire at War—Signals Communication in the South African War 1899–1902," *Military History Journal* 11(1) (South African Military History Society), http://web.archive.org/ web/20090214073219/http://rapidttp.com/milhist/ vol111jh.html.

109. Forley, Ian, "How to Calculate the Distance to the Horizon," Boat Safe, http://www.boatsafe. com/nauticalknowhow/distance.htm. This calculation ignores refraction.

110. "Albert J. Meyer Award," National Weather Service Forecast Office, Lake Charles, LA (May 29, 2003), http://web.archive.org/web/20050824062 727/http://www.srh.noaa.gov/lch/obs/cpm3.htm, citing Coe. Other sources on Boer War heliography include "The Second Boer War 1899–1902," Royal Signals Museum, http://web.archive.org/ web/20040603194124/http://www.royalsignals. army.org.uk/Displays%204Boerwar.htm, and Conan Doyle, Arthur, *The Great Boer War* (Smith, Elder, 1902). In Chap. VII, "The Battle of Ladysmith," Conan Doyle notes, "An attempt was made to convert a polished biscuit tin into a heliograph, but with poor success." See http://web.archive.org/ web/20150924073505/http://www.pinetreeweb. com/conan-doyle-chapter-07.htm.

111. Coe, 13.

112. Office of the Chief of Naval Operations, Naval History Division, Washington, "USS *McCall* II (DD-400)," Dictionary of American Naval Fighting Ships (DANFS), http://web.archive.org/ web/20170917061752/www.ibiblio.org/hyperwar/ USN/ships/dafs/DD/dd400.html.

113. Lambert, Reuben, "Alaska to Mexico and the Pacific to Denver, Can It Be Done?," Boy Scouts of America, http://web.archive.org/web/20030 206104543/http://www.ontargetbsa.org/alaska. htm.

114. Gold, Sean, "Best Signal Mirror for Rescue and Survival" (May 2, 2024), https://trueprepper. com/signal-mirrors-survival.

115. Johnson, Les, "Solar Sail Propulsion," 5 (2011), https://ntrs.nasa.gov/api/citations/20120016 691/downloads/20120016691.pdf.

116. Cooper, Iver, "S.S. Sunbeam," *Jim Baen's Universe,* 1(5) (2007).

117. Neufeld, supra.

118. Metzger, Robert A., and Geoffrey Landis, "Multibounce Laser-Based Sails," http://web.

archive.org/web/20121217062629/http://www.rametzger.com/nonfic-mblbs.htm.

119. Emissivity isn't a constant. Emissivity is the radiative output of the body, relative to a black body, and a black body's emission spectrum varies with temperature. The peak wavelength is inversely proportional to the temperature, and the emissivity of a material will vary with its temperature.

120. Matloff, Geoffrey, "Graphene: The Ultimate Interstellar Solar Sail Material?," *J. British Interplanetary Society*, 65: 378–81 (2012), equation (7), https://www.researchgate.net/publication/258659775_Graphene_the_Ultimate_Interstellar_Solar_Sail_Material.

121. Author's calculation per Matloff equation (7), with absorptivity/emissivity of 3.

122. Landis, Geoffrey A., "Small Laser-Propelled Interstellar Probe," Paper IAA-95-IAA.4.1.102, *46th International Astronautical Congress* (October 1995, Oslo, Norway), http://web.archive.org/web/20240701020039/http://www.aleph.se/Trans/Tech/Space/laser.txt.

123. Landis, Table 1. Matweb reports a reflectivity of 90% for tungsten light and emissivity of 0.05 for polished aluminum. https://www.matweb.com/search/DataSheet.aspx?MatGUID=0cd1edf33ac145ee93a0aa6fc666c0e0.

124. Landis, Table 2.

125. Landis, Geoffrey A, "Advanced Solar- and Laser-Pushed Lightsail Concepts, Final Report" (Ohio Aerospace Institute, May 31, 1999), 1.24, https://www.niac.usra.edu/files/studies/final_report/4Landis.pdf. Matweb reports emissivity of 0.61 at 650 nanometers and reflection of 50% visible, 55% UV and 98% IR. https://www.matweb.com/search/DataSheet.aspx?MatGUID=8a6a0df6122349b7bdc92662658d4a4f.

126. Matweb, "Niobium; Wrought," https://www.matweb.com/search/datasheet.aspx?matguid=5e2faf1388514c78a8b2290a210bd178.

127. Calculated from Filmetrics, "Refractive Index of Nb, Niobium," https://www.filmetrics.com/refractive-index-database/Nb/Niobium.

128. Kezerashvili, Roman Ya., "Solar Sail: Materials and Space Environmental Effects" (3rd International Symposium on Solar Sailing, 2013), https://arxiv.org/pdf/1307.7327.

129. Johnson, 7.

130. Mori, Osamu, et al., "First Solar Power Sail Demonstration by IKAROS," *Transactions Japan Society Aeronautical Space Sciences Aerospace Technology*, January 2010, doi:10.2322/tastj.8.To_4_25.

131. Japan Aerospace Exploration Agency, "Small Solar Power Sail Demonstrator 'IKAROS': Successful Attitude Control by Liquid Crystal Device" (July 23, 2010), https://www.jaxa.jp/press/2010/07/20100723_ikaros_e.html.

132. Oberth, Hermann, *Man into Space* (Harper & Bros., 1957), 97.

133. Oberth, 98–103.

134. Gernsback, Hugo, "Television of the Future," (December 1956), reproduced at http://web.

archive.org/web/20060811083210/http://www.twd.net/ird/forecast/1957tvfuture.html.

135. See Olsen, Carrie, "Orbital Velocity and Period Calculator," National Aeronautics and Space Administration (June 23, 1995), http://web.archive.org/web/20071226200343/http://liftoff.msfc.nasa.gov/academy/rocket_sci/orbmech/vel_calc.html.

136. Gernsback, Hugo, "Electronic Weather Control" (December 1963), http://web.archive.org/web/20060818104634/http://www.twd.net/ird/forecast/1964weather.html.

137. Lefcowitz, Eric, "Retrofuture: Space Mirrors and the Possibility of Perpetual Days," http://web.archive.org/web/20080518050245/http://www.retrofuture.com/weather.html.

138. Gernsback (1956), supra.

139. Crawford, Frank S., Jr., *Waves,* Berkeley Physics Course, Vol. 3 (McGraw-Hill, 1958), 213–14.

140. Lewis, Danny, "How a Russian Space Mirror Briefly Lit Up the Night," *Smithsonian* (January 21, 2016), https://www.smithsonianmag.com/smart-news/how-russian-space-mirror-briefly-lit-night-180957894.

141. Goldberg, Carey, "Russians Proclaim Mirror Test a Success as Way to Light World," *Houston Chronicle*, February 5, 1993, http://web.archive.org/web/20041210130031/http://www.chron.com/content/interactive/space/missions/mir/news/1993/19930205.html.

142. Lewis, supra.

143. Lefcowitz, supra.

144. McNally, David, "Znamya-2.5: One to Watch," *A&G News*, 39:4.4 (August 1998), https://academic.oup.com/astrogeo/article/39/4/4.4/188930.

Chapter 5

1. della Porta, Book XVII, chapter 4.

2. Id.

3. Hankins, Thomas L., and Robert J. Silverman, *Instruments and the Imagination* (Princeton University Press, 2014 [1995]), 43–49; De Roo, Henc, "Christiaan Huygens: The True Inventor of the Magic Lantern" (May 16, 2021), https://www.luikerwaal.com/newframe_uk.htm?/huygens_uk.htm.

4. Clayton, Peter, and Martin Price, eds., *The Seven Wonders of the Ancient World* (Routledge, 1988), Chap. 7.

5. Id.

6. Grout, James, "Pharos: The Lighthouse at Alexandria" (2024), https://penelope.uchicago.edu/encyclopaedia_romana/greece/paganism/pharos.html#:~:text=Statius%20compares%20its%20light%20to,the%20curvature%20of%20the%20earth.

7. Chugg, Andrew Michael, *The Pharos Lighthouse in Alexandria: Second Sun and Seventh Wonder of Antiquity* (Taylor & Francis, 2024), Chapter 9.

8. Proudman Oceanographic Observatory, "Insight into Marine Science," http://web.archive.org/web/20061210045543/http://www.pol.ac.uk/home/insight/hutch.html.

9. Trinity House, "Projecting the Light," http://web.archive.org/web/20041016205233/http://www.trinityhouse.co.uk/html/ie3.htm.

10. See photo light.jpg from Trinity House.

11. From the description by lighthouse engineer Robert Stevenson in 1801, quoted at Møbnn, Bill, "Wirral Lighthouses: Bidston Hill" (August 21, 2001), http://web.archive.org/web/20050824085153/http://ourworld.compuserve.com/homepages/m0bnn/Page4.htm.

12. Id.

13. Trinity House, "Lowestoft," http://web.archive.org/web/20041019012842/http://www.trinityhouse.co.uk/html/tlh32.htm.

14. Pepper, Terry and Sue, "The Lewis Lamp," *The Lighthouses of the Western Great Lakes* (March 26, 2003), http://web.archive.org/web/20031005075402/http://www.terrypepper.com/lights/fresnel/lewis-lamp.htm.

15. Rhein, Michael J., *Anatomy of the Lighthouse* (Barnes & Noble Books, 2000), 148.

16. Pepper (2003), supra; and Dolphin, Debbie, "Lewis Patent Lamps and Spherical Reflectors" (May 20, 2002), http://web.archive.org/web/20030131171310/http://home.attbi.com/~deb1/mass/LewisLamps.html.

17. Dolphin, supra.

18. Cardoza, Rod, "Evolution of the Sextant," West Sea Company, http://web.archive.org/web/20140906065305/http://www.westsea.com/tsg3/octlocker/octchart.htm.

19. Cardoza, supra.

20. O'Connor, J.J., and E.F. Robertson, "English Attack on the Longitude Problem," MacTutor (April 1997), http://web.archive.org/web/20210505161722/https://mathshistory.st-andrews.ac.uk/HistTopics/Longitude2.

21. O'Connor, supra; Calvert, J.B., "The Longitude" (July 7, 2000), http://web.archive.org/web/20070410194534/http://www.du.edu/~jcalvert/astro/longitud.htm.

22. Cardoza, supra, and Saunders and Cooke, "The Octant," http://web.archive.org/web/20051110054710/http://www.saundersandcooke.com/octant.html.

23. Staal, Julius D.W., *Patterns in the Sky* (McDonald & Woodward, 1988), 257–58.

24. O'Connor, J.J., and E.F. Robertson, "Longitude and the Académie Royale," MacTutor, http://web.archive.org/web/20190314033501/http://www-groups.dcs.st-and.ac.uk/~history/HistTopics/Longitude1.html.

25. Hong Kong Marine Department, "General Sight Reduction," http://web.archive.org/web/20030627200625/http://www.info.gov.hk/mardep/javascpt/sight.htm.

26. Allen, Dan, "Sextants" (November 3, 2002), http://dkallen.org/Sextants.htm.

27. O'Connor, J.J., and E.F. Robertson, "Nevil Maskelyne," MacTutor (March 2014), https://mathshistory.st-andrews.ac.uk/Biographies/Maskelyne.

28. Reed Navigation, "Lunars in the Nautical Almanac," http://reednavigation.com/lunars/na.html.

29. Id.; Wepster, Steven, "Precomputed Lunar Distance Tables" (December 23, 2012), https://webspace.science.uu.nl/~wepst101/ld/ldtab.html.

30. Huxtable, George, "About Lunars" (2002), http://fer3.com/arc/imgx/About-Lunars.pdf.

31. Siranah, Erik, "Lunar Distance" (January 7, 2020), http://www.siranah.de/html/sail008i.htm.

32. Marine Department Hong Kong, "General Sight Reduction," http://web.archive.org/web/20030214230809/http://www.info.gov.hk/mardep/javascpt/sight.htm.

33. Benson, Guy Meriwether, et al., "Observations of Latitude and Longitude at All Remarkable Points," in *Exploring the West from Monticello: A Perspective in Maps from Columbus to Lewis and Clark*, http://web.archive.org/web/20020402053906/http://www.lib.virginia.edu/exhibits/lewis_clark/ch5.html.

34. Gregory, 172.

35. Ruffell, W.L., "The Gun: Sights and Laying—Rangefinding" (1996), http://web.archive.org/web/20101129122721/http://riv.co.nz/rnza/hist/art90f.htm.

36. Diner, Daniel B., and Derek H. Fender, *Human Engineering in Stereoscopic Viewing Devices* (Springer, 2013), 15.

37. "History of the Camera," http://web.archive.org/web/20051127203829/http://www.bergen.org/AAST/Projects/Engineering_Graphics/_EG2000/camera/History.html.

38. Nickles, David, "Telegraph Diplomats: The United States' Relations with France in 1848 and 1870," *Technology and Culture*, 40(1): 1–25 (1999), http://web.archive.org/web/20020827092219/http://muse.jhu.edu/demo/tech/40.1nickles.html.

39. Livesey, R.J., "Letters: The 'Optical Lever,'" *J. British Astronomical Association*, 106: 5 (1996), http://web.archive.org/web/20020830204331/http://www.star.ucl.ac.uk/~hwm/octltrs.htm, "Johann Christian Poggendorff," http://web.archive.org/web/20061215224311/http://chem.ch.huji.ac.il/~eugeniik/history/poggendorff.html; Sella, Andrea, "Poggendorf's Mirror" (August 10, 2022), https://www.chemistryworld.com/opinion/poggendorfs-mirror/4015983.article.

40. Munro, J., *Heroes of the Telegraph* (1997), Etext #979, Project Gutenberg, https://www.gutenberg.org/files/979/979-h/979-h.htm; Calvert, J.B., "The Electromagnetic Telegraph," "iv. Invention of the Telegraph" (December 26, 2008), http://web.archive.org/web/20170511011033/http://mysite.du.edu/~jcalvert/tel/morse/morse.htm.

41. See also Antunes, Ermelinda Ramos, "Thomson Galvanometer," Catalogue, Ingenuity and Art exhibit, Physics Museum, University of Coimbra, http://web.archive.org/web/20030508181310/http://www.fis.uc.pt/museu/142ing.htm.

42. This description of the mirror galvanometer is based most closely on the one appearing in Munro, supra. It should be appreciated that there were many variations on this basic design. Other descriptions appear at Antunes, supra, and Duke, Charles, "Thomson's Mirror Galvanometer," Grinnell College Physics Historical Museum, http://web.archive.org/web/20110524111317/http://web.grinnell.edu/physics/PMuseum/MirrorGalv.html.

43. Smithsonian Institution Libraries, "Bold and Cautious," in "The Underwater Web: Cabling the Seas," http://web.archive.org/web/20090730015554/http://www.sil.si.edu/Exhibitions/Underwater-Web/uw-bold-and-cautious-02.htm.

44. Calvert, "The Electromagnetic Telegraph," section xxii, supra.

45. Munro, supra. See also Bridges, T.C., The Young Folk's Book of Invention (Little, Brown, 1926), Chap. 10, "Submarine Telegraphy," http://web.archive.org/web/20160504132915/http://www.usgennet.org/usa/topic/preservation/science/inventions/chpt10.htm.

46. Nebeker, Jakob, "Science and Technology: Lord Kelvin's Atlantic Cable," http://web.archive.org/web/20061101000000*/http://www.ieee.org/organizations/history_center/cht_papers/Nebeker.pdf; see also Smith, B. Webster, "Copper in Electrical Engineering: The Atlantic Cable," in Sixty Centuries of Copper, http://web.archive.org/web/20030817103033/http://60centuries.copper.org/electrical/electrical10.html.

47. McEwen, Neal, "Submarine Cable Telegraphy," in "The Telegraph Office," http://web.archive.org/web/20030626231358/http://fohnix.metronet.com/~nmcewen/tel_off.html.

48. Gooday, Graeme, The Morals of Measurement: Accuracy, Irony, and Trust in Late Victorian Electrical Practice (Cambridge University Press, 2004), 161.

49. Laser F/X, "How Laser Shows Work—Scanning System" (2008), https://www.laserfx.com/Works/Works3S.html.

50. Id.

51. Bowers, Brian, Sir Charles Wheatstone FRS: 1802–1875 (Institution of Electrical Engineers, 2001), 58–66.

52. Chinesta, Francisco, and Christine Evain, The History of Physics in Small Bites (Publibook, 2010), 118–19; Nolte, David D., Interference: The History of Optical Interferometry and the Scientists Who Tamed Light (Oxford University Press, 2023), 120–21; Bowers, 66–68.

53. For video of a praxinoscope in action, see "Praxinoscope, Charles Reynaud, 1877," North Carolina School of Science and Mathematics, http://web.archive.org/web/20171218110128/http://courses.ncssm.edu/gallery/collections/toys/html/exhibit11.htm.

54. Natsinas, Theodoros, "Reynaud, Charles-Émile," http://web.archive.org/web/20040409161905/http://www-personal.umich.edu/~natsinas/REYNAUD2bio.html.

55. "Comments from Leonardo's Notebooks," ArtCafe, http://web.archive.org/web/20031002220354/http://www.artcafe.net/artcenter/artfocus/leonardo.htm.

56. Calter, Paul, "Brunelleschi's Peepshow & the Origins of Perspective" (1998), http://web.archive.org/web/20200217161212/http://www.dartmouth.edu/~matc/math5.geometry/unit11/unit11.html.

57. Pliny, Natural History, Book 35, chapter 40.

58. Gregory, 19.

59. Champ, Heather, "The Mirror Project: Adventures in Reflective Surfaces," https://www.mirrorproject.com; Id., "Themes," http://web.archive.org/web/20080509143325/http://www.mirrorproject.com/themes; Long, Marion, "The Mirror Project," The Oprah Magazine 135–37 (September 2002), http://web.archive.org/web/20021017015902/http://www.harrumph.com/o-1.html.

60. Hoffman, Volker, "Brunelleschi's Invention of Linear Perspective: The Fixation and Simulation of the Optical View," 4th International Laboratory for the History of Science, Art, Science and Techniques of Drafting in the Renaissance (May 24–June 1, 2001, Florence and Vinci, Italy), http://web.archive.org/web/20060519001941/http://galileo.imss.firenze.it/news/intlabor/ehoffman1.html.

61. "Brunelleschi and the Origin of Linear Perspective," adapted from Joseph W. Dauben, "The Art of Renaissance Science," http://web.archive.org/web/20070817001204/http://www.kap.pdx.edu/trow/winter01/perspective.

62. Id.

63. Descamps, Jean Baptiste, La Vie des Peintres Flamands, Alemmands et Hollandois..., Vol. 2 (Jombert, 1754), 218–19.

64. Steadman, Philip, "Gerrit Dou and the Concave Mirror," in Lefèvre, Wolfgang, ed., Inside the Camera Obscura—Optics and Art Under the Spell of the Projected Image, Preprint 333 (Max Planck-Institut, 2007), 227–28.

65. Steadman, Philip, Vermeer's Camera: Uncovering the Truth Behind the Masterpieces (Oxford University Press, 2002), 5.

66. Quigley, Martin, Magic Shadows: The Story of the Origin of Motion Pictures (DigiCat, 2022), 1663.

67. Della Porta, Book XVII, Chapter 6.

68. Janson, Jonathan, "Vermeer and the Camera Obscura: Part I," http://www.essentialvermeer.com/camera_obscura/co_one.html.

69. Id.

70. Peres, Michael R., The Focal Encyclopedia of Photography (Taylor & Francis, 2012 [2007]), 51. Johann Zahn was the illustrator for a 1685–1686 publication by Sturm; see Steadman (2002), 178.

71. Delsaute, Jean-Luc, "The Camera Obscura and Painting in the Sixteenth and Seventeenth Centuries," Studies in the History of Art, 55: 110–23 (1998).

72. Jansen, supra.

73. Camuffo, Dario, et al., "The Little Ice Age in Italy from Documentary Proxies and Early Instrumental Records," Méditerranée, 122: 17–30 (2014).

74. Fiorentini, Erna, "Camera Obscura vs.

Camera Lucida: Distinguishing Early Nineteenth Century Modes of Seeing," 2006, Preprint 307, Max Planck-Institut https://www.mpiwg-berlin.mpg.de/sites/default/files/Preprints/P307.pdf.

75. Authier, André, *Early Days of X-ray Crystallography* (Oxford University Press, 2013), xcix.

76. Wollaston, British Patent 2993 (December 4, 1806).

77. Wollaston, W.J., "Description of the Camera Lucida," *J. Natural Philosophy, Chemistry and the Arts (Nicholson's)*, 17: 1–5 (1807).

78. Id.

79. Unpolarized light assumed. Rissanen, Joona, "Fresnel Reflection and Transmission Calculator," https://www.lasercalculator.com/fresnel-reflection-and-transmission-calculator.

80. Wollaston, William Hyde, British Patent 2993 (December 4, 1806), Specification in Repertory of Arts, Manufactures and Agriculture, Second Series, 1807, February; 57: 161–64; compare with figures available at Neolucida, https://neolucida.com/s/1806_WilliamHydeWollaston_Camera_Lucida_Patent_Specification.pdf.

81. Wollaston, 1807, supra; Rissanen, supra.

82. Steadman, Philip, "DMJ—Brunel's Camera Lucida" (August 1, 2024), https://drawingmatter.org/dmj-brunels-camera-lucida.

83. Fiorentini, supra.

84. Basil Hall letter in Dolland, George, *Description of the Camera Lucida* (np, 1830).

85. Terquem, A., "On the Employment of a Silvered Glass as a Camera Lucida," *London, Edinburgh & Dublin Philosophical Magazine & J. Science*, 3(21): 541–43 (1877).

86. Fiorentini, supra.

87. Beale, Lionel Smith, *How to Work with the Microscope* (J. Churchill, 1857), 20–21.

88. William Benjamin, and William Henry Dallinger, *The Microscope and Its Revelations*, Volume 1 (J&A Churchill, 1901), 279.

89. Carpenter, 281.

90. Carpenter, 284.

91. Hockney, David, *Secret Knowledge: Rediscovering the Lost Techniques of the Old Masters* (Viking: 2001), 23, 33, 60, 65.

92. Hockney, 118.

93. Hockney, 74–76.

94. Hockney, 103.

95. Lüthy, Christoph, "Hockney's Secret Knowledge, Vanvitelli's Camera Obscura," *Early Science & Medicine*, 10(2): 315–39 (2005).

96. Wikipedia, "Hockney–Falco Thesis," https://en.wikipedia.org/wiki/Hockney%E2%80%93Falco_thesis.

97. Gorman, David, "Art, Optics and History: New Light on the Hockney Thesis," *Leonardo*, 36(4): 295–301 (2003).

98. Steadman (2002), 27.

99. Steadman (2002), 24.

100. Jenison, Tim, "Vermeer's Paintings Might Be 350 Year-Old Color Photographs," https://boingboing.net/2014/06/10/vermeers-paintings-might-be.html.

101. Jenison, supra.

102. Janson, Jonathan, "An Interview with Philip Steadman" (April 25, 2003), https://www.essentialvermeer.com/interviews_newsletter/steadman_interview.html.

103. Kemp, Martin, *The Science of Art* (Yale University Press, 1992), quoted in Jusko, Don, "The Artists' Tools_of_Technique," http://web.archive.org/web/20050702083915/http://www.mauigateway.com/~donjusko/1artists.htm.

104. Hamilton, Gail (pseud.) [Mary Abigail Dodge (1833–1896)], *Gala Days* (Ticknor & Fields, 1863), 112, http://web.archive.org/web/20120204014456/http://www.merrycoz.org/voices/galadays/GALA04.HTM.

105. Howells, William Dean, *My Mark Twain: Reminiscences and Criticisms* (Harper & Brothers, 1910), Chap. X.

106. Leopard, Gordon, "Seeing Through Past Eyes—The Claude Glass" (July 26, 2023), https://gordonlepard.com/2023/07/26/seeing-through-past-eyes-the-claude-glass.

107. "Frogend Dweller," "Claude Glass and Romanticising the Landscape" (2016, Feb. 28) https://frogenddweller.wordpress.com/2016/02/28/claude-glass-and-romanticising-the-landscape.

108. Ray, Sidney, *Applied Photographic Optics* (Taylor & Francis, 2002), 557.

109. Della Porta, Book XVII, Chapter 6, "If you cannot draw...."

110. Book XVII, Chapter 1, "How letters may be cast out...."

111. Gorman, Michael John, "Projecting Nature in Early-Modern Europe," in Lefèvre (2007), 43.

112. Euler, Leonhard (transl. Henry Hunter), *Letters of Euler... to a German Princess*, Vol. II (Murray, 1802), Letter LXXXI (January 8, 1762), 322–26.

113. Farrar, John, *An Experimental Treatise on Optics* (Hilliard, 1826), 207.

114. "The Auxanoscope," *Cassell's Family Magazine*, 14?: 636–37 (1888).

115. Weynants, Thomas, "Dead Medium: The Fantasmagorie, Part Three," https://gebseng.com/media_archeology/dead_media_project/notes/41/419.html.

116. Hopkins, George Milton, *Experimental Science* (Munn, 1898), 773.

117. Gamwell, Lynn, *Exploring the Invisible* (Princeton University Press, 2020), 92–95.

118. Johnson, Jr., Charles, *Science for the Curious Photographer* (Taylor & Francis, 2017), 231.

119. Woodward, David, "Camera," USP 16,700 (February 24, 1857).

120. Gale, Moses, "Device for Adjusting Reflectors," USP 44,717 (October 18, 1864).

121. Zhang, Song, *High-Speed 3D Imaging with Digital Fringe Projection Techniques* (CRC Press, 2018), 33–34.

122. Id.

123. CLIR, "3. Disc Structure," https://www.clir.org/pubs/reports/pub121/sec3/.

124. Seneca (4 BCE–65 CE), *Natural Questions*

I, 16, reprinted at http://www.fordham.edu/halsall/pwh/seneca-nq1-16.html (Loeb translation). For the translation by Tho. Lodge (1614), see http://penelope.uchicago.edu/relmed/natqu116.html.

125. Schiffer, 14.

126. Polak, 129. See also Roche, Serge, *Mirrors* (Duckworth, 1957).

127. Wills, 16.

128. Id., 16; Diamond, Freda, *The Story of Glass* (Harcourt, Brace, 1953), 139.

129. "Versailles: Hall of Mirrors," Paris Guide, Smartweb, http://web.archive.org/web/20020212135604/www.smartweb.fr/versailles/a/pe/galeriedesglacesp.htm.

130. This is shown in a painting by Sir William Orpen, *The Signing of Peace in the Hall of Mirrors, Versailles, 28th June 1919* (Imperial War Museum, 152 cm x 127 cm, oil on canvas), reproduced at BBC, "History Trail: Wars & Conflict," http://web.archive.org/web/20060306174739/http://www.bbc.co.uk/history/lj/warslj/art_versailles.shtml.

131. "State Apartments (Grands Appartements) and Hall of Mirrors," http://web.archive.org/web/20030805092146/http://www.chateauversailles.fr/en/111.asp.

132. DeJean, Joan, *The Essence of Style* (Free Press, 2007), 192.

133. Velde, B., "Seventeenth-Century Varec Glass from the Great Hall of Mirrors at Versailles," in Janssens, Koen, *Modern Methods for Analysing Archaeological and Historical Glass Volume 1* (Wiley, 2013), 569.

134. Mississippi State University, "The Building," Teacher's Guide, http://web.archive.org/web/20071224054859/http://splendors-versailles.org/TeachersGuide/Building/index.middleFrame.html.

135. Barter, James, *The Palace of Versailles* (Lucent, 1999), 42–43.

136. Misniks, Christian, *Romantic Castles of the Fairy-Tale King* (Linderbichl Verlag, 2001), 53.

137. Mallingham, Margot, "The Netherlands," in *Dutch Tiles: Notes from a Neophite* (October 28, 1999), http://web.archive.org/web/20120717041955/http://www.tiles.org/pages/tilesite/mallingham/dutch1.htm.

138. Turek, Leslie, "Part 14—Touring Vienna by Bus," http://web.archive.org/web/20170711112121/http://www.leslie-turek.com/trip14.html.

139. Wills, 37.

140. "Aina Mahal," IndiaMart, http://web.archive.org/web/20021218000829/http://travel.indiamart.com/gujarat/monuments/aina-mahal.html.

141. "Lahore," Punjabilok, http://web.archive.org/web/20080827235731/http://www.punjabilok.com/pakistan/lahore_pak.htm.

142. Quoted in Hecht, 81.

143. Originally printed in *Nature*, September 23, 1880, pp. 500–503; reprinted at http://web.archive.org/web/20120416132527/http://histv2.free.fr/bell/bell8.htm.

144. Idem.

145. Library of Congress, "Inventor and Scientist," http://web.archive.org/web/20201027231856/https://www.loc.gov/collections/alexander-graham-bell-papers/articles-and-essays/inventor-and-scientist.

146. Bellis, Mary, "Alexander Graham Bell's Photophone Was an Invention Ahead of Its Time," ThoughtCo (March 7, 2019), https://www.thoughtco.com/alexander-graham-bells-photophone-1992318.

147. "Free Space Optics Pioneered by Alexander Graham Bell," Terabeam (2001), http://web.archive.org/web/20020207061329/http://www.terabeam.com/sol/car_fso.shtml.

148. O'Connor, J.J., and E.F. Robertson, "Jules Antoine Lissajous," MacTutor (December 2008) https://mathshistory.st-andrews.ac.uk/Biographies/Lissajous.

149. Matthews, John L., "Physiological Effects of Reflective, Colored and Polarizing Opthalmic Filters" (USAF School of Aviation Medicine, August 1949), 7.

150. 'Sunglass Lenses Serve Also as Mirrors," *Popular Science* (February 1937), 56.

151. "Mirror Glasses," *Life*, 103–4 (March 22, 1948).

152. Matthews, iii.

Chapter 6

1. Cohen, 262.

2. Wang, Jing-Guang, "Optics in China Based on Three Ancient Books," in Chen, Cheng-Yih, ed., *Science and Technology in Chinese Civilisation—Proceedings of the Workshop Held at the University of California* (World Scientific, 1987), 148–49.

3. Murray, Julia K., and Suzanne E. Cahill, "Recent Advances in Understanding the Mystery of Ancient Chinese 'Magic Mirrors,'" *Chinese Science*, 8: 1–8 (1987).

4. Chu, A. Kwang-Hua, "Comments on 'Oriental Magic Mirrors and the Laplacian Image,'" *arXiv*, December 15, 2005, arXiv:physics/0512139v1.

5. Teoh, Eden Kang Min, et al., "Investigation of Surface and Subsurface Profile, Techniques of Measurement, and Replication of the Chinese Magic Mirror," *International Conference on Optics in Precision Engineering and Nanotechnology*, 2013, https://doi.org/10.1117/12.2021112.

6. Mak, Se-yuen, and Din-yan Yip, "Secrets of the Chinese Magic Mirror Replica," *Physics Education*, 36: 102–7 (2001).

7. Riesz, Ferenc, "Visual Approach to the Imaging of Magic Mirrors (Makyohs)," *Results in Optics*, 2023, https://doi.org/10.1016/j.rio.2023.100477.

8. Schumacher, Mark, "Beginner's Guide Magic Mirrors China and Japan," 2023, https://www.onmarkproductions.com/JAHF/japanese-magic-mirrors-reference-guide.pdf.

9. Moond, Jodie, and Paul Carry, "The Magic Mirror Maker," *Kyoto J.* (February 4, 2014),

https://kyotojournal.org/renewal/the-magic-mirror-maker.

10. Pausanias, Description of Greece, 8.37.1, Perseus Project, Tufts University, http://www.perseus.tufts.edu/hopper/text?doc=Paus%2e+8%2e37%2e7&redirect=true

11. Bur, Tatiana, "Mirrors and Religious Aura in the Graeco-Roman World," in Gerolemou, 115.

12. Cohen, 267–68.

13. *Natural Magick*, Book XVII, chapter 2.

14. Christopher, Milbourne, and Maurine Christopher, *The Illustrated History of Magic* (Heinemann, 2nd ed., 1996), 61, 63.

15. Christopher, Milbourne, and Maurine Christopher, *The Illustrated History of Magic* (Pearson Education, 1996), 157, 159. Tobin had previously incorporated the mirrors into the Proteus cabinet that he built for Professor John Henry Pepper.

16. Sprott, Julius Clinton, "6. Light," in *Physics Demonstrations: A Sourcebook for Teachers of Physics* (1996), http://sprott.physics.wisc.edu/demobook/chapter6.htm.

17. Hopkins, Albert A., *Magic: Stage Illusions and Scientific Diversions...* (Sampson Low, 1897), 72–74.

18. Christopher, 160.

19. Hopkins, 82–83.

20. Christopher, 260.

21. Christopher, 160.

22. Hopkins, 75–77.

23. Tissandier, Gaston, *Popular Scientific Recreations in Natural Philosophy...* (W.H. Steele, 1882), 135.

24. Marion, 204.

25. Hopkins, 88.

26. Hopkins 79–80.

27. Hopkins, 523–25.

28. "The Secret History of the Magic Lantern Show," http://web.archive.org/web/20060305175236/http://www.heard.supanet.com/index2.html, and "History of the University of Westminster," http://web.archive.org/web/20040605192145/http://www.wmin.ac.uk/static/history.asp.

29. Sprott, supra.

30. Burdekin, Russell, "Pepper's Ghost at the Opera," Theatre Notebook (January 2015), https://www.researchgate.net/publication/295113791_Pepper%27S_Ghost_at_the_Opera?enrichId=rgreq-922d4ebe4e6bf69a70e75fa865cbdfd4-XXX&enrichSource=Y292ZXJQYWdlOzI5NTExMzc5MTtBUzo1MTYzMjI1MDcxOTAyNzJAMTUwMDExMjI4NjU3Nw%3D%3D&el=1_x_2&_esc=publicationCoverPdf.

31. Swiss, Jamy Ian, "The Science Behind the Ghost: A Brief History of Pepper's Ghost by Jim Steinmeyer" (Book Review), *Genii* (August 1999), https://www.vanishingincmagic.com/magic-book-reviews/the-science-behind-the-ghost-a-brief-history-of-peppers-ghost.

32. Burdekin, supra.

33. Steinmeyer, supra.

34. Ferguson, Doug, "The Haunted Hotel,"

Phantasmechanics, http://web.archive.org/web/20190428074512/http://www.phantasmechanics.com/hotel.html, paraphrasing Randi, James, *Conjuring* (St. Martin's Press, 1992), 26–27. Randi says that inventorship was also asserted by "Poole & Young, a Mr. Gompertz, and a magician named Silvester."

35. Swiss, supra; Steinmeyer, Jim, *Hiding the Elephant: How Magicians Invented the Impossible and Learned to Disappear* (Carroll & Graf, 2003), chapter 2.

36. "The 'Pepper's Ghost' Illusion, 1863," http://web.archive.org/web/20030505012412/http://www.multiliteracy.com/persist/10.html.

37. Castle, Terry, *The Female Thermometer: 18th-Century Culture and the Invention of the Uncanny* (Oxford University Press, 1995), 151, quoted in The Dead Media Project, http://web.archive.org/web/20170401035632/http://www.deadmedia.org/notes/15/152.html.

38. Steinmeyer, supra.

39. Dircks, Henry, *The Ghost!...* (Spon, 1863), 22–23, https://archive.org/details/ghostasproduced00dircgoog/mode/2up?view=theater.

40. Burdekin, supra.

41. Hopkins, 57–61.

42. Naughton, Pepper's Ghost, supra.

43. Evans, Henry Ridgely, *History of Conjuring and Magic* (International Brotherhood of Magicians, 1928), 192.

44. Steinmeyer, supra.

45. Pepper, John Henry, *The True History of the Ghost...* (Cassell, 1890), 24–25, https://www.gutenberg.org/cache/epub/72672/pg72672-images.html.

46. Naughton, Russell, "John Henry Pepper, 'Professor': 1821–1900," Adventures in Cybersound, Australian Centre for the Moving Image, http://web.archive.org/web/20071102190236/http://www.acmi.net.au/aic/pepper_bio.html.

47. Rettig, Hillary, "Million Dollar Monsters," Technocopia, http://web.archive.org/web/20020806183401/http://www.technocopia.com/fun-19991015-million.html.

48. Chef Mayhem (pseud.), "A Backstage Look at the Haunted Mansion: The Ghostly Grand Ballroom," Doombuggies, http://web.archive.org/web/20031010031335/http://www.doombuggies.com/tour_ballroom.htm. Randi says that this effect is "much closer to the original Dircks design."

49. Pepper, John, and James Walker, "Apparatus for Producing Optical Illusions," USP 221,605 (November 11, 1879).

50. Hopkins, 532–34.

51. Hopkins, 86–87.

52. Christopher, 176–77.

53. Brosnan, John, *Movie Magic: The Story of Special Effects in the Cinema* (St. Martin's Press, 1974), 23.

54. Brosnan, supra.

55. Hadlow, Steve, "Dr. No (1962)" (July 1999), http://web.archive.org/web/20020221204507/http://members.netscapeonline.co.uk/bondsupp007/movie/drno.htm.

56. Brosnan, 47.

57. Brosnan, 52.

58. Brosnan, 127–28.

59. Brosnan, 27, 94; NOVA Online, "Special Effects," http://web.archive.org/web/20221003125 050/http://www.pbs.org/wgbh/nova/specialfx2/1920.html.

60. Loew, Katharina, "Magic Mirrors: The Schüfftan Process," in North, Dan, et al., eds., *Special Effects: New Histories/Theories/Contexts* (Palgrave, 2015), 62–77, https://www.academia.edu/21890480/Magic_Mirrors_The_Sch%25C3%25BCfftan_Process.

61. Loew, supra.

62. Loew, supra.

63. Wu, chapter 3 note 16, supra.

64. Tolansky, S., *Curiosities of Light Rays and Light Waves* (American Elsevier Publ. Co., 1965), 21.

65. Jones, Alexander, 149, 173–74.

66. Nix, supra, XII, 345.

67. Porta, John Baptist, *Natural Magick* (1658 English edition), Book XVII ("Of Strange Glasses"), Chapters 2 and 3, https://web.archive.org/web/20180414154058/http://www.mindserpent.com/American_History/books/Porta/jportac17.html#bk17II. The twenty-book edition was first published in Latin in 1589.

68. Brewster, David, *The Kaleidoscope; Its History, Theory, and Construction, with Its Application to the Fine and Useful Arts...* (Hotten, 2nd ed. 1870), 167–73.

69. Brewster, David, "Optical Instrument Called the Kaleidoscope...," British Patent 4136 (August 22. 1817).

70. Baker, Cozy, *Through the Kaleidoscope* (Beechcliff Books, 1987), 13.

71. "The 2 B's—Brewster & Bush," http://web.archive.org/web/20140221154815/http://www.brewstersociety.com/2bs.html.

72. U.K. Patent No. 4,136 (August 30, 1817).

73. Quoted in Baker, 15.

74. Davlins, "F.A.Q.'s About ... Mirror Systems" (taken from Cozy Baker's book, *Kaleidoscope Renaissance*), http://web.archive.org/web/20111030083711/kaleido.com/mirrors.htm.

75. Davlins, supra; Walker, Jearl, "The Physics of Kaleidoscopes," in Baker, Cozy, *Through the Kaleidoscope... and Beyond* (Beechcliff Books, 1987), 172–73.

76. Bush, Charles G., USP 143,271, issued September 30, 1873, reissued November 11, 1873, as Re. 5,649.

77. Bush, USP 151,006 (May 19, 1874).

78. Bush, USP 151,005, "Improvement in Kaleidoscopes" (May 19, 1874).

79. Woodbury, Walter E., *Photographic Amusements* (Scovill & Adams, 1896), 8–9.

80. Optigone, "Giant Mirage: Makes Amazing Interactive 3-D Holographic Exhibit," https://optigone.com/index.php/giant-mirage, contains a video showing guests reacting to a 22-inch-diameter device. The physics are described at

Exploratorium, "Parabolas: It's All Done with Mirrors," https://www.exploratorium.edu/snacks/parabolas.

81. Dzierba, Alex R., Supplementary Note #4: Using Matrices for Geometric Optics, Course P360 (Spring 2000), Indiana University, http://www.dzre.com/alex/HonorsPhysics/Calendars/Note4.pdf. See also Adhya, Sriya, and John W. Noe, "A Complete Ray-Trace Analysis of the Mirage Toy," *Proc. of SPIE*, 9665: 966518–1 (June 3, 2007), https://www.spiedigitallibrary.org/conference-proceedings-of-spie/9665/1/A-complete-ray-trace-analysis-of-the-Mirage-toy/10.1117/12.2207520.full#_=_.

82. Lindberg, David C., *Roger Bacon and the Origins of Perspectiva in the Middle Ages* (Clarendon Press, 1996), corresponding to *Perspectiva* III.3.3.

83. Cohen, 262.

84. della Porta, supra, Book XVII chapter 2.

85. Hooker, John, "Method for Obtaining True or Positive Reflections," USP 370,623 (September 27, 1887).

86. Taggert, Anthony J., "True Image Mirror," USP 5,625,501 (April 29, 2014).

87. *Natural History* of Pliny, Book XXXIII, Chap. 45, page VI-126 of Bostock transl.

88. della Porta, Book XVII, Chaps. 1 and 9.

89. Stanton, Jeffrey, "Coney Island—Second Steeplechase1908–1964" (May 20, 1998), http://web.archive.org/web/20060928052457/http://naid.sppsr.ucla.edu/coneyisland/articles/steeplechase2.htm.

90. Adrian Fisher Maze Design, "A Short History of Mazes," http://web.archive.org/web/20070314054233/http://www.mazemaker.com/history_mazes.htm.

91. Castan, Gustav, "Mirror Maze," USP 545,678 (issued September 3, 1895; application filed January 6, 1891).

92. Perry Lionel, "Optical Illusion," USP 500,607 (issued July 4, 1893; application filed April 8, 1893).

93. Palm, G. Von Prittwitz, "Mirror Maze," USP 507,159 (issued October 24, 1893; application filed September 6, 1893).

94. "Mirror Maze in 'Alhambra' Style," Glacier Garden, http://web.archive.org/web/2007102 6091150/http://www.gletschergarten.ch/en/spiegel.html.

95. "The Magical Mirror Maze," Wookey Hole Caves (August 13, 2002), http://web.archive.org/web/20040605081437/http://www.wookey.co.uk/mirror.htm.

96. Quoted in Heverin, Aaron T., "The Midway!" (December 4, 1998), http://web.archive.org/web/20021028022949/http://intotem.buffnet.net/bhw/panamex/midway/midway.htm.

97. Stanton, supra.

98. Hancox, Clive, "History of 'Philo' and Stotesbury Mansion," http://web.archive.org/web/20081023045513/http://www.partyspace.com/facilitypages/philopatrian/history.html.

99. Stanton, Jeffrey, "Pacific Ocean Park

(1958–1967)" (April 6, 1998), http://web.archive. org/web/20060901131251/http://naid.sppsr.ucla. edu/venice/articles/pop.htm.

100. Hagopian, Kevin Jack, "The Lady from Shanghai," *10 Shades of Noir*, issue #2 http://web. archive.org/web/20240301104520/http://www. imagesjournal.com/issue02/infocus/shanghai.htm.

101. From documentary accompanying Bruce Lee, *Enter the Dragon*, Special Widescreen Edition, Warner Home Video, No. 15921 (1973, 1998) (VHS).

102. Quoted in Baltrušaitis, Jurgis, *Anamorphic Art* (Harry N. Abrams, 1969; English transl. 1976), 131.

103. Baltrušaitis, Fig. 118.

104. Baltrušaitis, Fig. 115.

105. Baltrušaitis, 85.

106. Baltrušaitis, Fig. 80. Another conical anamorph by Bettini is shown in Fig. 120.

107. Ucke, Christian, "Connecting Past and Future," Plenary Lecture, The International Conference 'Turning the Challenge into Opportunities' August 19–23, 1999, Guilin/China; http://www. ucke.de/christian/physik/ftp/lectures/origprocgui lin.PDF.

108. Baltrušaitis, fig. 34.

109. Baltrušaitis, fig. 122, and pp. 155–7. See also "Anamorphic Art," http://web.archive.org/ web/20060427054555/http://www.counton.org/ explorer/anamorphic.

110. Kent, Philip, Art of Anamorphosis Software https://www.anamorphosis.com/software. html.

111. Baltruišaitis, 159.

112. Baltruišaitis, 163–69.

113. Baltruišaitis, Fig. 129.

114. See generally Bedini, Silvio, *The Pope's Elephant* (Penguin Press, 2000).

115. Leeman, Fred, *Hidden Images* (Harry N. Abrams, 1975), 82.

116. Woeste, Louis Bernard, "Myriad Reflectors," USP 1,214,863 (February 6, 1917).

117. Hopcroft, Kevin, "History of Disco Lighting," http://web.archive.org/web/20020802070041/ http://www.njd.co.uk/hodl.html.

118. At 6:42, see Alexandria248, "Berlin: Symphony of the Metropolis" (posted August 14, 2011), https://www.youtube.com/watch?v=OSNNaq Vcauc.

119. "MD+F Laser Instructions (Classic Laser Chess)," http://web.archive.org/web/20150727221 024/http://www.laserchess.org/instructions_lc. html.

120. "Mirror Mania," https://boardgamegeek. com/boardgame/18677/mirror-mania.

121. "Mirror Mania," https://boardgamegeek. com/boardgame/18678/mirror-mania.

122. "Khet," https://boardgamegeek.com/ boardgame/16991/khet-the-laser-game.

123. "Laser Chess," https://boardgamegeek. com/boardgame/91034/laser-chess.

124. "Laser Battle," https://boardgamegeek. com/boardgame/24245/laser-battle.

125. 7,264,242.

126. Oblon, "Innovention Toys, LLC v. MGA Entertainment, Inc.," https://www.oblon.com/ news/innovention-toys-llc-v-mga-entertainment-inc; Finnegan, "MGA Entertainment's Infringement of Laser Board Game Patent Did Not Warrant Trebled Damages," https://www.finnegan.com/en/ firm/news/mga-entertainment-s-infringement-of-laser-board-game-patent-did.html.

127. "Mirror, Mirror," https://boardgamegeek. com/boardgame/94140/mirror-mirror.

128. Government of British Columbia, Ministry of Energy and Mines, "Common Rock-Forming Minerals," November 2, 2000, http://web.archive. org/web/20050504043321/http://www.em.gov. bc.ca/Mining/Geolsurv/Publications/InfoCirc/ Ic1987-5/rockmin.htm.The micas are actually a group of minerals, which include muscovite and biotite micas.

129. Andreae, Christopher, "Mirror, Mirror, in the Painting," *Christian Science Monitor* (February 22, 1999) (book review of Miller, *On Reflection*), http://web.archive.org/web/20021017034304/ http://www.csmonitor.com/durable/1999/02/22/ p18s1.htm.

130. Warren, Susan, "Nature's Glitter, the Mineral Mica, Ends Up in Auto Paints, Lipsticks," *Wall Street Journal* (October 16, 2000), 1, https://www. wsj.com/articles/SB971650617743185944.

Chapter 7

1. Preston, Thomas, *The Theory of Light* (Macmillan, 1928), 19.

2. Wood, Robert W., *Physical Optics* (Macmillan, 3rd ed., 1934), 163.

3. Bickerstaff, R. Paul, "The Michelson–Morley Experiment," in *Claustrophobic Physics: An Introduction to the Theory of Relativity and Poincaré Symmetry*, University of Idaho (January 12, 1999), http://web.archive.org/web/20020826013418/ http://www.phys.uidaho.edu/~pbickers/Courses/ 310/Notes/book/node46.html.

4. Harvey, Bruce, "The Michelson-Morley Experiment," *Alternative Physics* (1997), http://web. archive.org/web/20100122103751/http://users. powernet.co.uk/bearsoft/MickM.html.

5. A.A. Michelson, *American J. Science*, 122, 120 (1881), quoted in Hyperphysics, "A Bit of History: Michelson," http://web.archive.org/ web/20240413193557/http://hyperphysics.phy-astr.gsu.edu/hbase/relativ/mmhist.html.

6. Turner, Arthur F., "Metallic Mirror and Method of Making Same," USP 2,519,722 (August 22, 1950), citing Cartwright and Turner, "Minutes of the Washington, D.C. Meeting, April 27–29, 1939," *Physical Review*, 55: 1109–47 (1939).

7. Or an odd integer multiple of the quarter wavelength.

8. Orfanidis, Sophocles J., *Electromagnetic Waves and Antennas* (August 1, 2016), Example 6.3.1, https://eceweb1.rutgers.edu/~orfanidi/ewa.

9. Wolchover, Natalie, "To Make the Perfect

Mirror, Physicists Confront the Mystery of Glass" (April 2, 2020), https://www.quantamagazine.org/-to-make-the-perfect-mirror-physicists-confront-the-mystery-of-glass-20200402.

10. Brennesholtz, Matthew S., and Edward H. Stupp, *Projection Displays* (Wiley, 2008), 92–93.

11. Landis, table 4.

12. Kare, Jordin G., "High-Acceleration Micro-Scale Laser Sails for Interstellar Propulsion" (February 15, 2002), https://www.niac.usra.edu/files/studies/final_report/597Kare.pdf.

13. Turner, supra.

14. Paschotta, Rüdiger, "Reflection Spectrum of Tilted Dielectric Mirror," *Photonics Spotlight* (November 2, 2006), https://www.rp-photonics.com/spotlight_2006_11_02.html.

15. Joannopoulos et al., USP 6,130,780, "High Omnidirectional Reflector" (October 10, 2000).

16. Gaughan, Richard, "New Coatings Break Reflectivity Barriers," *Technology News*, http://web.archive.org/web/20020425110928/http://www.photonics.com/Content/Feb99/techBarriers.html. The stated wavelength band is in the infrared.

17. Cf. Schechter, Bruce, "M.I.T. Scientists Turn Simple Idea Into 'Perfect Mirror,'" *The New York Times on the Web*, National Science section (December 15, 1998), http://web.archive.org/web/20041024113824/http://home.earthlink.net/~jpdowling/mit.html; with Joannopoulos, J.D., et al., *Photonic Crystals: Molding the Flow of Light* (Princeton University Press, 1995).

18. Deopura, M., et al., "Dielectric Omnidirectional Visible Reflector," *Optics Letters*, 26: 1197 (August 1, 2001), https://www.semanticscholar.org/paper/Dielectric-omnidirectional-visible-reflector.-Deopura-Ullal/5ee467a512eef2ea38e2919f2e88a96c28d79a23.

19. Joannopoulos, John D., et al., *Photonic Crystals: Molding the Flow of Light* (Princeton University Press, 1990), 4–5.

20. O'Brien, J., et al., "Lasers Incorporating 2D Photonic Bandgap Mirrors," *Electronics Letters*, 32(24): 2243–44 (November 21, 1996).

21. Forward, Robert L., "Einstein's Legacy," *OMNI magazine* (March 1979), 54, http://web.archive.org/web/20040419204859/http://home.achilles.net/~jtalbot/history/einstein.html.

22. The original paper was Gordon, J.P., et al., *Physical Reviews*, 95: 282 (1954). For a history, see Talbot, J., "Microwave Laser," http://web.archive.org/web/20040420013310/http://home.achilles.net/~jtalbot/history/ammonia.html.

23. Maiman, T.H., "Stimulated Optical Radiation in Ruby," *Nature*, 187 (4736): 493–94 (1960).

24. Bertolotti, Mario, *The History of the Laser* (CRC, 2004), 243.

25. Bertolotti, 243–44.

26. See generally Goldwasser, Samuel M., "Sam's Laser FAQ, A Practical Guide to Lasers for Experimenters and Hobbyists Version 5.98" (2001), http://web.archive.org/web/20020209163131/http://www.misty.com/people/don/lasersam.htm.

27. Harbison, James P., and Robert E. Nahory,

Lasers: Harnessing the Atom's Light (Scientific American Library, 1998), 63.

28. LeGrand, Y. (transl. S.G. El Hage), *Physiological Optics* (Springer Berlin Heidelberg, 2013), 290.

29. Weber, Michael F., et al., "Giant Birefringent Optics in Multilayer Polymer Mirrors," *Science*, 287: 2451–56 (2000).

30. Shen, Yichen et al., "Optical Broadband Angular Selectivity," *Science*, 343 (6178): 1499–1501 (2014), http://www.mit.edu/~soljacic/angular-selectivity_Science.pdf.

31. Zhang, 34.

32. Zhang, 36.

Appendix

1. Alternative Energy Tutorials, "Solar Irradiance" (2024), https://www.alternative-energy-tutorials.com/solar-power/solar-irradiance.html. Dostrovsky says exoatmospheric solar intensity=1,353 W/m2 and provides formulas for calculating the surface intensity as a function of latitude, solar declination, and local time. See "Appendix I: The Calculation of Solar Energy Collected," in I. Dostrovsky, *Integration of Solar Energy in Multinational Networks*, http://web.archive.org/web/20060712044629/http://magnet.consortia.org.il/ConSolar/SunDaySymp/Dost/Dostrov6.html.

2. Koudouris, Giannis, et al., "Investigation on the Stochastic Nature of the Solar Radiation Process, *Energy Procedia*, 125: 398–404 (2017), Fig. 2.

3. For the exact formula for a parabolic dish, see Lovegrove, Keith, and Wes Stein, *Concentrating Solar Power Technology: Principles, Developments, and Applications* (Woodhead, 2020), 29. Also, the dimensions have the following relationship: $4FD = R2$, where F is the focal length, D the depth of the dish, and R the radius of the rim. Any two of these define the third, and R and D together determine the surface area of the dish (Wikipedia, "Paraboloid").

4. Lovegrove, 27–28.

5. I believe this assumes a spherical receiver, whereas the ship deck or hull is flat or slightly curved.

6. Lovegrove, 31–32. This assumes a "flat receiver" that is "large enough to accept reflected spots from the entire mirror surface."

7. Goswami, Dharendra Yogi, et al., *Principles of Solar Engineering* (Taylor & Francis, 2000), 146.

8. "Ancient Sieges," *The Penny Magazine* No. 214 (August 1, 1835), reprinted at http://web.archive.org/web/20110114081911/http://www.history.rochester.edu/pennymag/214/as.htm.

9. See "NOAA Solar Calculator," https://gml.noaa.gov/grad/solcalc.

10. Gustafs, "Light Reflectance Values," https://gustafs.com/knowledge-hub/surfaces/ligt-reflection-values.

11. Bonsor, Kevin, "How Wildfires Work," www.howstuffworks.com/wildfire1.htm.

12. LookChem, "Pine Tar Chemical Properties," https://www.lookchem.com/pine-tar.

13. Kaye, Theodore P., "Pine Tar; History and Uses," San Francisco Maritime National Park Association, https://www.maritime.org/conf/conf-kaye-tar.php.

14. https://en.climate-data.org/europe/italy/siracusa/siracusa-764495.

15. Beall, F.C., "Specific Heat of Wood—Further Research Required to Obtain Meaningful Data," U.S. Forest Service Note FPL-0184 (USDA, February 1968).

16. Hasburgh, Laura E., and Charles R. Boardman, "Measuring the Specific Heat Capacity of Wood during Pyrolysis," in *Obtaining Data for Fire Growth Models* (ASTM, 2023), http://dx.doi.org/10.1520/STP164220210098.

17. "Thermal Conductivity," http://hyperphysics.phy-astr.gsu.edu/hbase/Tables/thrcn.html.

18. "Infrared Emissivity Table," https://www.thermoworks.com/emissivity-table.

Bibliography

ARTFL Encyclopedia Project (University of Chicago). Autumn 2022 digital edition of Diderot, Denis, and Jean le Rond d'Alembert, *Encyclopédie, ou dictionnaire raisonné des sciences, des arts et des métiers*. Le Breton, 1771.

Baltrušaitis, Jurgis. *Anamorphic Art*. Harry N. Abrams, 1969; English transl., 1976.

Bertolotti, Mario. *The History of the Laser*. CRC, 2004.

Brewster, David. *The Kaleidoscope; Its History, Theory, and Construction, with Its Application to the Fine and Useful Arts*. Hotten, 2nd ed., 1870.

Brosnan, John. *Movie Magic: The Story of Special Effects in the Cinema*. St. Martin's Press, 1974.

Buffon, Comte de. *Mémoires de l'Académie Royale des Sciences pour 1747* (1752).

Buffon, Comte de [Leclerc, Georges-Louis]. *Oeuvres complètes de Buffon avec les supplémens*, Suppl., Vol. I, Sixième tome (1835).

Butti, Ken, and John Perlin. *A Golden Thread: 2500 Years of Solar Architecture and Technology*. Van Nostrand Reinhold, 1980.

Christopher, Milbourne, and Maurine Christopher. *The Illustrated History of Magic*. Heinemann, 2nd ed., 1996.

Coe, Lewis. *The Telegraph: A History of Morse's Invention and Its Predecessors in the United States*. McFarland, 1993.

Cohen, Morris R., and I.E. Drabkin. *A Source Book in Greek Science*. Harvard University Press, 1948.

Della Porta, Giambattista. *Seventeenth Book of Natural Magic* (1658 English edition). http://web.archive.org/web/20030803063243/http://members.tscnet.com/pages/omard1/jportac17.html; facsimile, https://www.loc.gov/item/09023451.

Dircks, Henry. *The Ghost! as Produced in the Spectre Drama, Popularly Illustrating the Marvellous Optical Illusions Obtained by the Apparatus Called the Dircksian Phantasmagoria: Being a Full Account of Its History, Construction, and Various Adaptations*. Spon, 1863.

Ellerman, Ferdinand. "Silvering the 100-Inch Hooker Telescope." Leaflet 52 (1933, May). https://adsabs.harvard.edu/full/1933ASPL....2....5E.

Ellis, William S. *Glass: From the First Mirror to Fiber Optics, The Story of the Substance That Changed the World*. Avon Books, 1998.

Feist, Ulrike. "The Reflection Sundial at Palazzo Spada in Rome: The Mirror as Instrument, Symbol and Metaphor." In *The Mirror in Medieval and Early Modern Culture: Specular Reflections*, edited by Nancy M. Frelick. Brepols, 2016.

Ford, Brian J. *Single Lens: The Story of the Simple Microscope*. Harper & Row, 1985.

Friedman, Joseph Solomon. *History of Color Photography*. American Photographic Publishing, 1947.

Gibbon, Edward. *Decline and Fall of the Roman Empire*, Vol. IV. Methuen, 1898.

Goldberg, Benjamin. *The Mirror and Man*. University of Virginia Press, 1985.

Gros-Galliner, Gabriella. *Glass: A Guide for Collectors*. Stein & Day, 1970.

Hawthorne, John G., and Cyril Stanley Smith, transl. *On Divers Arts: The Treatise of Theophilus*. University of Chicago Press, 1963.

Hecht, Jeff. *City of Light: The Story of Fiber Optics*. Oxford University Press, 1999.

Holzmann, Gerald J., and Björn Pehrson. "The Early History of Data Networks: Mirrors and Flags." http://web.archive.org/web/20020402051025/http://www.it.kth.se/docs/early_net/ch-2-1.2.html.

Hopkins, Albert A. *Magic: Stage Illusions and Scientific Diversions Including Trick Photography*. Sampson Low, 1897.

Ives, Frederic. *Krömsköp Color Photography*. Photochromoscope Syndicate Ltd., 1898.

King, Henry C. *The History of the Telescope*. Dover, 2003 [1955].

Krehl, Peter O.K. *History of Shock Waves, Explosions and Impact: A Chronological and Biographical Reference*. Springer Berlin Heidelberg, 2008.

Landis, Geoffrey A. "Small Laser-Propelled Interstellar Probe," Paper IAA-95-IAA.4.1.102, *46th International Astronautical Congress* (October 1995, Oslo, Norway). http://web.archive.org/web/20240701020039/http://www.aleph.se/Trans/Tech/Space/laser.txt.

Learner, Richard. *Astronomy Through the Telescope*. Van Nostrand Reinhold, 1981.

Loew, Katharina. "Magic Mirrors: The Schüfftan Process." In *Special Effects: New Histories/Theories/Contexts*, edited by Dan North et al. Palgrave, 2015.

Lovegrove, Keith, and Wes Stein. *Concentrating Solar Power Technology: Principles, Developments, and Applications.* Woodhead, 2020.

MacFarlane, Alan, and Gerry Martin. *Glass: A World History.* University of Chicago Press, 2002.

Marion, Fulgence. *The Wonders of Optics.* Scribner, 1871.

McCray, W. Patrick. *Glassmaking in Renaissance Venice.* Ashgate, 1999.

Mehlman, Felice. *Phaidon Guide to Glass.* Prentice Hall, 1982.

Melchior-Bonnet, Sabine. *The Mirror: A History.* Routledge, 2001.

Newman, Jay Hartley, and Lee Scott. *The Mirror Book: Using Reflective Surfaces in Art, Craft, and Design.* Crown, 1978.

Newton, Isaac. *Opticks or a Treatise of the Reflections, Refractions, Inflections and Colours of Light.* William Innys, 4th ed., corrected, 1730.

Pendergrast, Mark. *Mirror, Mirror.* Basic Books, 2003.

Pepper, John Henry. *The True History of the Ghost and All About Metempsychosis.* Cassell, 1890. https://www.gutenberg.org/cache/epub/72672/-pg72672-images.html.

Pilkington, L.A.B. "Review Lecture. The Float Glass Process." *Proceedings Royal Society London, Series A* 314, no. 1516 (1969): 1–25.

[Pliny the Elder, *Naturalis Historia*]. Bostock, John, transl. *Pliny the Elder, The Natural History,* Vol. VI. Henry G. Bohn, 1857.

Polak, Ada. *Glass: Its Tradition and Its Makers.* G.P. Putnam's Sons, 1975.

Prenderghast, Gerald. *Repeating and Multi-Fire Weapons: A History from the Zhuge Crossbow Through the AK-47.* McFarland, 2018.

Ray, Sidney. *Applied Photographic Optics.* Taylor & Francis, 2002.

Saxby, Graham. *The Science of Imaging.* CRC Press, 2016.

Schiffer, Herbert F. *The Mirror Book: English, American and European.* Schiffer Publ. Ltd., 1983.

Sircus, W. "Milestones in the Evolution of Endoscopy: A Short History." *J. Royal College Physicians Edinburgh* 33, no. 2 (2003): 124–134.

Smith, Cyril Stanley, and Martha Teach Gnudi. *The Pirotechnia of Vannoccio Biringuccio: The Classic Sixteenth-Century Treatise on Metals and Metallurgy.* Dover, 1990 [1959].

Steinmeyer, Jim. *Hiding the Elephant: How Magicians Invented the Impossible and Learned to Disappear.* Carroll & Graf, 2003.

Tissandier, Gaston. *Popular Scientific Recreations in Natural Philosophy, Astronomy, Geology, Chemistry.* W.H. Steele, 1882.

[Vitruvius, *De Architectura*]. *The Architecture of Marcus Vitruvius Pollio in Ten Books,* translated by Joseph Gwilt (1826). https://lexundria.com/vitr/0/gw.

Waldman, Gary. *Introduction to Light: The Physics of Light, Vision, and Color.* Dover, 2002 [1983].

Wall, Edward John. *The History of Three-Color Photography, Part 1.* American Photographic Publishing, 1925.

White, Michael. *Newton: The Last Sorcerer.* Perseus Books, 1997.

Wills, Geoffrey. *English Looking Glasses: A Study of the Glass, Frames, and Makers, 1670–1820.* Country Life, 1965.

Wilson, R.N. *Reflecting Telescope Optics,* vol. I. Springer-Verlag, 1996.

Wilson, R.N. *Reflecting Telescope Optics II: Manufacture, Testing, Alignment, Modern Techniques.* Springer, 1996.

Wrixon, Fred B. *Codes, Ciphers & Other Cryptic & Clandestine Communication.* Black Dog & Leventhal, 1998.

Zerwick, Chloe. *A Short History of Glass.* Corning Museum of Glass, 1980.

Zhang, Song. *High-Speed 3D Imaging with Digital Fringe Projection Techniques.* CRC Press, 2018.

Author's Preexisting Work

As noted in the front matter, portions of this book were previously published in the *Grantville Gazette,* which ceased publication in July 2022. That online magazine presented fiction set in and nonfiction relating to the fictional literary universe created by the late Eric Flint's alternate-history sci-fi novel *1632.* The nonfiction works listed below considered the limited knowledge and resources that would have been available to the characters in that novel and its sequels. Naturally, the fictional aspects have been omitted from the present work!

Portions of the discussion of solar sailing were previously published in a nonfiction article in *Jim Baen's Universe,* an online fantasy and science fiction magazine that ceased publication in April 2010.

Cooper, Iver P., "In Vitro Veritas: Glassmaking After the Ring of Fire," *Grantville Gazette* 5 (Sept. 2005).

Cooper, Iver P., "S.S. Sunbeam," *Jim Baen's Universe* (Feb. 2007).

Cooper, Iver P., "Seeing the Heavens," *Grantville Gazette* 16 (Mar. 2008).

Index